Aquatic Ecotoxicology: Fundamental Concepts and Methodologies

Volume I

Editors

Alain Boudou, D.Sc.
Professor
Laboratory of Fundamental
Ecology and Ecotoxicology
University of Bordeaux I
Talence, France

Francis Ribeyre, D.Sc.
Engineer
Laboratory of Fundamental
Ecology and Ecotoxicology
University of Bordeaux I
Talence, France

CRC Press, Inc.
Boca Raton, Florida

Library of Congress Cataloging-in-Publication Data

Aquatic ecotoxicology.

Includes bibliographies and index.
1. Aquatic ecology. 2. Water — Pollution — Toxicology.
3. Water — Pollution — Environmental aspects.
4. Environmental monitoring. I. Boudou, Alain,
1949-   . II. Ribeyre, Francis.
QH541.5.W3A68 1989   574.5′263   88-24169
ISBN 0-8493-4828-5 (v. 1)
ISBN 0-8493-4829-3 (v. 2)

Direct all inquiries to CRC Press, Inc., 2000 Corporate Blvd., N.W., Boca Raton, Florida, 33431.

© 1989 by CRC Press, Inc.

International Standard Book Number 0-8493-4828-5 (v.1)
International Standard Book Number 0-8493-4829-3 (v.2)

Library of Congress Card Number 88-24169
Printed in the United States

FOREWORD

The volumes entitled *Aquatic Ecotoxicology: Fundamental Concepts and Methodologies,* edited by A. Boudou and F. Ribeyre, are a welcome addition to the ecotoxicological literature.

Ecotoxicology is one of the essential disciplines which provide the scientific basis for environmental protection policy. In the European Community, such a policy has been in force since 1973. As of June 1987, 57 directives have been adopted and translated into national law in each of the 12 member states. Thirteen of these pertain exclusively to the aquatic environment and many more are relevant to it. Indeed, the initial thrust of the European Community action plan for the environment was directed to the protection of fresh water, coastal waters, and ground water. A dual approach was adopted, combining the concept of quality objectives with that of emmission standards. Quite often the scientific basis for setting water quality objectives was found to be insufficient. Later, the marketing of new chemicals was regulated in the European Community as well as in the U.S. and elsewhere. The notification file was to contain certain information on aquatic ecotoxicology, hence the need to develop a series of simple, reliable tests.

All this spurred the initiation of research programs coordinated by the Commission of the European Communities and involving many laboratories in the member states as well as the European Community Joint Research Centre.

Research priorities covered the familiar sequence starting with dispersion in the medium, transfer, and fate, and health and ecological effects. Particular emphasis was placed on methodology, concerning the chemical analysis of micropollutants and ecotoxicological testing.

These topics remain relevant and are discussed in detail in this book by a number of North American and European authors, including the editors, who have themselves produced valuable contributions to the field in the last 15 years.

Ecologists and administrators have now come to recognize the inadequacy of simple single species tests for predicting sublethal effects of harmful substances on the complex environment. Therefore, the development of more adequate approaches for laboratory, semifield, and field tests to mimic effects at population, community, and ecosystem levels of biological organization is subject to ongoing research. This book reviews some of these new concepts in aquatic ecotoxicology.

Undoubtedly, this two-volume text will be of great use not only to research scientists and students, but to those who have to prepare regulations to protect the aquatic environment.

Ph. Bourdeau
Commission of the European Communities
Brussels, Belgium

PREFACE

Although its roots can be traced all the way back to the Middle Ages, aquatic ecotoxicology is only now beginning to emerge as a bona fide scientific discipline. One of the first environmental issues to attract public attention was the eutrophication of natural waters which subsequently engendered a water pollution management strategy aimed primarily at the protection of water for domestic purposes, fisheries resource, agriculture, and industry. Toxic contaminants in water did not emerge as an environmental issue until the 1970s, and have since led to the shift in the focus of environmental management towards the conservation of natural ecosystems for recreational and aesthetic purposes. Ecotoxicology is the direct outgrowth of this recent interest in the effects of contaminants on whole ecosystems which have become of primary management concern. Its continuing growth has been assured by the unprecedented demand for predictive testing occasioned by various legislations pertaining to Water Quality Criteria, Clean Water Acts, and Water Quality Standards, pretesting and registration of new pesticides and chemcials, the Environmental Impact Assessments under the Toxic Substances Control Acts, and many other legislative actions. The increasing use of ecotoxicological methods in environmental hazard assessment has led to numerous studies on the ambient levels and speciation of toxic contaminants in water, their accumulation in the biota, and the functional response of biological species and populations to specific contaminants. So far, these studies have failed to yield a coherent and consistent picture of how contaminants affect total aquatic ecosystems.

The close analogy between aquatic ecotoxicology and human medicine has been pointed out by several researchers (see, in particular, D. J. Rapport, *Adv. Environ. Sci. Technol.*, 16, 315, 1984). The concept of "ecosystem medicine" indeed embodies the holistic frame-work for research in the emerging specialty. Thus, aquatic toxicology regards an ecosystem as a de facto living entity (hence the aphorism "Lake Erie is dead") which is responsive to stress due to contaminant overload. Numerous case studies suggest that the common symptoms of ecosystem distress are surprisingly similar and often include an increase in primary productivity, changes in composition of the biota to favor more opportunistic species, prevalence of diseases (especially tumors), as well as reductions in species diversity and size of dominant fauna and flora. These classic distress signs and many more have been identified in the Great Lakes (see M. Evans, *Adv. Environ. Sci. Technol.*, in press, 1987). An important challenge to the practitioners of ecosystem medicine must be to relate the symptoms of ecosystem stress to the causative environmental agent. Advancement in this area has been particularly hampered by our understanding of ecosystem dynamics, which is still very primitive. As Dr. J. R. Vallentyne (The Algal Bowl — Lakes and Man, Miscellaneous Special Publications 22, Department of Fisheries and Oceans, Canada, 1974) aptly observed, "Our knowledge of ecosystems today is equivalent to that of the human body in the latter part of the 18th century."

We are even less advanced in our understanding of the sequelae of ecosystem distress. The mammalian response to stress can be characterized by three stages: (1) the alarm reaction, (2) the stage of resistance where adjustments and accommodations can be made, and (3) the stage of exhaustion leading ultimately to death. The main objective of ecotoxicology is to detect pollution-related problems during the reversible stages or before exhaustion occurs.

The earliest symptomology of ecosystem distress may be subtle and transient physiological or behavioral changes in the sensitive organisms or groups and as such may go undetected. In the Great Lakes, for example, some of the recognized early warning signs of dysfunctional changes include (1) diminished mayfly population, (2) elevated contaminant accumulation in tern and herring gull eggs, and (3) impaired reproductive success of local populations and/or colonies of piscivorous avian and mammalian species. The reproductive failure may be brought about through declines in egg production, eggshell thinning, embryo toxicity, altered parental behavior, congenital abnormalities, biochemical and histopathological dysfunction and aberrant behavior of the recently hatched young, or any number of other deleterious changes.

As in clinical chemistry where biochemical changes are widely used in disease diagnosis, chematological indicators are coming increasingly into play (especially for fish populations) in the assessment of ecosystem health. Among the physiological parameters that have been used to asssess the health of fish populations in the Great Lakes are changes in serum protein, transaminases, blood glucose, cortisol, and ammonia as well as changes in specific antibodies. The hematological parameters in use include erythrocyte and leukocyte counts and mean corpuscular hemaglobin concentration as well as the morphological and numerical characteristics of blood cells in the peripheral vascular system and in the hematopoietic centers of the spleen and kidney. Changes in some enzyme activity presage sublethal effects of contaminants on the zooplankton. The indicator enzymes that suggest toxic effects on the Great Lakes communities include acetylcholinesterase, ALA-D, adenosine triphosphatase, leucine aminonaphthylaminase, allantoinase, and the group of mixed function oxidase enzymes. Chronic effects of sublethal quantities of contaminants can also be manifested in steroidogenesis, endocrine stress response, and other ways. This area of biochemical indicators of ecosystem dysfunction remains a fertile field of study.

The "target organ" concept in human toxicology is equally applicable in ecosystem distress. A fundamental principle in clinical and ecosystem medicine is that certain parts of the system are crucial to the functioning and well-being of the whole. Stress in the sensitive parts will result in pronounced dysfunctional changes in the system as a whole. As noted above, the destabilization of target populations in stressed ecosystems is well documented, for example, in the Great Lakes. Nearshore zones and the subadjacent marshes, bogs and wetlands are often the critical contaminant receptors ("target organs") of concern; stress in such a habitat for fish spawning can exercise a significant impact on the composition of the biota in the lake.

As is to be expected, the distress syndrome becomes manifest when the ecosystem is already far advanced in its breakdown process. Changes in vital signs generally occur after the systemic transformations have advanced to such a degree that the self-cleansing processes can no longer maintain the ecological functions within the normal range. It needs to be noted that most of the government initiatives in controlling water pollution are preventive in intent. At the moment, there is no scientific basis for determining whether the removal of the sources of stress will result in the recovery of ecosystems over-burdened with contaminants.

The discussion above clearly shows that aquatic ecotoxicology is still an emerging science with an exciting future ahead of it. The large volume of disciplinary, deparmentalized, medium-by-medium, and pollutant-by-pollutant studies have only yielded some clues on the principal distress syndrome of severely distressed environments. A holistic framework must now be developed to (1) further elucidate the vital signs of ecosystem health, (2) diagnose the early warning symptoms of ecosystem stress, (3) assess the sensitivity and long-term response of ecosystems to low doses of contaminants, and (4) formulate the proper treatment protocols for ecosystem rehabilitation.

This volume thus is timely in every respect. It provides an overview of the fundamental concepts that have evolved so far in both the theoretical and methodological aspects of the field. The collation and evaluation of the available information should lead to the development of some unifying principles on whole ecosystem response to different levels of contaminant insult. It emphasizes the essential linkage of aquatic toxicology and environmental chemistry and hence provides the groundwork for aquatic hazard assessment and environmental fate modeling. This volume should serve as a focal point in the future development of aquatic ecotoxicology as a key scientific discipline.

J. O. Nriagu
National Water Research Institute
Burlington, Ontario, Canada

INTRODUCTION

During the last five years, several scientific books have been devoted to the ecotoxicology of aquatic systems. To complement these different publications, we felt it was desirable to produce a synthesis, bringing together the fundamental concepts and the main methodologies currently being developed in this discipline.

Research into ecotoxicology can be classified into three fundamental concerns: abiotic factors, which characterize the physicochemistry of environments; biotic factors, relating to biological structures and functions; and contamination factors, which define the modes of pollution of ecosystems. All actions and interactions between these factors, subjected to continual variations in space and in time, lead to extremely complex ecotoxicological mechanisms. With such a complex situation to consider, it becomes vitally important that specific methodologies should be devised, both from the point of view of fundamental research, and for the potential uses which may result.

The vastness of the subject made it imperative that we restrict this work to some extent: we deal with freshwater ecosystems only. This theme is illustrated by many examples of research on the main types of contaminants, apart from radionuclides, which are usually dealt with separately. However, despite these limitations, we still cannot presume to have produced an exhaustive analysis of the subject.

As the bases of ecotoxicology lie in the analysis of the structure and the functioning of natural systems, the first two chapters of Volume I, Part I, are devoted to a synthesis of the current state of knowledge in relation to two main types of freshwater ecosystems: running water (rivers) and still water (lakes). In relation to this ecological basis, the main concepts of ecotoxicology are developed in Chapter 3, giving particular emphasis to the mechanisms which bring about the transfers of contaminants between the different compartments, and also the effects this produces at each biological integration level.

Almost all studies carried out in this discipline call for quantitative knowledge of the localization of toxic products within the systems being studied, in order to assess the contamination levels, to follow their evolution, and to establish links between the resulting effects. The problems of contaminant quantification, from the collecting of samples to the different dosing methods, are described in Chapter 4, using trace metals and metalloids as examples. The other three chapters in Part II deal with the evolution of pollutants in aquatic biotopes: chemical speciation of trace metals (Chapter 5), adsorption of trace inorganic and organic contaminants by solid particulate matter (Chapter 6), and geochemistry and bioavailability of trace metals in sediments (Chapter 7).

The most significant research methodologies currently being developed in aquatic ecotoxicology are presented in Part III and in Part I of Volume II, the former being devoted to *in situ* studies and the latter to experimental approaches in the laboratory.

After describing the essential mechanisms of contamination of the hydrosphere and its effects on ecosystems (Chapter 8), three case studies were selected to illustrate particular features of field research, the methodologies used and the type of results produced (Chapters 9.1, 9.2, and 9.3).

In an intermediate position between the "ecosystem" level and laboratory models are enclosures and artificial streams, which are presented in Chapters 10.1, 10.2, and 11. They offer the possibility of compromise between "representativity", owing to their situation in a natural environment, and "simplicity", as they are limited to a certain extent and offer the possibility of intervention.

Among the main research methodologies developed in the laboratory (Volume II, Part I), we made a distinction according to the biological supports used:

1.   Ecotoxicological models. Their chief objective is to show the effect of interspecific relationships on the transfer of contaminants and their effects: linear transfer models

(experimental trophic chains) and interactive models (experimental ecosystems or microcosms) (Chapters 1.1 and 1.2).

2.   Monospecific approaches. Using ''tools'' borrowed from toxicology, physiology, biochemistry, and other sciences, this approach gives access to the fundamental mechanisms of bioaccumulation and the dysfunctions that can be induced at organism, cell, or even molecule level. This approach can also be useful in perfecting toxicological tests to estimate the possible risks from using new molecules, from modifying the environments, etc. (Chapters 2.1 to 2.4.).

The last chapter of Volume II, Part I, is a presentation of the genetic effects of the contamination of aquatic systems. This is a particularly important aspect of ecotoxicology, concerning as it does both the immediate effects of toxicants at individual and population levels and, above all, the evolution of species through hereditary transmission.

In Part II, Chapter 4 consists of a comparative analysis of the principal methodologies used in aquatic ecotoxicology, drawing conclusions about the relative merits and limitations of each.

Chapter 5 is devoted to mathematical modelization, a synthetic method with which it is possible to describe, and even predict, the fate of contaminants in natural systems. The devising of the different mathematical models is illustrated by examples, accompanied by a critical analysis of their structure and use, and a description of their contribution in the decision-making processes (artificial intelligence).

After considering the fundamental bases of ecotoxicology and describing the major established methodologies, we felt it was important to devote the last chapter of the book to areas of possible application. Implementing tools for diagnosis and creating methods of intervention or prevention concerning the risks involved gives rise to many problems related to the complexity of phenomena and the means available to tackle them.

In the conclusion, we discuss the future of ecotoxicology as a science. This must, of necessity, be based on an increase in fundamental research in order to understand the mechanisms involved and to use the acquired knowledge in considered applications; it must also focus on the very close complementarity between the different methodologies, strengthening the multidisciplinarity of the approach but also addressing problems that the different methods have in common.

A. Boudou
F. Ribeyre

THE EDITORS

Alain Boudou, D.Sc. is Senior Lecturer at the University of Bordeaux I. After obtaining his Master's degree at the Science Faculty in Toulouse, he was awarded his Aggregation (Teacher's Diploma) in Biological Science in 1971. He received his Doctorate (thèse d'Etat) in 1982 at the University of Bordeaux. He lectures in fundamental ecology and ecotoxicology.

Francis Ribeyre, D.Sc. is an engineer at the Fundamental Ecology and Ecotoxicology Department at the University of Bordeaux I. After his Master's degree in Electronics in 1973, he received his Doctorate (thèse d'Etat) in Biological Science in 1988 at the University of Bordeaux. He lectures part-time in ecotoxicology and data analysis.

Since 1974, Drs. Boudou and Ribeyre have set up an ecotoxicology research team as part of the University of Bordeaux I. The main theme of their work, based mainly on experimental studies, is the processes of bioaccumulation and contaminant transfers in freshwater ecosystems.

They have authored or coauthored more than 60 scientific publications in the field of aquatic ecotoxicology. Their research is financed by various national and international bodies, notably the Ministère Français de l'Enseignement Supérieur et de la Recherche, Ministère de l'Environnement, Centre National de la Recherche Scientifique, Etablissement Public Régional d'Aquitaine, Commission of the European Communities.

They are members of the following associations: Société d'Ecotoxicologie Fondamentale et Appliquée, Société de Biologie, Association Française de Limnologie, Société Française d'Ecologie, International Society for Ecotoxicology and Environmental Safety.

CONTRIBUTORS

Volume I

Rod Allan, D.Sc.
Director
Lakes Research Branch
National Water Research Institute
Canada Centre for Inland Waters
Burlington, Ontario, Canada

M. Astruc, D.Sc.
Professor
Department of Chemistry
University of Pau
Pau, France

David Bennett
Senior Scientist
Environmental and Biochemical
  Toxicology Division
Sittingbourne Research Centre
Shell Research Ltd.
Kent, England

Alain Boudou, D.Sc.
Professor
Laboratory of Fundamental Ecology and
  Ecotoxicology
University of Bordeaux I
Talence, France

Alain Bourg, Ph.D., D.Sc.
Water Resources Department
French Geological Survey
Bureau de Recherches Géologiques et
  Minières
Orleans, France

Peter G. C. Campbell, Ph.D.
Professor
Institut National de la Recherche
  Scientifique — Eau
Université du Québec
Ste. Foy, Québec, Canada

Jacques Capblancq, D.Sc.
Professor
Laboratory of Hydrobiology
University of Toulouse III
Toulouse, France

Norman O. Crossland
Principal Scientist
Environmental and Biochemical
  Toxicology Division
Sittingbourne Research Centre
Shell Research Ltd.
Kent, England

Henri Décamps
Director, Center for Ecology of
  Renewable Resources
Centre National de la Recherche
  Scientifique
Toulouse, France

O. Décamps, D.Sc.
Laboratory for Botany and Forestry
University of Toulouse III
Toulouse, France

Robert J. Kosinski, Ph.D.
Associate Professor
Biology Program
Clemson University
Clemson, South Carolina

Akira Kudo, Ph.D., D.Eng.
Senior Research Officer and Visiting
  Professor
Division of Biological Sciences
National Research Council
Ottawa, Ontario, Canada

Francois A. J. Ramade, D.Sc.
Professor and Director
Laboratory of Zoology and Ecology
University of Paris-Sud
Orsay, France

Oscar Ravera
Senior Scientist and Professor
Environmental Institute
Commission of the European
  Communities
Ispra, Italy

Francis Ribeyre, D.Sc.
Engineer
Laboratory of Fundamental Ecology and
 Ecotoxicology
University of Bordeaux I
Talence, France

W. Salomons
Delft Hydraulics Laboratory
Delft, Netherlands

Markus Stoeppler, Ph.D.
Section Head
Institute of Chemistry
Nuclear Research Center (KFA) Juelich
Juelich, Federal Republic of Germany

Andre Tessier, Ph.D.
Professor
Institut National de la Recherche
 Scientifique — Eau
Université du Québec
Ste. Foy, Québec, Canada

ACKNOWLEDGMENTS

We should like to offer our most profound thanks to all the contributors whose participation has made this work possible. As the size of the book had to be restricted to some extent, we were obliged to limit the number of chapters selected to cover each topic; we must therefore offer our apologies to all those other specialists whom we would have liked to see participate and who could have enriched this approach to the fundamental concepts of aquatic ecotoxicology and the main associated methodologies. We should also like to thank the editorial staff at CRC Press for their understanding and help and for the confidence they have continually shown throughout the preparation of this collective work. Madame V. Serre, our secretary, also deserves recognition. We gratefully acknowledge D. Geffre, L. Graham, H. Koziol, and M. Perrin (Département de Langues Vivantes Pratiques, Université de Bordeaux II) for the English translation.

TABLE OF CONTENTS

Volume I

Freshwater Ecosystems: Ecological and Ecotoxicological Bases

Chapter 1

RIVER ECOSYSTEMS: ECOLOGICAL CONCEPTS AND DYNAMICS

H. Décamps and O. Décamps

TABLE OF CONTENTS

I. INTRODUCTION

A theory of running-water ecosystem functioning has developed in recent years. This theory finds its roots in a series of papers, the most notable of which is by Hynes[1] in 1975. Taking into account the unity of a river and its valley, to consider them within their catchment, is, in fact, essential for an understanding of the dynamics of running-water ecosystems. The concepts of river continuum[2] and of spiraling of nutrients[3] were largely developed on this notion as was the work on the fish yield of African rivers[4] and the recent syntheses on river ecology.[5,6]

The evolution of ideas on running-water ecosystems has developed very rapidly since 1970, reflecting a great interest in this area. Books have followed one another.[7-20] The International Large Alluvial River Symposium held in Ontario, Canada in September 1986 underlined the importance of a concentrated effort to "provide an understanding of the management of large rivers for fish production, to produce estimators of fish yields, to promote the development of inventory and assessment techniques and productivity models, to identify areas needing further study, to improve communication and liaison among managers with large river interest". At approximately the same time in France, the first "Journées de l'Environnement" organized by the Centre National de la Recherche Scientifique clearly posed the problem of rivers as means of communication and economic development, energy sources, living systems, and recreational areas. Their careful management, taking into consideration the sometimes conflicting demands on them, is a major preoccupation of both scientists and environmental managers.

The results obtained by such symposia demonstrate that a predictive ecology of fluvial ecosystems is being developed and is clearly needed. In this chapter we will try to elaborate on several of its essential characteristics.

II. DIVERSITY AND COMPLEXITY OF RIVERS

Rivers are distributed over the whole world under extremely varied climatic conditions. They differ in their lengths, their flows, and their sediment loads (Table 1). Their forms reveal a complexity that is reflected in the fluvial dynamics. Figure 1 shows the principal geomorphological determinants of several fluvial forms. The extent to which a river meanders depends on the slope of the channel, the sediment load, and the degree of river regulation.[21] The resulting types of rivers distinguish themselves from each other by their meandering and their stability.

For larger rivers, the primary characteristic to consider for understanding ecosystem dynamics is the variety of interactions between the channel and the floodplain. The importance of these interactions has long been recognized. A simple diagram shows that they act in the three spatial dimensions: longitudinal, vertical, and cross-sectional (Figure 2). A fourth dimension, time, also has an effect, during flooding, for example. These different dimensions regulate the set of interactions that act on the terrestrial and aquatic environments and that organize the structure and function of the fluvial community. All of this suggests that the water residence time is critical in explaining floodplain ecology. Water residence time varies as a function of the geomorphological characteristics and determines the influx and efflux of nutritive substances and the resulting aquatic plant and animal production.

On the basis of the catchment as a whole, the recognition of the different stream orders permits a consideration of the dendritic network.[22] Streams of the first order, lacking tributaries, form the second-order streams when they join, which in turn create the third-order streams, and so on. The grouping of streams of the same order allows a characterization of the catchments. In general, the number of streams increases with decreasing stream order in the following form:

$$N_s = a \cdot b^s$$

Table 1
LARGE RIVERS OF THE WORLD
RANKED BY LENGTH, MEAN ANNUAL
DISCHARGE AT MOUTH, AND
AVERAGE SUSPENDED LOAD

Length (km)

Nile	Africa	6650
Amazon	South America	6437
Mississippi/Missouri	North America	6020
Yangtze	Asia	5980
Yenisey	Asia	5540
Ob	Asia	5410
Hwang-Ho	Asia	4845
Zaire	Africa	4700
Amur	Asia	4444
Lena	Asia	4400

Mean Annual Discharge at Mouth (1000 m³/s)

Amazon	South America	212.5
Zaire	Africa	39.7
Yangtze	Asia	21.8
Brahmaputra	Asia	19.8
Ganges	Asia	18.7
Yenisey	Asia	17.4
Mississippi/Missouri	North America	17.3
Orinoco	South America	17.0
Lena	Asia	15.5
Paraná	South America	14.9

Average Annual Suspended Load

	Total t × 106	Load/discharge
Ganges	1625.6	86.9
Brahmaputra	812.9	41.0
Yangtze	560.8	25.7
Indus	488.7	87.3
Amazon	406.4	1.9
Mississippi	349.5	20.
Nile	123.8	44.2

From Welcomme, R. L., *River Fisheries,* Tech. paper 262,
Food and Agriculture Organization, Rome, 1985, 1. With
permission.

in which N_s is the number of streams of order s. Similarly, for the length (L_s) of a stream of order s:

$$L_s = x \cdot y^s$$

Consequently, a large number of small tributaries constitutes an important part of the total length of the watercourses of a given region (Table 2).

Moreover, a relationship exists between river length and drainage basin which, according to Welcomme,[4] takes the following form for the 50 largest rivers in the world:

$$L = 1.7084 \, A^{0.5418} \qquad (r^2 = 0.70)$$

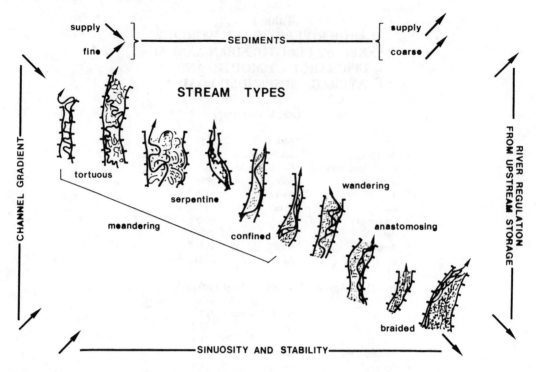

FIGURE 1.   Mollard's continuum of stream channel and floodplain types,[57] with governing factors (top and side margins) and general response (bottom). (Modified from Reference 21.)

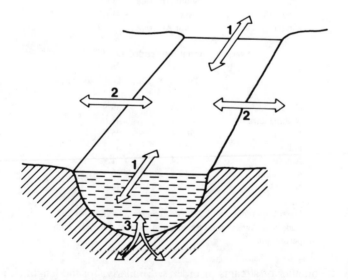

FIGURE 2.   Fluvial ecosystem with (1) longitudinal (headwater — riverine — estuarine), (2) lateral (riverine — riparian — floodplain), (3) vertical (riverine — groundwater) interactive pathways. (Modified from Reference 6.)

where L (in km) is the length of the main channel and A (in km^2) is the area. The coefficient and exponent of this relationship change somewhat when the different continents and climatic zones are considered individually.

One way of studying systems as complex as rivers is to consider them as systems organized hierarchically in space and in time.[23] Different levels have been distinguished by Frissell et

Table 2
NUMBER AND LENGTH OF THE
VARIOUS STRAHLER STREAM
ORDERS IN THE U.S.

Order	Number	Length	Percent cumulated
1	1,570,000	1.6	48.4
2	350,000	3.7	73.2
3	80,000	8.5	86.3
4	18,000	19.2	92.9
5	4,200	44.9	96.6
6	950	102.4	98.6
7	200	235.2	99.3
8	41	540.8	99.5
9	8	1,243.2	99.9
10	1	2,880.0	100.0

From Welcomme, R. L., *River Fisheries,* Tech. paper 262, Food and Agriculture Organization, Rome, 1985, 1. With permission.

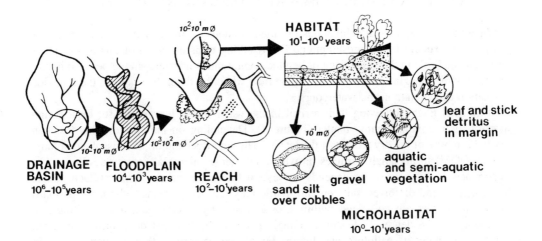

FIGURE 3. Hierarchical organization of a fifth- to sixth-order reach system as a part of floodplain and drainage basin systems, its habitat, and microhabitat subsystems. The diameters of the circles indicate the different spatial scale. The time scales are given in parentheses. (Modified from Reference 24.)

al.[24]: microhabitat, pool/riffle, reach, and stream segment (Figure 3). At each of these levels, a specific spatial-temporal scale can be recognized. For example, geological phenomena are infrequent, but have a major effect on the stream segments, while more frequent phenomena affect the microhabitat subsystems. In this hierarchy, each subsystem makes up the environment of subsystems of lower levels and exerts control over them.

To consider the processes that govern the dynamics of flowing-water ecosystems, we will consider them at three hierarchical levels:

1. The microhabitat scale allows a consideration of microbial processing and nutrient dynamics.
2. The stream scale relates to the river continuum concept.
3. The landscape scale is needed to consider the dynamics of the fluvial corridor, that is to say, the floodplain as a whole.

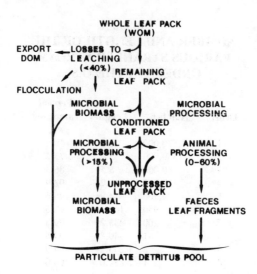

FIGURE 4.   Simplified diagram of leaf processing in streams. (Modified from Reference 54.)

III. MICROBIAL PROCESSING AND NUTRIENT DYNAMICS

Whatever its origin — allochthonous or autochthonous — most of the organic matter is processed by stream fauna consumption. The resulting transformation of this matter plays a fundamental role in the structuring of benthic communities in rivers. In temperate climate, the development begins in the autumn, at the time when the fallen leaves from the riverside vegetation accumulate in the watercourses.

In the first hours following their arrival in the water, these leaves release, by leaching most of their soluble organic matter (Figure 4). Depending on the species, this loss can reach 25 to 30% of the dry weight of the leaves. Through flocculation, the dissolved organic matter (DOM) forms small particles on which an intense microbial activity develops. The importance of this processing depends on several factors in the environment such as turbulence, pH, and dissolved salt content. While they are being leached, the large particles of organic matter are colonized within 1 or 2 weeks by microorganisms, with the rate depending on the degree of leaching. Fungi appear to be the first decomposers of plant detritus, attacking its cellulose and lignin. Fungi and bacteria within the detritus obtain nitrogen from the water and convert it into protein, thus enriching the food available for animals. At the same time, large particles of organic matter, leached and colonized by microorganisms, are reduced to finer particles both by mechanical demolition and by the action of detritivore animals and microbic metabolism. This whole process is influenced by factors such as turbulence, temperature, availability of nitrogen and phosphorus, and calcium content. It depends equally on the tree species available and the lignin content of the leaves.

Different size ranges of plant fragments have been classified.[25] Important transport of whole organic material (WOM >16 mm) and of large particulate organic matter (LPOM: 1 to 16 mm) occurs during high water. However, several factors can influence the quantity of LPOM, medium particulate organic matter (MPOM: 250 μm to 1 mm), and small particulate organic matter (SPOM: 75 to 250 μm). These factors are stream size, morphology, watershed topography, weather, decomposition rate, and time since the last rain. The differences existing between streams or between different sections of the same stream make prediction based solely on flow difficult. The fine particulate organic matter (FPOM: 0.45 to 0.75 μm) has various origins: wood, compacted fecal pellets, diatoms and desmids,[26] scoured benthic algae after rains,[27] and plankton.[28] Other sources are shredder feeding by invertebrates and the flocculation and precipitation of DOM.[29]

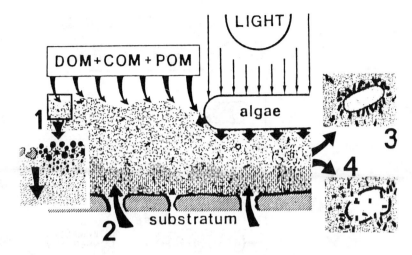

FIGURE 5. Schematic representation of river epilithon with light and organic energy pathways. (1) Microbial fragmentation and diffusion of surface-adsorbed DOM, colloidal organic matter (COM), and POM; (2) pathway from groundwater through the substratum; (3 and 4) details of the polysaccharide matrix showing exoenzyme concentration near a bacterium and releasing of enzymes from cell lysis. (Modified from Reference 31.)

DOM <0.45 μm constitutes the most important energy source in streams. It originates primarily from groundwater,[30] but also partly from the leaching of materials that have fallen directly into the water. The epilithon, the organic layer that covers the bottom of streams (Figure 5), plays a key role in the transformation of DOM. Large quantities of slime, probably secreted by bacteria and algae on the streambed, form a polysaccharide matrix of fibers producing enzymes capable of degrading organic materials to small molecular weight products.[31] The FPOM, trapped by slime and fungal mycelia at the bottom, is progressively broken down in the epilithon and colonized by bacteria and fungi. The standing stock of algae of the epilithon is largely determined by light and maintained in a state of rapid turnover by scouring and grazing. As pointed out by Winterbourn,[32] the biochemical nature of DOM is poorly understood with its different fulvic and humic acids, phenols, lipids, carbohydrate proteins, and amino acids. Soil exudates transported by groundwater are probably one of the major sources of DOM. The biologically active fraction of DOM seems to be small in comparison to refractory compounds that dominate DOM.[33] In Figure 6 is shown the complexity of the relationships that exist between autotrophy and heterotrophy at the epilithon level. These two processes are mixed, interactive, and interdependent. Their relationship changes over both space and time: this is a fundamental characteristic of flowing-water ecosystems. This will be seen again at other levels of integration.

Research the last 20 years has shown that the particulate matter on the riverbank is the principal source of energy for most headwaters. The latter act as transformers of organic matter produced by the terrestrial environment. This transformation also occurs within large rivers.[34] The organic matter is constantly transported downstream by the current. There is no long-term, on-site recycling, but rather a more or less continuous transfer during which the detrital pool is ingested and egested repeatedly during the downstream movement.[35] These various organic-matter transformations are primarily controlled by living organisms. Hyphomycetal fungi, with their mycelium, penetrate the vegetal substrate and digest cellulose and lignin, thereby raising the levels of phosphorus, nitrogen, and proteins. Among the invertebrates, the shredders attack the largest fragments, the collectors the most tender tissues, and the grazers the surface microflora. As well, the shredders increase the availability of nutrients to other organisms. This resource sharing is completed by a repeated utilization of fecal material released by the different invertebrate species. This fecal material, its abundance

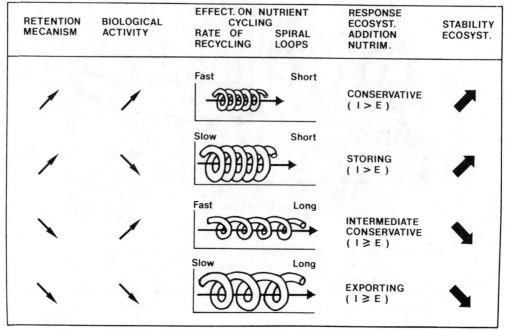

RETENTION MECANISM	BIOLOGICAL ACTIVITY	EFFECT. ON NUTRIENT CYCLING		RESPONSE ECOSYST. ADDITION NUTRIM.	STABILITY ECOSYST.
		RATE OF RECYCLING	SPIRAL LOOPS		

I : Import , E : Export

FIGURE 6.   Characterization of nutrient dynamics in streams differing in their retention mechanisms and biological activity. Effects on nutrient cycling and response of stream ecosystem to addition of nutrients. Hypothesized consequence on ecosystem stability. (Modified from Reference 40.)

resulting from a low efficiency of invertebrate assimilation, represents a high-quality nutrient source for microorganism colonization. Large quantities of wood (bark, branches, trunks) are also used, but their decomposition is much slower, varying from 1 year for twigs to 10 to 20 years for branches 2 cm in diameter to several centuries for thicker trunks. Chauvet et al.[36] have examined differences in decomposition of poplar lignin in a river and its shoreline zone and found this latter environment among the most active for the ligninolyse when considering different natural environments.

These successive transformations of organic matter occur during downstream movement. This has given rise to the concept of nutrient spiraling in flowing waters.[3,37,38] The idea of spiraling emphasizes the originality of the phenomenon of nutrient transport together with its recycling (Figure 6). This downstream movement within the ecological system distinguishes flowing waters from closed systems or from open ones, such as lakes, for example. In running-water ecosystems the recycling occurs simultaneously and proportionally to the resources moving downstream, thus demonstrating its importance relative to the distance over which a complete cycle occurs. In fact, this spiral of nutrients in running water combines two processes: the on-the-spot utilization of drifting organic material by benthic organisms and, on the other hand, the reutilization of this organic material downstream. Moreover, as Elwood et al.[3] stress, the concept of spiraling allows a consideration of the dynamics of nutrients and carbon in lotic ecosystems. The dynamics are characterized by a close link between hydrological and ecological processes in the control of nutrient recycling and oxidation of organic carbon.

Thus, in the case of a relatively conservative element, for example, phosphorus, originating on the catchment, the length of a spiral can be defined. The length represents the average distance covered on its way downstream by a nutrient atom during a cycle. A rapid recycling, resulting from high biological activity associated with a slowing of the downstream move-

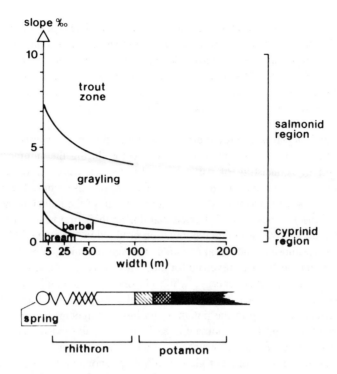

FIGURE 7.   Schematic representation of the different stream zones with the fish zones (above) according to Huet[41] and the corresponding rhithron-potamon (below) according to Illies and Botosaneanu.[42]

ment, results in a shortening of the spiral loops. Restated, the shorter the cycle length, the larger the number of times a nutrient can be used in a stream section. Newbold et al.[37] showed that the recycling distance of carbon was related to the index of fluvial metabolism, defined by Fisher and Likens[39] as the relationship between the amount of carbon respired within the ecosystem to the amount brought into the system. This notion was used by Minshall et al.[40] to characterize the nutrient dynamics in different American streams (Figure 6). The retention mechanisms and the biological activity, in fact, determine the metabolic conditions of running water through the rate of recycling and the distance between spiral loops. This, in turn, allows the prediction of the response of a particular lotic ecosystem to the addition of nutrients and its resulting stability.

IV. THE RIVER CONTINUUM

Numerous authors have attempted to formalize the ecological succession from the source to the mouth of rivers by the study of the zonation of the fauna. Two important European papers on this deal with the fish[41] and the invertebrate fauna[42] (Figure 7).

It is clear that two main systems succeed each other along the watercourse from upstream (the rhithron) to downstream (the potamon). The rhithron system has well-oxygenated, fast-flowing waters and an eroding streambed. In temperate regions the water temperatures are on average less than 20°C, and the fauna consists of current-adapted, cold-water stenotherms, and plankton is virtually absent. Portions of the potamon system are important sites of oxygen deficiency. Here the current is normally slower and the stream/riverbed is primarily sandy and muddy. The potamon fauna is eurythermic and consists of lake-dwelling species with plankton developing during the summer.

In the upstream portion (the rhithron), the vegetal debris of riparian origin constitutes the principal energy source. During fragmentation this debris, colonized by bacteria and fungi, is utilized repeatedly by the invertebrate community. Leached DOM aggregates into particles also colonized by microorganisms. Aquatic vegetation (macrophytes and periphytic algae) is utilized as well, either directly or during decomposition. The consumer community contributes to this by fractioning the detritus stock into progressively finer particles and by enriching the stock with its waste products and remains. These cycles are repeated along the river course, with continual enrichment during any alternation between rapid and slow water zones.

Fine detritus and DOM of rhithronic origin also act as the energy source in the potamon. Because of slower currents, phytoplankton may develop during the summer, as well as a zooplankton normally consisting of parthenogenetic species whose development is more rapid than that of sexually reproducing species. This is the case where the residence time is of the order of weeks, such as in dead arms, slow portions of the main channel, or dammed rivers. Studies of the Thames River[43] showed that during their first year of life, roaches and bleak contributed about 70% of the total population production, obtained 53% of this energy from zooplankton (primarily Cladocera) and 46% from chironomids, allochthonous material, and detritus. The relationship was reversed for older individuals: the zooplankton constituted no more than 2% of consumed energy, while chironomids, allochthonous material, and detritus contributed 70%. In fact, the detrital material constituted the food for a succession of filtering invertebrates living in the potamon: mollusks, oligochaetes, and insect larvae.

This division into rhithron and potamon, each in turn subdivided, has largly inspired research on the structure of running-water ecosystems. More recently, the river continuum concept[2] has contributed to a coherent view of all the phenomena that succeed one another along the length of a river (Figure 8). According to this concept, geophysical variables present a continuous gradient from headwaters to river mouths, with the biotic communities succeeding each other along this gradient in such a way that energy losses are reduced to a minimum. This tendency toward a reduction of energy loss results in a sharing of available resources over space and time. Diverse communities succeed one another the length of this continuum, from headwaters (orders 1 to 3) to medium-sized streams (orders 4 to 6) to large rivers (orders >6). The headwaters are characterized by a strong influence from the riparian vegetation. This riparian vegetation is the source of allochthonous nutrients and, in turn, reduces autochthonous plant production by shading. In medium-sized streams the bed is enlarged, the allochthonous inputs are relatively less important, and aquatic communities tend to be increasingly based on autochthonous plant production. Organic allochthonous material, which comprises a large part of LPOM in upstream waters, fragments little by little into FPOM, molecular components, amino acids, sugars, etc. Invertebrate benthic communities reflect these transformations and are, in turn, dominated by shredders, grazers, and collectors. Fish communities undergo concomitant transformations, with a dominance of insectivores, little by little joined by grazers and predators.

This approach to the river continuum concept permits useful generalizations about the contribution of organic matter, about the structure of the invertebrate community, and about resource sharing along the length of rivers. This concept remains the object of study.[44,45] It leads to a better understanding of rivers as integrated systems and allows the formulation of hypotheses at the ecosystem level.

Two significant additions to the river continuum concept merit mention: the serial discontinuity concept[46] and the drainage network perspective.[24,26,47]

As Figure 9 indicates, in a pristine stream the continuum river concept describes how a given parameter of the ecosystem varies as water flows downstream. The serial discontinuity concept attempts to predict the effect of a dam or a series of dams on a parameter. The utility of this model is to allow the quantification of disturbances (such as dams or a series of dams). It can be equally applied to other situations where the stream/river continuum has been artificially disturbed.

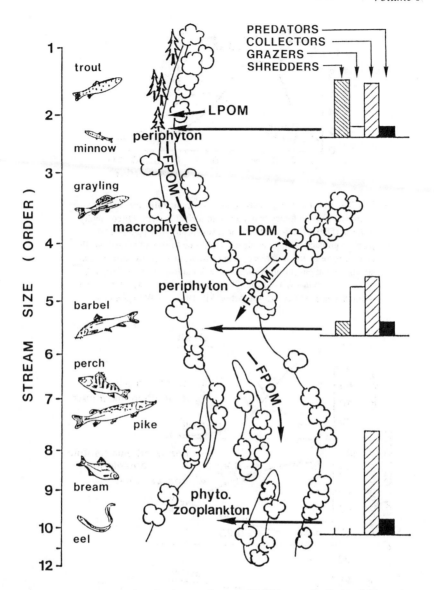

FIGURE 8. Representation of a river continuum, with P/R = production/respiration ratio. (Modified from Reference 2.)

Based on the ideas proposed by Vannote et al.[2] and Minshall et al.[40] Naiman and colleagues[26,47] considered the totality of the networks in the drainage area. They then established quantitative predictions concerning the origin and fate of organic matter and nutrients in the aquatic landscape. This study, conducted on the Moisie River, Quebec, between 1979 and 1985, demonstrated (Figure 10) that as a stream size increased from the first to the ninth order, there was a decrease in total carbon inputs (i.e., precipitation, throughfall, primary production, and allochthonous materials). This was followed by a gradual increase due to greater primary production in streams greater than order 6. The standing stock of carbon decreased exponentially downstream, and total carbon outputs (i.e., respiration, leaching, methane evasion, and insect emergence) increased slightly downstream. When placed in a watershed perspective, the data show that for the Moisie River watershed, total carbon inputs were evenly distributed throughout the drainage network, with most carbon stored in rivers with orders 7 to 9.

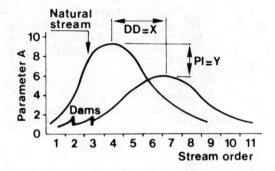

FIGURE 9.   The serial discontinuity concept: influence of an impoundment on ecological parameters in a river system. Discontinuity distance (DD) is the downstream or upstream shift of a parameter (e.g., primary production) a given distance (X) due to stream regulation. PI is a measure of the difference in the parameter intensity attributed to stream regulation. (From Ward, J. V. and Stanford, J. A., in *Dynamics of Lotic Ecosystems,* Fontaine, T. D. and Bartell, S. M., Eds., Ann Arbor Science Publishers, Ann Arbor, MI, 1983, 29. With permission.)

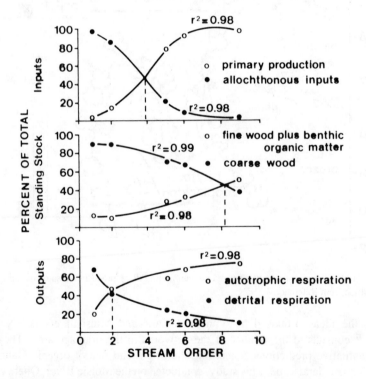

FIGURE 10.   Percentage of total organic carbon inputs, outputs, and standing stock as a function of stream order. (From Naiman, R. J., Melillo, J. M., Lock, M. A., Ford, T. E., and Reice, S. R., *Ecology,* 68(5), 1147, 1987. With permission.)

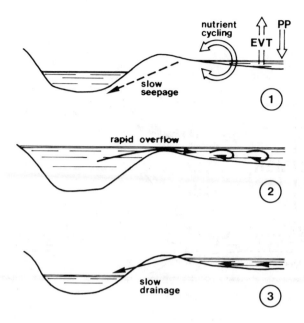

FIGURE 11. Lateral interactions in floodplain. (1) Normal channel-contained flow; (2) overbank flow during flood; (3) post-flood drainage and impoundment. (From Brinson, M. M., Bradshaw, H. D., and Holmes, R. N., in *Dynamics of Lotic Ecosystems*, Fontaine, T. D. and Bartell, S. M., Eds., Ann Arbor Science Publishers, Ann Arbor, MI, 1983, 199. With permission.)

V. DYNAMICS OF FLUVIAL LANDSCAPES

In the case of large rivers, the interactions between the channel and the floodplain determine the dynamics of the fluvial landscape.[48,49] When flooding occurs, the floodplains receive a deposit of sediments and nutrients. These floods also load the water and aid in the recharging of aquifers. The various fish species utilize the floodplain for reproduction. With the exception of those rivers diked or without floodplain, the resulting interactions during the hydrologic phases are similar to those described for the lower Mississippi valley (Figure 11).

At the fluvial landscape scale, Welcomme[4,11] clearly showed the importance of the catchment and the floodplain areas in the prediction of fish yields. The study of about 20 African rivers exploited by a fishery yielded a good correlation between the catch (C) of fish (in tons) and the catchment area (A) of the river system (in km²) or the length (L) of the main channel (in km).

$$C = 0.03 \ A^{0.97} \qquad (r = 0.91)$$
$$C = 0.032 \ L^{1.98} \qquad (r = 0.90)$$

When rivers with large floodplains (>2% of catchment) were included, the relation became

$$C = 0.44 \ A^{0.90}$$

Such simple formulas have also been used outside Africa to predict fish catch in rivers with extensive floodplains and even in those with relatively smaller floodplains.[4,11]

The significance of the floodplain area in the determination of the production of fluvial

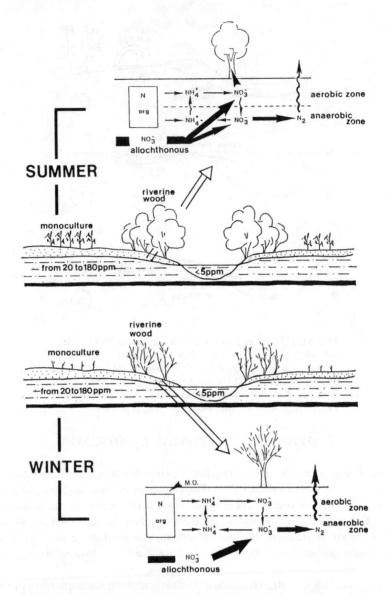

FIGURE 12. Model of the cycling of nitrogen in riverine forest on the alluvial plane
of the Garonne River. (From Pinay, G. and Décamps, H., *Regulated Rivers*, 2, 507,
1988. With permission.)

systems is equally apparent upon a consideration of interyear fluctuations of the fish stock.
These fluctuations are directly linked to the variability of the hydrologic regime. The analysis
of, for example, the central delta of the Niger River indicates a highly significant correlation
between the fisheries of 1 year and the flow regime of the previous one, or better yet, of
the previous 2 years.[4]

The interactions between the river channel and the floodplain area occur via the filtering
action of the riparian woods. Therefore, these woods have an essential role in the fluxes of
nutrients from the terrestrial to the aquatic environments as suggested in Figure 11. Pinay[50]
has shown the impact of a riparian forest in the Garonne valley on the coupling of the land-
water nitrogen flux (Figure 12). The riverine forests, a remainder of the original alluvial
forest, form a safety barrier between the ground- and surface waters. The disappearance of
nitrates from the groundwater during its passage through the shoreline forest led to a major

reduction in concentrations between the aquifer of the agricultural areas and the river water. This reduction is the result of a microbial denitrification of which the rates measured — up to 50 mg $N_2/m^2/d$ — are much higher than those obtained in other natural ecosystems.

On the banks of large rivers, the alluvial vegetation is variously linked to the river itself and to the water table. Because sediments are deposited on the floodplain and on gravel bars, the riparian vegetation becomes less under the influence of the stream and allows the development of pioneer communities of grasses and willows on recently deposited shoreline material that grade into the willow communities, only flooded during highwater, to the hardwoods at yet higher levels.[51] With the river capable of changing its course on the floodplain, the resulting erosion of the banks can return the plant succession to an earlier stage. This reversibility of successions can end after centuries when the new shore can no longer be submerged. This can be the result of an important sediment accumulation, but is, above all, due to the deepening of riverbed, accompanied by concomitant lowering of the water table. The latter is the basis for the irreversible, but normally slow successions observable on the alluvial terraces in the Garonne valley. However, the rate can be dramatically increased by man's activities, which produce a more rapid deepening of the riverbed by dams storing upstream sediments and by embankments increasing the erosive forces of the river. As well, dams can block the cycle of reversible successions by isolating the river channel from the floodplain. The river thereby becomes isolated from its terrestrial environment. The shoreline vegetation copes with these imposed barriers and forms communities maintained in a particular stage. Forest dynamics constitute, therefore, a biotic indicator of the various types of natural or artificial evolution along the fluvial corridor.

Another consequence of human activity in the floodplain is the increasing fragmentation of the original alluvial forest. Décamps et al.[52] have shown some possible consequences of this on the nesting communities of shore-dwelling birds. The conservation of a continuous ribbon of riparian woodlands is an important requirement for the maintenance of a community of nesting birds along the alluvial corridor. The presence of these ribbons of continuity is one characteristic of large rivers in pristine conditions,[53] and one that permits longitudinal migrations of species along the length of the water courses.

While it is evident that there has been much progress made in the last 20 or so years in understanding the functioning of running-water systems, much remains to be done, in particular with reference to large rivers. To the great physical complexity and diversity of large rivers can be added complex temporal and spatial changes resulting from human impact (and the much greater physical difficulty in sampling such rivers in comparison to the small streams that have received the most attention). Clearly, more research is needed to link large rivers to their terrestrial environment at the level of the alluvial valley and to determine the extent to which research findings in small streams can be applied to much larger systems.

ACKNOWLEDGMENTS

We gratefully acknowledge the assistance of Dr. J. Kalff in reviewing the manuscript and his family in the translation of the paper, as well as Dr. J. Sedell for his helpful comments.

REFERENCES

1. **Hynes, H. B. N.,** The stream and its valley, *Verh. Int. Ver. Limnol.,* 19, 1, 1975.
2. **Vannote, R. L., Minshall, G. W., Cummins, K. W., Sedell, J. R., and Cushing, C. E.,** The river continuum concept, *Can. J. Fish. Aquat. Sci.,* 37, 130, 1980.

3. **Elwood, J. W., Newbold, J. D., O'Neill, R. V., and Van Winkle, W.,** Resource spiraling: an operational paradigm for analyzing lotic ecosystems, in *The Dynamics of Lotic Ecosystems,* Fontaine, T. D. and Bartell, S. M., Ed., Ann Arbor Science Publishers, Ann Arbor, MI, 1983, 3.

4. **Welcomme, R. L.,** *River Fisheries,* Tech. paper 262, Food and Agriculture Organization, Rome, 1985, 1.

5. **Ward, J. V. and Stanford, J. A.,** The ecology of regulated streams: past accomplishments and directions for future research, in *Advances in Regulated Streams Ecology,* Craig, J. F. and Kemper, J. B., Eds., Plenum Press, New York, 1987, 391.

6. **Ward, J. V. and Stanford, J. A.,** Riverine Ecosystems: the influence of man on catchment dynamics and fish ecology, presented at the Int. Large Alluvial River Symp., Delawana, Ontario, September 1986, 1.

7. **Hynes, H. B. N.,** *The Ecology of Running Waters,* Liverpool University Press, Liverpool, 1970, 1.

8. **Whitton, B. A., Ed.,** *River Ecology,* Blackwell Scientific, London, 1975, 1.

9. **Whitton, B. A., Ed.,** *Ecology of European Rivers,* Blackwell Scientific, London, 1984, 1.

10. **Ward, J. V. and Stanford, J. A., Eds.,** *The Ecology of Regulated Streams,* Plenum Press, New York, 1979, 1.

11. **Welcomme, R. L.,** *Fisheries Ecology of Floodplain Rivers,* Longman, London, 1979, 1.

12. **Lock, M. A. and Williams, D. D., Eds.,** *Perspectives in Running Water Ecology,* Plenum Press, New York, 1981, 1.

13. **Fontaine, T. D. and Bartell, S. M., Eds.,** *The Dynamics of Lotic Ecosystems,* Ann Arbor Science Publishers, Ann Arbor, MI, 1983, 1.

14. **Barnes, J. R. and Minshall, G. W., Eds.,** *Stream Ecology: Application and Testing of General Ecological Theory,* Plenum Press, New York, 1983, 1.

15. **Lillehammer, A. and Saltveit, S. J., Eds.,** *Regulated Rivers,* Universitets Forlaget, Oslo, 1984, 1.

16. **Petts, G. E.,** *Impounded Rivers. Perspectives for Ecological Management,* Wiley, Chichester, 1984, 1.

17. **Sioli, H., Ed.,** *The Amazon. Limnology and Landscape Ecology of a Mighty Tropical River and Its Basin,* Junk, Dordrecht, Netherlands, 1984, 1.

18. **Gore, J. A., Ed.,** *The Restoration of Rivers and Streams. Theories and Experience,* Butterworths, London, 1985, 1.

19. **Penka, M., Vyskot, M., and Klimo, E., Eds.,** *Floodplain Forest Ecosystem,* Elsevier, Amsterdam, 1985, 1.

20. **Davies, B. R. and Walker, K. F.,** *The Ecology of River Systems,* Junk, Dordrecht, Netherlands, 1986, 1.

21. **Kellerhals, R. and Church, M.,** The morphology of large rivers: characterization and management, presented at the Int. Large Alluvial River Symp. Delawana, Ontario, September 1986, 1.

22. **Strahler, A. N.,** Quantitative analysis of watershed geomorphology, *Trans. Am. Geophys. Union,* 38, 913, 1957.

23. **Allen, T. F. H. and Starr, T. B.,** *Hierarchy: Perspective for Ecological Complexity,* University of Chicago Press, Chicago, 1982, 1.

24. **Frissell, C. A., Liss, W. J., Warren, C. E., and Hurley, M. D.,** A hierarchical framework for stream habitat classification: viewing streams in a watershed context, *Environ. Manage.,* 10, 199, 1986.

25. **Dance, K. W.,** Seasonal aspects of transport of organic and inorganic matter in streams, in *Perspectives in Running Water Ecology,* Lock, M. A. and Williams, D. D., Eds., Plenum Press, New York, 1981, 69.

26. **Naiman, R. J.,** The influence of stream size on the food quality of seston, *Can. J. Zool.,* 61, 1995, 1983.

27. **Lush, D. L. and Hynes, H. B. N.,** Particulate and dissolved organic matter in a small partly forested Ontario stream, *Hydrobiologia,* 60, 177, 1978.

28. **Liaw, W. K. and MacCrimmon, H. R.,** Assessment of particulate organic matter in river water, *Int. Rev. Gesamten Hydrobiol.,* 62, 445, 1977.

29. **Naiman, R. J. and Sedell, J. R.,** Characterization of particulate organic matter transported by some Cascade Mountain streams, *J. Fish. Res. Board Can.,* 36, 17, 1979.

30. **Hynes, H. B. N.,** Groundwater and stream ecology, *Hydrobiologia,* 100, 93, 1983.

31. **Lock, M. A.,** River epilithon — a light and organic energy transducer, in *Perspectives in Running Water Ecology,* Lock, M. A. and Williams, D. D., Eds., Plenum Press, New York, 1981, 3.

32. **Winterbourn, M. J.,** Recent advances in our understanding of stream ecosystems, in *Ecosystem Theory and Application,* Polunin, N., Ed., Wiley, Chichester, 1986, 240.

33. **Dahm, C. N.,** Pathways and mechanisms for removal of dissolved organic carbon from leaf leachate in streams, *Can. J. Fish. Aquat. Sci.,* 38, 68, 1981.

34. **Chauvet, E.,** Changes in the chemical composition of alder, poplar and willow leaves during decomposition in river, *Hydrobiologia,* 148, 35, 1987.

35. **Webster, J. R. and Patten, B. C.,** Effects of watershed perturbation on stream potassium and calcium dynamics, *Ecol. Mongr.,* 49, 51, 1979.

36. **Chauvet, E., Fustec, E., and Gas, G.,** Etude expérimentale de la dégradation des lignines de peuplier dans un sol alluvial et dans l'eau d'un fleuve, *C.R. Acad. Sci. Paris,* 3, 87, 1986.

37. **Newbold, J. D., Mulholland, P. J., Elwood, J. W., and O'Neill, R. V.,** Organic carbon spiralling in stream ecosystems, *Oikos,* 38, 266, 1982.
38. **Webster, J. R.,** Analysis of Potassium and Calcium Dynamics in Stream Ecosystems on Three Southern Appalachian Watersheds of Contrasting Vegetation, Ph.D. thesis, University of Georgia, Athens, 1975.
39. **Fisher, S. G. and Likens, G. E.,** Energy flow in Bear Brook New Hampshire: an integrative approach to stream ecosystem metabolism, *Ecol. Monogr.,* 43, 421, 1973.
40. **Minshall, G. W., Petersen, R. C., Cummins, K. W., Bott, T. L., Sedell, J. R., Cushing, C. E., and Vannote, R. L.,** Interbiome comparison of stream ecosystem dynamics, *Ecol. Monogr.,* 51, 1, 1983.
41. **Huet, M.,** Aperçu des relations entre la pente et les populations piscicoles des eaux courantes, *Schweiz. Z. Hydrol.,* 11, 333, 1949.
42. **Illies, J. and Botosaneanu, L.,** Problèmes et méthodes de la classification et de la zonation écologiques des eaux courantes, considérées surtout du point de vue faunistique, *Mitt. Int. Ver. Theor. Angew. Limnol.,* 12, 1, 1963.
43. **Mann, K. H., Britton, R. H., Kowalczewski, A., Lack, T. J., Mathews, C. P., and McDonalds, I.,** Productivity and energy flow at all trophic levels in the River Thames, England, in *Proc. IBP — UNESCO Symp. Productivity Problems in Freshwater,* Kajak, Z. and Hillbricht-Ilkowska, A., Eds., Polish Scientific, Warsaw, 1972, 579.
44. **Minshall, G. W., Cummins, K. W., Petersen, R. C., Cushing, C. E., Bruns, D. A., Sedell, J. R., and Vannote, R. L.,** Developments in stream ecosystem theory, *Can. J. Fish. Aquat. Sci.,* 42, 1045, 1985.
45. **Statzner, B. and Higler, B.,** Questions and comments on the river continuum concept, *Can. J. Fish. Aquat. Sci.,* 42, 1038, 1985.
46. **Ward, J. V. and Stanford, J. A.,** The serial discontinuity concept of lotic ecosystems, in *Dynamics of Lotic Ecosystems,* Fontaine, T. D. and Bartell, S. M., Eds., Ann Arbor Science Publishers, Ann Arbor, MI, 1983, 29.
47. **Naiman, R. J., Melillo, J. M., Lock, M. A., Ford, T. E., and Reice, S. R.,** Longitudinal patterns of ecosystem processes and community structure in a subarctic river continuum, *Ecology,* 68(5), 1147, 1987.
48. **Décamps, H.,** Towards a landscape ecology of river valleys, in *Trends in Ecological Research for the 1980s,* Cooley, J. H. and Golley, R. B., Eds., Plenum Press, New York, 1984, 163.
49. **Amoros, C., Bravard, J. P., Pautou, G., and Roux, A. L.,** Methodological research applied to the ecological management of water resources of fluvial systems, *Regulated Rivers,* 1, 17, 1986.
50. **Pinay, G. and Décamps, H.,** The role of riparian woods in regulating nutrient fluxes beween the alluvial aguifer and surface water: a conceptual model, *Regulated Rivers,* 2, 507, 1988.
51. **Pautou, G. and Décamps, H.,** Ecological interactions between the alluvial forests and hydrology of the upper Rhône, *Arch. Hydrobiol.,* 104, 13, 1985.
52. **Décamps, H., Joachim, J., and Lauga, J.,** The importance for birds of the riparian woodlands within the alluvial corridor of the river Garonne, S. W. France, *Regulated Rivers,* 1, 301, 1987.
53. **Cummins, K. W., Minshall, G. W., Sedell, J. R., Cushing, C. E., and Petersen, R. C.,** Stream ecosystem theory, *Verh. Int. Ver. Limnol.,* 22, 1818, 1984.
54. **Bird, G. A. and Kaushik, N. K.,** Coarse particulate organic matter in streams, in *Perspectives in Running Water Ecology,* Lock, M. A. and Williams, D. D., Eds., Plenum Press, New York, 1981, 41.
55. **Brinson, M. M., Bradshaw, H. D., and Holmes, R. N.,** Significance of floodplain sediments in nutrient exchange between a stream and its floodplain, in *Dynamics of Lotic Ecosystems,* Fontaine, T. D. and Bartell, S. M., Eds., Ann Arbor Science Publishers, Ann Arbor, MI, 1983, 199.
56. **Leopold, L. B., Wolman, M. G., and Miller, J. P.,** *Fluvial Processes in Geomorphology,* W. H. Freeman, San Francisco, 1964, 1.
57. **Mollard, J. D.,** Air photo interpretation of fluvial features, in *Proc. 7th Can. Hydrology Symp.,* Department of the Environment, Edmonton, 1973, 341.

Chapter 2

SPECIAL FEATURES OF LAKE ECOSYSTEMS

J. Capblancq

TABLE OF CONTENTS

I. GENERAL CHARACTERISTICS OF LAKE SYSTEM ECOLOGY

Lakes have long been considered as prototypes of ecosystems, i.e., more or less closed units, well defined physically, and characterized by a certain functional autonomy.[1,2] As for any ecosystem, overall function is based on the dependence and interdependence of the various communities of living organisms and their interactions with the abiotic environment. The overall functional metabolism can be divided into two main parts:

1. Photosynthetic production of organic material by autotrophic organisms
2. Consumption and degradation (respiration) of this organic material by heterotrophs (animals, microbes) with the concomitant recycling (mineralization) of nutrients

These processes of production and biodegradation are regulated by a set of morphodynamic and climatic variables (depth, residence time of water, sun and wind energy) which impose constraints on the chemical and physical properties of lake water (temperature, transparency, gas content, dissolved salts, etc.). These properties are themselves undergoing continuous change in response to seasonal variations of climatic variables and to feedback effects of the metabolic activity of biota.

This general outline of the operation of a lake ecosystem (Figure 1) differs from that of a terrestrial ecosystem in three main characteristics.

First, in contrast to the terrestrial plant production of organic material where large masses of supporting tissue are formed (cellulose, lignin), primary production in lakes is mainly carried out by microphytes (algal phytoplankton). Since this phytoplankton is consumed and degraded as fast as it is produced, the stock of organisms is rapidly renewed, and nutrients are recycled relatively quickly. This can lead to permanent changes in living communities as well as in their physicochemical environment within a relatively short time span.

Second, in closed terrestrial plant systems, most of the incident light energy is absorbed by the photosynthetic tissues. In an aquatic ecosystem, the medium itself absorbs much of the incoming light. There is less light available, therefore, for photosynthesis, and aquatic primary production is often light limited.

Third, water in equilibrium with the atmosphere contains around 25 times less oxygen than an equivalent volume of air, and the molecular diffusion of O_2 is 10^4 times lower in water than in air. The amounts of dissolved oxygen and the transfer capabilities of O_2 from the atmosphere therefore limit the capacity to oxidize and mineralize the organic matter produced in the system or imported from the watershed.

However, lakes are open systems continuously supplied by inorganic and organic matter, either dissolved or suspended in the water flowing in from the water basin. There is also continuous exchange with the atmosphere (gas and heat) and with sediments (Figure 1). A lake is, thus, intimately coupled with its watershed; it can be considered as a part of the catchment ecosystem where it operates as a retention zone for components transported from the land due to the greatly reduced transportation power of the water. The soil vegetation complex, representing the major compartment of transfer between these two systems, has undergone a radical change over the last few decades, due mainly to modern agricultural techniques, urbanization, and industrialization. These processes have led to the marked changes observed in natural productivity of lake ecosystems.

The retention properties which differ from one lake to another, depending on hydrological and morphometric properties of the lake basin, do not influence all materials uniformly. In a supposedly homogeneous lake, the water retention time (Tw) is expressed as a ratio of the inflow (Q = m^3/t) and the lake volume (V = m^3)

$$Tw = V/Q$$

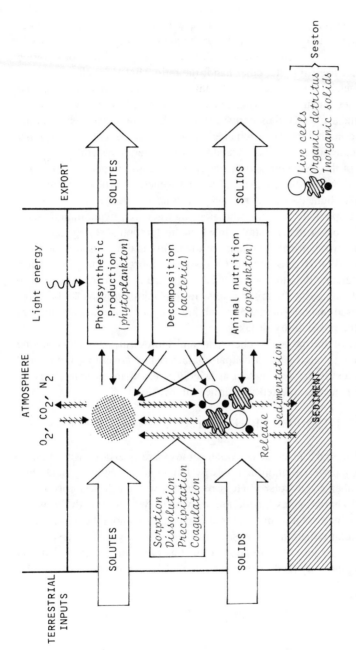

FIGURE 1. Schematic representation of a lake ecosystem showing the main physicochemical (left) and biological (right) processes operating in the production and transformation of dissolved and suspended matter in lake water.

That of an element E can be defined with respect to the rate of input as

$$TE = |E| \cdot V/|E_{in}| \cdot Q$$

where $|E|$ and $|E_{in}|$ are the concentrations in the lake water and the input water, respectively. The value of TE falls with increasing reactivity, i.e., the capacity to precipitate, to coagulate, to be adsorbed on inert particles, to be incorporated into the biomass, or to sediment. Thus, the retention times of relatively unreactive solutes such as sodium, potassium, or chloride ions are close to that of water itself. On the other hand, the ratio TE/Tw for elements such as phosphorus and heavy metals, which are strongly adsorbed on suspended material and then sedimented, is much less than unity. Thus, lake sediment acts as a permanent or temporary storage compartment for materials derived from both the drainage basin and the metabolism of the aquatic communities. The exchanges which take place with the overlying water, therefore, play an important role in the regulation of the overall system.

The different reactivities of the inflowing substances cause them to segregate in the lake. This segregation is enhanced by the development of a heterogeneous vertical structure. In relatively deep lakes, gravity, light transmission, and heat transport lead to the formation of layers (trophogenic and tropholytic zones, epilimnion, thermocline, hypolimnion, sediments). Exchange of materials between these strata, in which the duration of storage increases from top to bottom, is rarely symmetrical. The main nutrient flux is downward in the form of particulate organic matter (POM), and upward for dissolved inorganic materials. Since the rate of sedimentation is greater than the rate of diffusion, there is a transfer of chemical energy and nutrients from the upper to the lower layers, from plankton toward the benthos, and from water toward the sediments. This transfer inhibits recycling of nutrients and has a controlling influence on the productivity of lakes.

The vertical stratification of the physical, chemical, and biological parameters arising from the stagnation of water is one of the principal characteristics governing the overall operation of a lacustrine ecosystem. This aspect is widely stressed in most treatises on limnology,[3-10] to which the reader is referred for further information.

II. VERTICAL DISTRIBUTION OF LIGHT ENERGY AND TURBULENCE

Heat and radiation exchange take place at the water atmosphere interface, with gain in energy from solar and atmospheric radiation and loss by radiation, evaporation, and conduction. It should be remembered that solar input is penetrant, whereas losses only affect the superficial layers. Wind also provides kinetic energy which by friction with the water surface leads to movements which are transmitted and dissipated throughout the water mass.

The energy provided by solar radiation is almost entirely absorbed by surface layers and transformed into heat. The coefficient of absorption (ϵ) of this radiation by water depends on wavelength (λ) so that the intensity (I) decreases with depth (Z) according to a hyperbolic function:

$$I_\lambda(Z) = I_\lambda(o) \cdot \exp(-\epsilon_\lambda \cdot Z)$$

Absorption of infrared radiation ($\lambda > 750$ nm) is high, over 90%/m for all types of water, leading to significant heating of the superficial layers. The absorption of visible light is much more variable and is markedly affected by the presence of suspended particles and colored substances which reflect, diffuse, and absorb light. The attenuation of light energy by these materials is more marked in the UV and in the blue and green regions of the spectrum. This attenuation is thus selective and produces intensity and spectral gradients which differ significantly from lake to lake, and even in the same lake at different times of the year. The depth of the euphotic zone (light intensity >1% of the surface intensity) ranges from some tens of centimeters (orange and red light are then the most penetrant) to several tens of meters (where green and blue light are most penetrant).

The thermal conductivity of water is relatively low ($1.43 \cdot 10^{-3}$ cal/cm/s/°C). In stationary water, penetration of light and propagation of heat by conduction leads to the formation of a vertical temperature gradient in a hyperbolic distribution. Since the density of water falls with increasing temperature above 4°C, a parallel density gradient is produced which enhances resistance to mixing.

The energy required for mixing is essentially derived from the wind by frictional effects on the water surface producing wind drift flows, convection cells (Langmuir vortices), and waves (which propagate downward in an orbital motion).[11] In addition, temporal variations in density, arising from diurnal heating and cooling, and the spatial variations arising from irregular surface heating (regions of shade or light) produce density currents. Movements caused by the wind and the density currents become disordered or turbulent, i.e., trajectories and velocities become distributed at random. The significance of turbulence stems from the fact that these random fluctuations are accompanied by transport in directions other than that of the mean flow with the concomitant dispersion of kinetic energy, temperature, solutes, and suspended particles.

In homogeneously dense water, the surface turbulence can be propagated to considerable depths. On the other hand, this propagation is considerably reduced where there are density gradients. In this case, gravity provides a resisting force to the relatively low energies which become dissipated in the turbulent eddies. The presence of a high density gradient (picnocline) acts as a barrier to the turbulence. The gradient behaves like a layer analogous to the lake bottom, but this layer is fluid, and, when turbulence increases, it can become sufficiently unstable to carry underlying layers along with it.

The convection currents, which occur when superficial water is cooled, increase in density and sink (night, cloudy conditions, or late summer), and the frictional wind stress which generates water motion can explain the formation of thermal strata in summer and their disappearance in autumn in relatively deep lakes in temperate zones. The typical summer stratification (see Figure 3) divides the water into two layers (epilimnion, hypolimnion) separated by a zone with a marked temperature gradient (metalimnion or thermocline) where vertical diffusion coefficients are considerably reduced (10^4 to 10^6 times less than in the epilimnion and 10 to 30 times less than in the hypolimnion). This reduction in conductivity attenuates heat flow and amplifies the thermal gradient. The thermocline stabilizes in summer between 5 and 20 m below the surface, depending on lake morphometry, its latitude, and its exposure to wind. The ecological impact is considerably enhanced if this zone coincides with the depth of the euphotic zone.

Thermal stratification can occur even in relatively shallow lakes. However, it is much less stable since mixing of water masses of different density requires less work as the distance between the centers of gravity falls. Under the influence of the wind and diurnal temperature variations, there is frequent mixing. In deeper lakes complete mixing occurs only once or twice a year (monomictic or dimictic lakes) or at irregular intervals between several successive years of stratification (oligomictic lakes).

III. BIOLOGICAL REGULATION OF THE CHEMICAL COMPOSITION OF LAKE WATER

The circulation of elements carried from the water basin is controlled not only by physical processes depending on turbulence (eddy diffusion, sedimentation, resuspension, or advection), but also by biological activity. Exchanges of matter between water, organisms, and

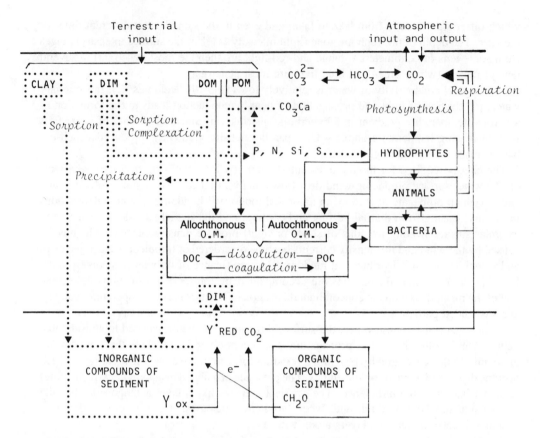

FIGURE 2. Simplified carbon cycle in a lake ecosystem and its interactions with the cycle of inorganic elements. Electrons are transferred from sedimented detrital organic carbon to electron acceptors in the absence of O_2 ($Y_{ox} \rightarrow Y_{red}$); these reduced compounds are generally more soluble and may diffuse out of the sediment. DIM, dissolved inorganic matter; DOC and POC: dissolved and particulate organic carbon, respectively.

sediments are intimately linked to the carbon cycle which impinges directly and indirectly on the cycles of numerous other elements (Figure 2).

Carbon recycling is carried out by photosynthesis and respiration which enables carbon, nitrogen, and phosphorus to be recycled in more or less fixed proportions corresponding to the mean composition of organic matter. Assimilation of CO_2 by photosynthesis leads to a concomitant uptake of inorganic nitrogen (NO_3 or NH_4), phosphates, and liberation of oxygen in a mean molar ratio for O/C/N/P around 276/106/16/1 in accordance with the equation:[12,13]

$$106CO_2 + 122H_2O + 16HNO_3 + HPO_4^{-2} + \text{Energy} \underset{R}{\overset{P}{\leftrightarrow}}$$

$$|(CH_2O)_{106}(NH_3)_{16}(H_3PO_4)| + 138O_2$$

Other elements (S, K, Fe, etc.) found in tissues should also be included as well as those elements constituting cell walls (SiO_2 of diatoms).

The variations in levels of CO_2 and O_2 are the most obvious chemical manifestations of this reaction, and measurement of these variations is used to evaluate the rates of production and biodegradation. The consumption and liberation of CO_2 leads to displacement of the carbonate/bicarbonate/carbonic acid equilibrium. This leads to pH changes in relation to the buffering capacity (alkalinity) of the water. The variations of pH and dissolved oxygen levels, in turn, regulate most of the biochemical and chemical reactions which affect water composition.

The organic matter produced in the euphotic zone is sedimented and degraded in the underlying layers. Thus, each atom of a nutrient element utilized by an organism is returned to solution in a deeper layer than that in which it was taken up. In this deeper layer, the reducing power of organic matter is transferred to the liquid milieu.

By analogy with pH, which measures the capacity of a solution to accept or give up protons, the capacity of a medium to accept (oxidizing) or give up (reducing) electrons can be expressed by the relative activity of electrons ($pE = -\log_{10} e^-$) or by the potential with a respect to a reference electrode (EH = redox potential with respect to the hydrogen electrode $= 0.059\ pE$).[12]

Water in equilibrium with atmospheric O_2 at 25°C and pH 7 has a well-defined potential ($pE = 13.6$ or $EH = 800$ mV) for which all elements are in the state of maximum oxidation: C as CO_2, HCO_3^- or CO_3^{2-}; N as NO_3^-; S as SO_4^{2-}; Fe as $FeOOH$ or Fe_2O_3; Mn as MnO_2. Oxidation of organic matter leads to reduction of these compounds in the order of decreasing potentials (Figure 3). O_2 is reduced first, but when its concentration falls below a certain threshold (around 4 ppm), nitrates are used as oxidant ($NO_3^- \rightarrow NO_2^- \rightarrow N_2$). When O_2 and NO_3^- are depleted, sulfates and CO_2 can be used as electron acceptors ($SO_4^{2-} \rightarrow SH^-$; $CO_2 \rightarrow CH_4$). These reactions are all catalyzed by different microorganisms (aerobic heterotrophic, denitrifying, sulfate-reducing, or methane-producing bacteria). The seasonal succession and depth distribution of these bacteria are related to the redox potentials of the reactions they catalyze. The changes in EH and pH regulate most of the chemical reactions determining the dissolution or precipitation of many metallic elements, especially iron and manganese.

When these reactions take place in conditions of low turbulence, the processes of biosynthesis and biodegradation lead to vertical gradients of oxygen, pH, and EH. These gradients, which determine the concentrations of many chemical elements as well as the form in which they exist, depend on the rates of photosynthetic production and sedimentation. These gradients are rarely uniform, but are usually characterized by zones of discontinuity (picnoclines).

The water/sediment interface represents a zone of discontinuity between the pore water in the sediment and the free water. The coefficients of diffusion between the pore water are close to the molecular diffusion coefficients and are 10^4 to 10^6 times lower than in the water above. Oxidation of the organic fraction of the sediments in a weakly dispersive medium leads to the formation of a marked redox potential gradient in the first few centimeters under the interface. Therefore, there is an increased concentration of reduced elements (Fe^{2+}, Mn^{2+}, NH_4, HS^-, organic acids) dissolved in the pore water (Figure 3). Phosphates and many metal ions (heavy metals) adsorbed or coprecipitated with hydroxides are liberated during the reduction of $Fe(OH)_3$ to Fe^{2+}. Liberation may also occur if there is a fall in pH with production of organic acids leading to dissolution of carbonates and ferric hydroxide. The pore water also becomes enriched in silica, the dissolution of which depends on the CO_2 concentration.

The reduced elements which are dissolved in the pore water diffuse toward the superficial layer of the sediments, where they are oxidized and precipitated either chemically or biologically (e.g., ferrobacteria) when the redox potential exceeds a certain threshold. A fall in potential at the water/sediment interface concomitant with exhaustion of dissolved oxygen causes these elements to diffuse from the pore water to the hypolimnic water (Figure 3). The rate of this process is mainly a function of the difference in concentration between these two compartments. It also depends on the turbulence at the interface and the mixing of superficial sediments by benthic invertebrates (bioturbation).

The thermocline represents a second zone of discontinuity in stratified lakes. The increases in density and viscosity of the water inhibit sedimentation and lead to the accumulation of particulate organic matter, both living and dead. Depending on the transparency of the water, there can be oxygen consumption leading to a metalimnic deficit or production of oxygen which then accumulates in the metalimnion.

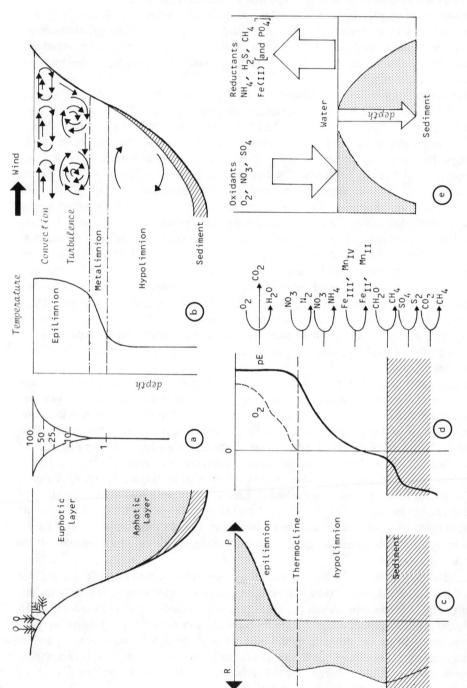

FIGURE 3.   Photosynthetic production (P) and heterotrophic destruction (R) of organic matter may become vertically segregated in a lake (c) due to vertical distribution of light (a) and temperature (b). Depletion of oxygen in the hypolimnetic layer in a very eutrophic lake is the first step in a sequence of microbially mediated redox processes which follow the fall in pE with increasing depth (d). Distribution of concentration in sediment pore water (shaded area) and direction of fluxes expected at the sediment water interface is shown (e).

The exchanges with the atmosphere and the constant redistribution of elements by eddy diffusion oppose the formation of gradients. In isothermic lakes and in the epilimnion of stratified lakes, these transport processes enable excess oxygen produced by photosynthesis to be eliminated or to compensate the overconsumption, maintaining levels close to saturation. In the hypolimnion, these compensatory processes are relatively limited, and oxidation of sedimented organic matter progressively depletes the oxygen. It should be stressed that the anoxia of the deep layers does not necessarily signify a disequilibrium. The evolution of dissolved O_2 in the hypolimnion also depends on the morphometric characteristics, the water transparency, and the stratification of the lake. The combined effect of these factors can be expressed by the ratios VH/VE (hypolimnion/epilimnion volumes) and Z_{eu}/Z_{mix} (euphotic/thermocline depths). For an identical input of organic matter the deficit in the hypolimnion will be higher in lakes with low VH/VE and Z_{eu}/Z_{mix}

IV. REGULATION OF PRIMARY PRODUCTION AND BIODEGRADATION

The major part of the carbon in a lake comes from a combination of the input from the water basin and the autochthonous primary production. The input is essentially made up of residues of terrestrial vegetation carried in from the water basin. These more or less degraded materials are relatively refractory, and they participate both directly in biological processes (bacterial degradation) and indirectly, by reducing light penetration as well as by their ability to form complexes with many ions.

The autochthonous primary production is carried out by phytoplankton and by benthic flora (macrophytes + periphyton). The latter tend to be limited to the littoral zone, while the major part of the production is carried out by phytoplankton distributed over the whole euphotic zone. The importance of phytoplankton in the metabolic activity of lacustrine ecosystems probably accounts for the large number of studies devoted to its ecology, and reviews by Hutchinson,[7] Harris,[14] Pourriot et al.,[15] and Reynolds[16] can be cited in this respect.

The biomass of phytoplankton produced in a lake depends on the amounts of nutrients available. This is the basis of the distinction between the eutrophic (nutrient-rich) and the oligotrophic lakes which was introduced early in limnology.[17] Photosynthetic production ranges from 4 g carbon per square meter per year in the most oligotrophic lakes to more than 800 g carbon per square meter in the most eutrophic lakes.[9,18]

The stoichiometry for the production equation indicates that the elements making up the plant biomass are used in a mean molar ratio C/N/P: 106/16/1. The plant biomass produced is thus limited by the least abundant element with respect to this ratio (limiting nutrient). Analysis of the N/P ratio in lake water shows: (1) that the majority of lakes in temperate zones are characterized by a N/P ratio >10 by weight (>22 in moles) and (2) that this ratio is close to 10 in eutrophic lakes.[19,20] It has also been shown that there are processes which can compensate for deficits in carbon, nitrogen, and silica. It has been found that:

1.   The rate of diffusion of atmospheric CO_2 is sufficient to sustain the carbon requirements of increased phytoplankton biomass in phosphate-enriched experimental lakes.[21]
2.   A low N/P ratio favors the development of nitrogen-fixing, filamentous blue-green algae. Biological N_2 fixation can then compensate for the limited nitrogen content of the water.
3.   The exhaustion of silica due to a large growth of diatoms leads to their replacement by nonsilica-containing algae (green algae, dinoflagellates, blue-green algae).

Overall, these observations indicate that in most cases phosphorus is the element controlling primary production in temperate lakes. Numerous comparative studies have shown convincingly the existence of relationships between (1) the annual loading of P (imported mass/m^2 water per year) and its concentration in the lake and (2) the phosphorus concentration of the lake and its plant biomass. These observations have led to phosphorus budget models[19,20,22] relating the loading (Lp) and the concentration (Pλ) in an equation that also includes the mean depth of the lake, the turnover rate of lake water, and the net rate of sedimentation of phosphorus. Since the biomass of phytoplankton is correlated with Pλ and the production increases with biomass, this equation has predictive value.

The increase in production of phytoplankton resulting from an increased input of limiting nutrient (eutrophication) has marked repercussions on the optical properties of the water. Phytoplankton has a filtering effect which alters the spectral characteristics of the light and increases its attenuation coefficient. This self-shading effect reduces the depth of the euphotic zone (Z_{eu}) and determines, in combination with the irradiance on the surface layer (Io) and the depth of the mixing zone (Z_{mix}), the light climate of the algae. Any reduction in Io or the ratio Z_{eu}/Z_{mix} (e.g., colored water or water with high content of suspended particles, deep isothermal lakes) increases the relative dark exposure of the algae and reduces net primary production. Conversely, any increase in Io or Z_{eu}/Z_{mix} (transparent, shallow or stratified lakes) increases net production. This means that the production of phytoplankton (g carbon per square meter per unit time) is regulated more by the available light in eutrophic and turbid water than in oligotrophic transparent lakes, where the limiting factor is the availability of nutrients.

The spring outburst (March to April) and the autumnal decline (October to November) which characterize the growth cycle of phytoplankton in temperate lakes correspond to a seasonal change in daily solar radiation. These two periods usually coincide with the appearance and disappearance of thermal stratification on which the two principal components of phytoplankton ecology depend: residence time of algae in the euphotic zone and the availability of nutrients. During the periods of circulation, the amounts of nutrients available and the renewal of these nutrients by exchange with sediments are increased. On the other hand, net photosynthetic production is reduced (low Z_{eu}/Z_{mix}). By limiting the circulation of algae between the surface and the thermocline, stratification increases the length of time in the euphotic zone, but it reduces the nutrient resources to the amounts available in the epilimnion alone. These seasonal variations in Io, Z_{eu}, and Z_{mix} can explain the phases of growth and decline of phytoplankton which are observed to follow an identical cycle from year to year in any given lake. The algal biomass remains relatively low in oligotrophic lakes where algal nanoplankton (<50 μm) predominate throughout the year. This biomass is much higher in eutrophic lakes which are characterized by blooms of algal netplankton (>50 μm) in summer.

Overall, variations in numbers, biomass, and specific composition of phytoplankton, which characterize the annual cycles or the shift from oligotrophy to eutrophy, can be formulated as an equilibrium between growth and loss processes as follows:

$$dB/dt = |\mu.—Ks— Kg—Kw—Kd|B$$

where B is the biomass (number of cells, biovolume, mass of carbon, etc.) and μ, Ks, Kg, Kw, and Kd are the rates of growth and loss by sedimentation, grazing, hydraulic washout, and death, respectively. These rates are influenced at any given time by independent variables (temperature, water renewal, duration and intensity of light at the surface, stratification, nutrient inputs) and variables dependent on the algal biomass (concentration of available nutrients, light attenuation, numbers of herbivorous zooplankton). They also depend on the morphological and physiological characteristics of the different algal species making up the phytoplankton.

In this way three things happen. First, the intrinsic rate of growth of algal populations (μ max) along with most physiological functions (photosynthesis, respiration, nutrient absorption) falls with increasing cell size.[16,23] This inverse relationship can be explained by the fact that cell volume increases faster than membrane surface area. It means that species of small size (nanoplankton) have higher specific rates of production (production/biomass) and multiplication than the larger species (netplankton). Second, the rate of sedimentation (Ks) increases with algal volume and specific gravity (Stokes law). Density is a function of cell structure (lipid and carbohydrate content; formation of siliceous frustules of varying size in diatoms; formation of mucilage or gaseous vacuoles) which differs from one species to another and with their physiological state. Third, the capacity of consumption of algae by herbivorous zooplankton and the size of the ingested particles fall with the size of the animal. The filtering organisms of the zooplankton and herbivorous micropredators mostly feed on bacteria and small algae (<10 to 15 μm for Rotifers, <50 to 60 μm for Cladocera and calanoid copepods).[15]

Growth, sedimentation, and grazing are therefore selective processes. The nanoplankton is highly productive; sediment, the slowest. On the other hand, it is highly exploited by the herbivorous zooplankton and, as a result, rarely produces significant biomass. The large algae (netplankton), less productive but less taken up by herbivores, are responsible for the formation of blooms when sedimentation losses are minimal (blooms of diatoms in well-mixed water, mobile forms such as dinoflagellates, or floating forms such as blue-green algae in eutrophic lakes during stratification).[24]

The autochthonous primary production along with the allochthonous input forms a pool of living POM, detrital POM, and dissolved organic matter (DOM). The relative proportions are approximately 1/10/100, although this ratio differs from lake to lake and is subject to seasonal variations. Exchange between compartments involves production, transformation, and degradation (secretion of DOM by organisms, lysis of dead cells and washing out of dissolved compounds, selective ingestion of particles and excretion by zooplankton, enzymatic hydrolysis of detrital POM and DOM, and mineralization by bacteria) via mechanisms which are still poorly understood. Heterotrophic organisms (bacteria, fungi, planktonic and benthic animals) participating in these processes simultaneously act as decomposers (enzymatic degradation and mineralization of organic matter) and producers (formation of biomass), although the ways in which these two functions are carried out are somewhat different.

Herbivorous animals and zooplankton ingest daily a mass of living and dead particles ranging from 0.4 times to twice their own weight, and they filter a volume of water equivalent to 10^4 to 10^6 times their body volume.[15] Their excreta in the form of nitrogenous compounds (ammonium salts, urea), phosphates, DOM, and fine particles of organic matter which disperse in the milieu (Rotifers, Cladocera) or in the form of fecal pellets which sediment rapidly (copepods), provide substrates for the growth of microorganisms. Herbivorous zooplankton filter a large fraction of the water volume per day, sometimes exceeding 100%,[25] from which they selectively extract living and detrital POM in relation to their size. It has been estimated that the fraction of the phytoplanktonic production grazed by zooplankton is around 30% in eutrophic lakes and 60 to 100% in oligotrophic lakes.[26] Given the repeated ingestion of particles of fecal and bacterial origin associated with this process, the importance of grazing in the degradation of POM can be readily appreciated.

The detrital organic matter (POM + DOM) is made up of a great diversity of chemical substrates that are decomposed by bacteria at rates ranging from 500 to 1000%/d for the most labile molecules (glucose, amino acids) to less than 1%/d or per week for more refractory dissolved compounds or particulate material. The concentration of these organic compounds in the water is inversely proportional to their rate of disappearance. The least degradable dissolved organic substances, therefore, predominate in the organic carbon content of lake water. Conversely, the soluble organic matter synthesized by phytoplankton and released by secretion or autolysis is decomposed as fast as it is formed. The concentration of such substances in the water is of the order of a few micrograms per liter.

The processes of decomposition are carried out by a complex ensemble of bacteria and animals which alter the size and chemical composition of organic matter, thereby making it accessible to attack from other species. On the one hand, the mechanical fragmentation of particles during digestion, bacterial enzymatic hydrolysis, and flushing out the POM gradually reduce it to smaller and smaller particles and DOM. On the other hand, the bacterial production from DOM and the formation of bacterial aggregates associated with flocculated organic matter together with the excretion of fecal pellets leads to the formation of larger particles which tend to sediment. Depending on the length of time (i.e., distance) during which POM remains in an oxygenated layer, the fraction decomposed in a body of water differs from lake to lake in relation to a wide range of physical (temperature, morphometry, stratification) and chemical factors (depending on the nature and mass of input of organic material).

In the least productive, deep, and well-oxygenated lakes, most of the phytoplanktonic production is degraded in the open water, and only small amounts sediment out. As the depth of the lake falls and the input of organic matter by production of plant materials that are relatively little consumed by the phytoplankton (netplankton, littoral macrophytes) increases, there is a rise in the proportion of decomposition taking place in the sediments (Figure 4). In temperate lakes, the spring outburst in phytoplankton which uses nutrients accumulated during the winter (inputs from the water basin and those released from sediments by mixing) is poorly exploited by zooplankton. There is, therefore, little recycling in the open water. Most of the nutrients are transferred to sediments where decomposition takes place. In summer, the growth of zooplankton and thermal stratification, which attenuates sedimentation, enhance recycling of nutrients in the open water (Figure 4). These recycling mechanisms, which play an essential part in the regulation of the phytoplanktonic biomass in oligotrophic lakes during the summer, become less significant when there is an increased nutrient input, i.e., high concentration and low flow (which leads to algal bloom) or low concentration and high flow (in this case there may be no bloom).

V. CONCLUSIONS

The study of lakes and reservoirs has now extended well beyond the purely descriptive and typological stage of early limnology and has now become an ecosystem-oriented aquatic science. This probably stems from the fact that the interactions between the biota and their physicochemical environment are more clear-cut than in many other ecological systems.

Over and above the purely academic level, the study of the interactions outlined in this review constitute an indispensable basis for the establishment of measures designed to conserve and improve water quality. Limnology is perhaps one of the few disciplines where pure and applied research are so closely connected.

Contemporary limnology is a much more experimental and quantitative subject than in the past. This is significant since many parameters including the flux between compartments can be measured, and the mechanisms responsible for the evolution of the overall system can be investigated, thus leading to the formulation of dynamic and functional models. These methods of systems analysis enable data from various approaches (from laboratory experiments on cultures to manipulation of the overall ecosystem) to be integrated and help us to understand the general processes involved and to interpret these complex phenomena in a more quantitative way.

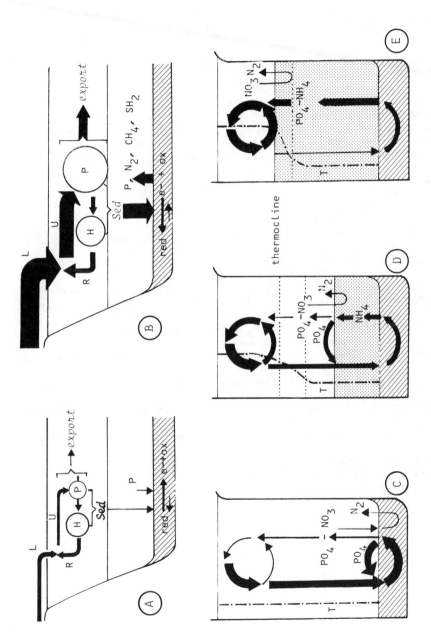

FIGURE 4. Representation of nutrient pathway in oligotrophic (A) and eutrophic (B) lakes, and in a very productive lake in spring (C), early summer (D), and later summer (E). Relatively more nutrient is recycled in the euphotic layers (fast cycle) in oligotrophic lakes and is sedimented in eutrophic ones. Nutrient loss from the epilimnetic fast cycle is reduced by the development of a steep thermocline (lower rate of sinking), and phosphorus internal loading from deoxygenated deep-water layers is increased in eutrophic lakes. Shaded areas represent O_2 depletion; dashed lines, temperature (T). P = phytoplankton; H = biomass of zooplankton + bacteria; U = uptake; R = recycling; L = external load.

REFERENCES

1. **Forbes, S. A.,** The lake as a microcosm, *Bull. Peoria Sci. Assoc.,* p. 77, 1887.
2. **Lindeman, R. L.,** The trophic-dynamic aspect of ecology, *Ecology,* 23, 399, 1942.
3. **Welch, P. S.,** *Limnology,* McGraw-Hill, New York, 1952, 538.
4. **Ruttner, F.,** *Fundamentals of Limnology* (transl.), University of Toronto Press, Toronto, 1963, 295.
5. **Dussart, B.,** *Limnologie. L'étude des Eaux Continentales,* Gauthier-Villars, Paris, 1966, 677.
6. **Hutchinson, G. E.,** *A Treatise on Limnology. I,* John Wiley & Sons, New York, 1957.
7. **Hutchinson, G. E.,** *A Treatise on Limnology.* II, *the Limnoplankton,* John Wiley & Sons, New York, 1967.
8. **Golterman, H. L.,** *Physiological Limnology.* Elsevier, Amsterdam, 1975.
9. **Wetzel, R. G.,** *Limnology,* W. B. Saunders, Philadelphia, 1975.
10. **Margalef, R.,** *Limnologia,* Omega, Barcelona, 1983.
11. **Mortimer, C. H.,** Lake hydrodynamics, *Mitt. Int. Ver. Limnol.,* 20, 124, 1974.
12. **Stum, W. and Morgan, J. J.,** *Aquatic Chemistry,* John Wiley & Sons, New York, 1981.
13. **Redfield, A. C., Ketchum, B., and Richards, F. A.,** The influence of organisms on the composition of sea water, in *The Sea,* Vol. 2, Hill, M. N., Ed., Wiley Interscience, New York, 1966, 26.
14. **Harris, G. P.,** Photosynthesis, productivity and growth: the physiological ecology of phytoplankton, *Ergeb. Limnol.,* 10, 1, 1978.
15. **Pourriot, R., Capblancq, J., Champ, P., and Meyer, J. A.,** *Ecologie du Plancton des Eaux Continentales,* Masson, Paris, 1982.
16. **Reynolds, C. S.,** *The Ecology of Freshwater Phytoplankton,* University Press, Cambridge, 1984.
17. **Thieneman, A.,** Seetypen, *Naturwissenschaften,* 18, 1, 1921.
18. **Westlake, D. F.,** Primary production, in *The Functioning of Freshwater Ecosystems,* Le Cren, E. D. and Lowe-McConnell, R. H., Eds., Cambridge University Press, London, 1980, 141.
19. *Eutrophisation des eaux. Méthodes de surveillance, d'évaluation et de lutte,* Organisation de Cooperation et de Développement Économique, Paris, 1982.
20. **Golterman, H. L. and Kouwe, F. A.,** Chemical budgets and nutrient pathway, in *The Functioning of Freshwater Ecosystems,* Le Cren, E. D. and Lowe-McConnell, R. H., Eds., Cambridge University Press, London, 1980, 85.
21. **Schindler, D. W., Brunskill, G. J., Emerson, S., Broeker, W. S., and Peng, T. H.,** Atmospheric carbon dioxide: its role in maintaining phytoplankton standing crops, *Science,* 177, 1192, 1972.
22. **Vollenweider, R. A.,** Input-output models with special reference to the phosphorus loading concept in limnology, *Schweiz. Z. Hydrol.,* 37, 53, 1975.
23. **Banse, K.,** Rates of growth, respiration and photosynthesis of univellular algae as related to cell size — a review, *J. Phycol.,* 12, 135, 1976.
24. **Kalff, J. and Knoechel, R.,** Phytoplankton and their dynamics in oligotrophic and eutrophic lakes, *Annu. Rev. Ecol. Syst.,* 9, 475, 1978.
25. **Haney, J. F.,** An in situ examination of the grazing activities of natural zooplankton communities, *Arch. Hydrobiol.,* 72, 87, 1973.
26. **Morgan, N. C.,** Secondary production, in *The Functioning of Freshwater Ecosystems,* Le Cren, E. D. and Lowe-McConnell, R. H., Eds., Cambridge University Press, London, 1980, 274.

Chapter 3

FUNDAMENTAL CONCEPTS IN AQUATIC ECOTOXICOLOGY

A. Boudou and F. Ribeyre

TABLE OF CONTENTS

I. INTRODUCTION

The term ''ecotoxicology'' was first coined by Truhaut in 1969[1] when it was recognized that the ever-increasing amounts of pollution in natural environments necessitated the establishment of a new science based on the study of the ecological effects of pollutants.

Several definitions of the term have been proposed in works of synthesis, but all are more or less restrictive, depending on the degree of specialization of the different authors.[2-6] The definition established by the Scientific Committee on Problems of the Environment (SCOPE)[2] will be used: ''Ecotoxicology is concerned with the toxic effects of chemical and physical agents on living organisms, especially on populations and communities within defined ecosystems; it includes the transfer pathways of those agents and their interactions with the environment.''

The main aim of ecotoxicology is based on the analysis of the transfer processes of contaminants within natural systems and on the study of the structural and functional effects to which they give rise. It highlights the three fundamental concerns in this discipline (Figure 1).

On the one hand, there are abiotic factors and biotic factors, which characterize natural systems and define the ecological base of any ecotoxicological approach. On the other, there are contamination factors, which include all physical, chemical, and, in some cases, biological agents which are likely to create perturbations, either temporary or more long lasting, in ecosystems or their subcompartments. Many different terms are employed, their definition and use not being, in general, too precise: for example, contaminant (''substance released by man's activities''),[4] pollutants (''all substances that occur in the environment and which have a deleterious effect on living organisms''),[4] toxicant (''agent or material capable of producing an adverse response or effect in a biological system, seriously injuring structure or function or producing death''),[6] xenobiotic (''new-made chemicals, not produced in nature'').[5]

Each of these three concerns is characterized by the extreme diversity of its elements, by factors which are almost constantly changing, in space and in time, and by a multitude of interrelationships. The result is hypercomplexity in ecotoxicological mechanisms, with contamination factors bringing yet another dimension to the fundamental difficulty which characterizes ecological studies.

To begin with, the area of investigation covered by ecotoxicology within the general context of analysis of the structure and functioning of ecosystems will be briefly described. Next, the main concepts of this discipline, using a synthesis of contamination mechanisms in freshwater environments, will be presented. This synthesis will be based on a sequence of different stages, beginning with the pollutants reaching the biotopes, their distribution throughout the aquatic phase and the sediments, the processes of bioaccumulation and transfer within the trophic networks, and, last, their effects on biocenotic structures and the ecological functioning of aquatic systems.

II. ECOLOGICAL BASES OF ECOTOXICOLOGY

Since the introduction of the concept of ecology by Haeckel in 1869, numerous definitions have been proposed. The one developed by Duvigneaud[7] has been chosen: ''Ecology is the science of complex, functional biological systems, called ecosystems.''

Ecosystems are, by definition, unitary structures limited in space which combine a physical and chemical environment — the biotope — with a community of living beings — the biocenose. In fact, the delimitation of an ecosystem is the result of a more or less arbitrary division within a higher level of integration — the ecosphere.[8]

Since the emergence of the first living beings on our planet about 3.6 billion years ago,

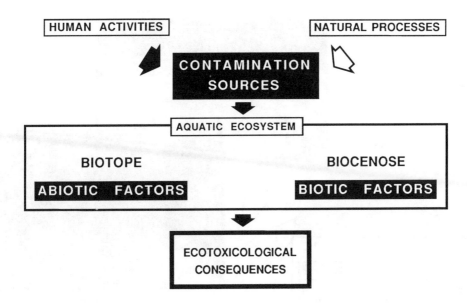

FIGURE 1. The three fundamental concerns in aquatic ecotoxicology.

the processes of evolution have produced an extraordinary diversification of species (microorganisms, plants, animals). These different phases of biologization have played an essential role in the evolution of the ecosphere, for example, the gradual formation of the protective layer of ozone in the lower atmosphere (average altitude 30 km), mainly from gaseous oxygen, a by-product of photosynthesis. These phenomena are still going on today and, indeed, are more pronounced if we take into account the pressure exerted by the human species and its multiple activities on the functioning of natural systems: "The evolution of the ecosphere and that of human societies are now indissolubly bound up together."[9]

Between the two levels of ecosphere and ecosystem, intermediate functional units can be defined, the ecocomplexes,[10] which correspond to an assembly of interdependent ecosystems which have been shaped during a common ecological history. The two preceding chapters (Chapters 1 and 2) illustrate this concept and its importance with regard to the ecological approach to continental aquatic systems: the river and its valley, and the lake and its terrestrial basin.

For the biologist, the living world is a reflection of a structural organization, in levels or stages of integration, from materials at molecular level (e.g., nucleic acids, proteins, lipids) to the biocenose or biosphere (Figure 2). However, as one ascends through the various stages, new mechanisms and properties emerge which in their complexity go far beyond being simply one added step above the structural and functional characteristics of the lower levels. Hence, a population is a group of individuals belonging to the same species, living in a defined biotope at a determined moment. The environment has a continuous effect, direct or indirect, on the structure, the functioning, and the evolution of population (Figure 3). The individual, the basic biological element, is the principal target on which ecological factors act; consequences may be observed at the biochemical, physiological, and/or ethological levels. Depending on the extent of these effects, they may influence demographic criteria and, in certain cases, the genetic structure of the population (Figure 4). Genetic diversity gives rise to a polymorphism on which the pressure of selection is felt. Thus, evolution depends both on the adaptability of each individual and on that of populations, determined by their genetic pool.[11] Any modification to one species may have an effect, directly or indirectly, on the structure of the genetic inheritance of the others by modifying the selective contexts to which they are subjected and, for the natural systems, their capacity for self-regulation and their evolutionary potentiality.[12]

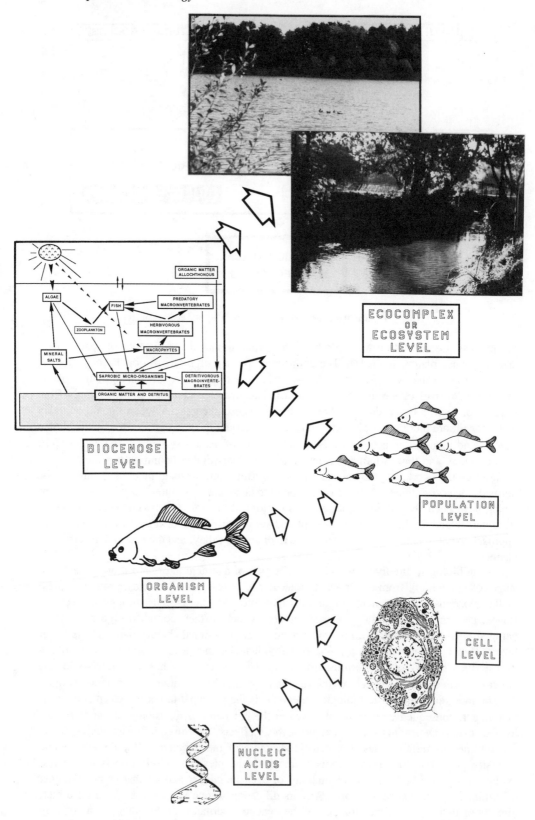

FIGURE 2.   Illustration of the biological integration levels in freshwater ecosystems.

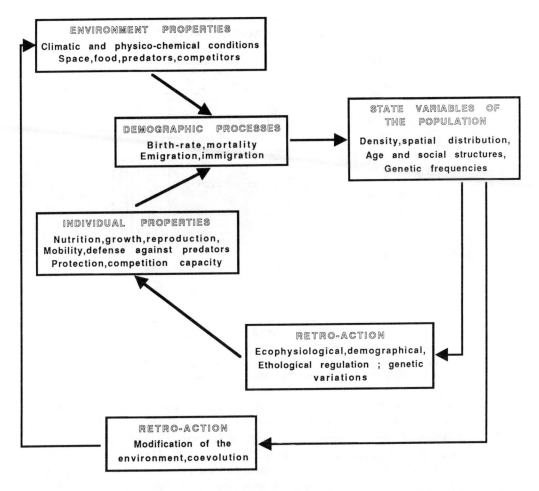

FIGURE 3. The population-environment system. (From Barbault, R., *Ecologie Générale,* Masson, Paris, 1983. With permission.)

A biocenose, or community, represents all populations living together in a given biotope and can be described by enumerating its specific composition. Formulating such an inventory produces many difficulties, some almost insuperable (sampling, determination). A functional diversity, however, provides a much richer source of information than a specific diversity. Generally speaking, the study of the organization of biocenoses reveals that these biological hyperstructures are fairly well ordered and coordinated, and the distribution of species is not simply at random nor independent from one another.

Among the relationships which ensure both cohesion and regulation in communities, alimentary links have pride of place. Within the community are trophic networks, extremely complex interlinked systems made up of linear transfer structures (food chains) which develop from autotrophic organisms or primary producers. The flow of energy passing through these networks is closely dependent on the amount of energy trapped by the autotrophs. If transfer rates are estimated by measuring productivity, this results in a very low rate of efficiency, whatever the type of ecosystem considered (average estimate: 10%).[13]

As well as classifying organisms according to their position in the trophic networks, it is also possible to define them according to their life form or life habit. Thus, in freshwater systems, the following types are traditionally classified separately: benthos (organisms living on or in the bottom sediments); periphyton (organisms, both plant and animal, attached to rooted plants or other surfaces above the bottom); plankton (floating organisms whose

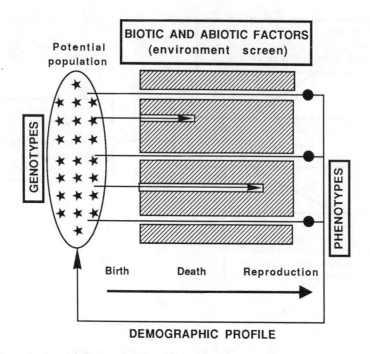

FIGURE 4.   The population against the environment screen. (From Barbault, R., *Ecologie Générale*, Masson, Paris, 1983. With permission.)

movements are more or less dependent on currents); nekton (swimming organisms able to navigate at will); and neuston (organisms resting or swimming on the surface).[13]

The distribution of species depends very closely on the physicochemical characteristics of the biotopes. Thus, the main ecological factors that control the distribution of organisms in freshwater systems are current speed and stability of water depth; light and temperature regimes; substratum condition and stability; dissolved oxygen concentration and water quality (e.g., acidity, hardness, nutrient concentrations).[14] Integration of all these factors gives rise to zonation phenomena: on a large scale, zonation of rivers or lakes; and on a much smaller scale, microhabitats (see Chapters 1 and 2).

The ecological study of the structure and functioning of ecosystems shows up fluctuations which are sometimes quite important, both in the abiotic and the biotic factors: "Ecosystems do not behave in constant and repeatable ways."[15] When we look at the evolution of the ecosphere from a spatiotemporal aspect, we see that from the time of appearance of the first living creatures to the emergence of the human species, billions of years went by with several very long periods which were particularly harmful to the processes of biologization — glaciation, volcanic activity, earthquakes, etc. — and led to enormous numbers of aquatic and terrestrial species being wiped out. These harmful periods alternated with more favorable phases in which a more diversified taxonomy could once again become established. On a much more limited scale, a very great number of abiotic and biotic factors within the aquatic ecosystems present some very large variations of a cyclical nature, e.g., nychthemeral, seasonal, and annual cycles. This is the case, for instance, in the effect of climatic factors on the hydrodynamic characteristics of rivers (flood and low water) or lakes where the summer thermic stratification can, depending on the level of eutrophication, lead to a sharp increase in the phenomenon of anoxia in the hypolimnion, with severe consequences for the benthos populations.

In addition to all these phenomena that are often described somewhat inaccurately as "natural" or "normal", man, too, has left his mark in varying degrees on these fluctuations since the first members of the genus *Homo* appeared on the earth in the Paleolithic age. During the entire prehistoric period, his interventions can be summarized as an almost permanent struggle against a hostile environment (e.g., rigors of climate, lack of food, dangers from other species). As he moved into the agropastoral age, he learned how to make tools and arms and, thus, increased his capabilities. For the first time he was able to exercise control over animals and plants. He became a shepherd and farmer and destroyed vast areas of forest. During the preindustrial age, his technical and energy resources remained relatively moderate, although his spirit of adventure led navigators to explore the world. For the last century, the age of the craftsman has been replaced by the industrial era: man controls energy and its various sources. This has increased his material possibilities to a tremendous degree, and he has applied them to a technology which has become more and more diversified and sophisticated, especially in the fields of modern agriculture, industry, urban environment, transport, and health. Man's action on his environment is mainly guided by the hope of a better life, a hope which still remains the major guideline behind most of his actions.

This evolution has been concomitant with the demographic explosion on the planet as a whole. It has taken only 12 years for the population of the world to increase from 4 to 5 billion, whereas to reach 1 billion, which happened at the beginning of the 19th century, had taken 2 million years.[16]

Human activity and its results have developed from elements drawn from within natural systems. These elements are used in a variety of ways which can be characterized as a progression. At one end of the scale is a total absence of any kind of transformation of the elements. At the other end is a very complex combination of actions using chemical synthesis, for example, which leads to the manufacture of new products, unknown in natural environments (xenobiotics). In all cases, man intervenes in biogeochemical cycles by increasing the fluxes between compartments, modifying the chemical forms of elements, and selectively enriching the biotopes according to the amounts of material he puts back in and the extent of the discharge areas. Nowadays, the impact of man on his environment is seen as being more and more determinant in its nature, because of the diversity and effectiveness of the means at his disposal, the rapidity and scale of their effects at the level of the continents, and even of the entire planet. The most convincing illustration of these means at our disposal is provided by the various scenarios which have recently been published portraying nuclear war. These combine a primary phase where living creatures are destroyed by radiation and its indirect effects (e.g., burns, traumatic shock) and a secondary phase of physical disturbances in the atmosphere ("nuclear winter").[16,17]

Referred to by the vague terms "pollution" or "contamination", the secondary effects of human activity were the basis for the emergence of ecotoxicology, a new discipline to complement a whole group of scientific fields participating in the study of environments and how they are populated: biology, genetics of populations, chemistry of the environment, etc. Their common objective, defined essentially from an anthropocentric viewpoint, is to improve our knowledge of these complex systems by means of methodologies resulting from a multidisciplinary approach. Among the principal goals of the research are a better use of natural resources and an increase in productivity, establishing protective and preventive measures.

Briefly, the contamination of continental aquatic ecosystems can be characterized by a set of essential processes which ensure transfers between the main abiotic and biotic compartments. By means of these transfers, toxic products are distributed throughout the system and over a period of time. Also, each contaminant has its own contamination scenario, which differs not only according to the system being considered, but also according to the state of the receiving environment.

A synthesis of the principal ecotoxicological mechanisms which characterize the contamination of freshwater systems is presented below.

III. CONTAMINATION SOURCES OF FRESHWATER ECOSYSTEMS

Pollution sources can be classified in various ways, depending on the criteria considered: origin (e.g., domestic, agricultural, industrial, mining); main components (e.g., organic, metallic, saline, heated); properties and their effects (e.g., putrescible, toxic, inert, colloidal).[14] It is important to stress the fact that the sources of pollution in freshwater ecosystems are not limited solely to discharge from chemical substances. Other human interventions contribute to the modifications (and, in some cases, considerable disruptions) in their biocenotic structure and functioning: river channel modifications, river impoundment, river discharge regulation, intercatchment transfers, etc.[14]

For Ramade,[3] there are three principal causes of contamination of the ecosphere:

1. The production of energy from combustible fossils or from the atom: this currently forms the basis of almost all human activity in industrialized countries and leads to many types of pollution, both direct and indirect: gaseous discharge into the atmosphere (e.g., nitrogen, sulfur, and carbon oxides; products of combustion of hydrocarbons), heated effluents, and thermal enrichment of the receiving biotopes, etc.
2. The chemical industry and its extraordinary expansion over the last few decades: different estimates mention more than half a million chemicals currently in use, this number increasing by 1000 to 2000 per year, although only about 10,000 or so are produced in amounts of between 500 kg and 1000 t.[2,14]
3. Modern agriculture: it exerts an ever-increasing pressure on the natural environment due to mechanization, the spreading of fertilizers, and protective treatments such as herbicides, insecticides, and fungicides. This represents a diffuse pollution against which intervention is limited, although new strategies have been established for the management of agroecosystems. Take the example of nitrogenous pollution of continental waters such as rivers, lakes, and underground water. This is a perfect illustration of both the size of the problem and the difficulties encountered when trying to set up concrete protection measures.[18]

Together with the criteria already described, the geographical extent of the discharge is a very important element in characterizing sources of pollution. If we take Miller's classification,[19] three main categories are defined:

1. Point spills: "When a significant amount of a chemical has entered an ecosystem at a point, in both space and time, and effects of contamination are expected in a well-defined more or less local area"
2. Chronic local releases: "Cases in which discharges have taken place over such periods of time, and in such quantities, that a larger region (a river system, for example), has been contaminated"
3. Widespread releases: "Release of a substance in sufficient quantity and over a wide enough area that there could result a noticeable pollution of a significant part of the entire earth's surface"

According to Butler,[2] there are something like 1000 substances manufactured in such quantities as to be capable of polluting the entire globe: for example, radioactive fallout; DDT; PCBs; freons and similar fluorinated hydrocarbons; production of carbon dioxide, oxides of sulfur and nitrogen and their end products; and acid rains.[19]

For a given aquatic ecosystem, the amounts of contaminant received may arrive from a variety of transfer pathways (from the atmospheric compartment or from the watershed) and are referred to by the global term "pollution input".

Precipitation accounts for the bulk of exchanges between the toxic products present in the atmosphere — gaseous compounds, aerosols, fine particles — and the continental aquatic biotopes. In certain cases, especially in lakes with a large surface area, contaminants are received directly at the water-air interface, though this is very closely related to climatic conditions. In Lake Michigan, for example, the dry deposition contributes 75% of the total loading for lead and 50% for zinc.[20] Atmospheric trace elements enter the surface microlayer (3 Å to about 3 mm thick), which is highly enriched in organic matter as compared with the bulk water and which also has a higher microbial activity. There are a lot fewer results available for freshwater systems than there are relating to the surface microlayer in the marine environment.[21]

Contamination sources that originate from the terrestrial basin are due to isolated waste from urban or industrial sources, for example, or to more diffuse sources such as water from surface runoff or soil drainage, groundwater, feeder streams (lakes), or tributaries (rivers)[22] as shown in Figure 5. An inventory taken in 1976 of the relative magnitude of various sources of pollution of the Thames estuary, in terms of the oxygen-demanding load, shows a very marked preponderance of sewage effluents (74% of the total load) compared with direct industrial discharges (9%), affluents (5 to 6%), freshwater from the upper Thames (7.5%), and storm water (3.0%).[23]

In Chapter 9.3, covering the contamination of the Laurentian Great Lakes by persistent toxic organic chemicals, the major sources of pollution of these limnetic ecosystems are dealt with at the level of land-water and atmosphere-water exchanges.

In very many cases, the conjunction of human activity and the environment gives rise only to an amplification of natural phenomena, which, having thus reached a high level or threshold, lead to more or less severe dysfunctions within the ecosystems. This is the case, for instance, when hot water is discharged into rivers or estuaries after the installation of electric power stations, or when supplies of biodegradable organic matter from urban or industrial effluents contribute to enrich the endogenous stock of inert matter. The ecological consequences of this contamination depend on the capacity of the receiver system for self-purification and the degree to which the pollution has spread. Last, there is the use made by man, directly or indirectly, of a very large number of chemical elements present in the ecosphere which are toxic for living beings. Some of these do not seem to play any role in vital processes (e.g., mercury, cadmium, lead); others, called hormetines,[24] are indispensable in very small quantities and can become dangerous, or even fatal, if the quantities absorbed go beyond a certain threshold (e.g., copper, cobalt, iron, fluorine). It is a very delicate problem to estimate the effect of human activity on the biogeochemical cycles of these elements within the ecosphere. Several experimental models have been set up for certain heavy metals from a theoretical base or from the results of experimentation (e.g., cadmium, Figure 6).[25] It should be pointed out, however, that the results proposed often diverge considerably. For mercury, estimates of the fractional global circulation which originates with man range from 5 to as much as 30%.[19]

Classically, besides this first set of contamination factors is the category of xenobiotics. Some of these substances are dispersed in the environment intentionally. Such is the case for organic insecticides (organochlorines, organophosphates, carbamates, synthetic pyrethroids) which are very widely used in agriculture or in programs set up to fight certain parasitic diseases. Others are introduced unintentionally, by accident, or as by-products of industrial activity or residues of other molecules. This is the case, for example, with PCBs.

The first step in a study of the contamination of an aquatic ecosystem — the inventory of the sources of pollution and their quantification — is very difficult to carry out when one considers not only the diversity of the waste products, but also their variation in space and time and their characteristics which are often diffuse and indirect. In order to be precise and exhaustive, such an approach would require a sampling strategy for each of the potential

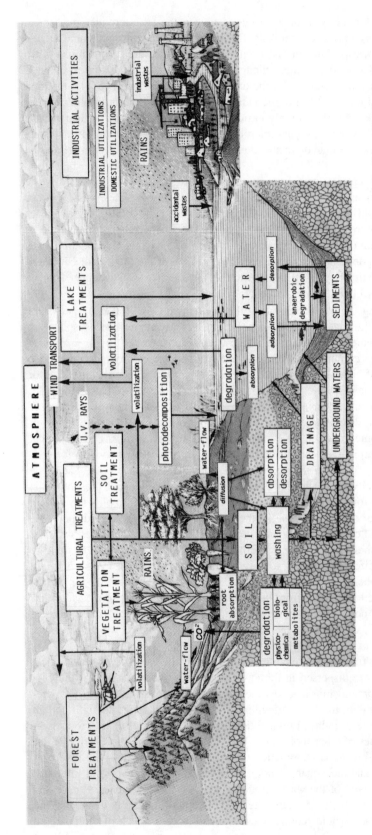

FIGURE 5.   Pesticide dispersion in the environment. (Adapted from Jamet, P., in *Les Produits Antiparasitaires à Usage Agricole: Conditions d'Utilisation et Toxicologie*, Fournier, E. and Bonderf, J., Eds., Lavoisier, Paris, 1983, 189.)

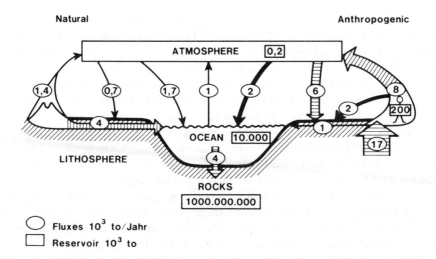

FIGURE 6. Present-day and prehistorical cadmium fluxes. (From Brunner, P H. and Baccini, P., in *Metals in the Hydrocycle,* Salomons, W. and Förstner, U., Eds., Springer-Verlag, Berlin, 1984. With permission.)

transfer routes described above, together with considerable analytical facilities. In fact, estimates relating to sources of pollution are more often the result of a reverse operation: the contamination levels measured in the different compartments of the system studied (water, sediments, organisms) dictate the choice of essential pollutants and the search for their origins and entry routes into the biotope.

IV. CONTAMINANT DISTRIBUTION AND CHEMICAL TRANSFORMATIONS IN AQUATIC BIOTOPES

A. Principal Ecotoxicological Processes in the "Water Column"

The processes that ensure the distribution of contaminants within aquatic biotopes are extremely complex. In the water column three major components play a part: (1) the pollution conditions, (2) the physicochemical properties of the pollutants, and (3) the wide array of competing reactions with the different elements present in the aquatic phase. The conditions of discharge, as already described, characterize the frequency and quantity of the supply of contaminant. When linked with certain characteristics of the biotope, such as the mixing of the water or the surface/depth ratio, the conditions have a considerable influence on the dispersal of toxic products. For any given substance, a whole group of physicochemical properties will come into play regarding its environmental behavior. Hence, depending on its degree of solubility in water or, conversely, in apolar solvents, dilution will be accordingly more or less rapid and homogeneous. Such is the case with certain low-density hydrophobic hydrocarbons which spread over the water-atmosphere interface, sometimes covering very wide areas. Contamination of the water column is generally delayed, involving, as it does, modifications to the structure of the surface film by means of oxidation reactions or by the formation of water-in-oil emulsions.[26]

Together with these solubility properties, which are directly or indirectly influenced by abiotic factors such as temperature and pH, many other physicochemical characteristics of pollutants condition their mobility and distribution throughout aquatic biotopes including dissociation constants, formation of chemical complexes, volatilization, and leaching and dissipation characteristics.[2]

In the case of metallic pollutants, "the behavior in aquatic systems is highly complex, due to a large number of possible interactions with ill-defined dissolved and particulate components and nonequilibrium conditions.''[25] From a single chemical form introduced initially into the natural environment, a whole set of chemical species may appear, each one having its characteristic chemical reactivity and lifetime, and each reacting differently with the matrices of the environment[27] (see Chapter 5). The study of the speciation of dissolved metals is currently based on a combination of several complementary methods, within speciation schemes: theoretical calculations (chemical equilibrium models,[28,29] see Figure 7); physical and chemical separation techniques (e.g., filtration, voltammetry); direct chemical methods, more particularly for the organometal species.[30] The greatest limitation for natural waters is the absence of ''appropriate methods to address the complexation of metals by organic ligands and the disequilibrium between redox couples.''[25]

Among the other contaminants found in continental aquatic ecosystems are organic substances, which, in general, have physicochemical properties different from those of metals and the behavior of which in the biotope is extremely variable, especially as it depends on molecule chemical stability or remanence. Consider pesticides, taking account of the extreme diversity of products used and the major role played by their physicochemical properties in their activities. There is a very wide range of readings for aqueous solubility, from a few μg/l (ppb) for some organochlorines like DDT, to 800,000 mg/l for some herbicides. As a general rule, their hydrosolubility is close to ppm (mg/l). When introduced into aquatic biotopes, they bond to suspended organic matter. The sorption phenomena are rapid and usually irreversible. For some, volatilization processes are considerable.[31]

The persistence of these toxic substances is classically expressed in terms of chemical half-life and is dependent on a number of abiotic and biotic mechanisms. Research carried out in this field more and more often shows that a knowledge of degradation products is as important, if not more so, as the quantification of the disappearance kinetics of the original chemical form. Indeed, certain metabolites can be seen to be very dangerous, with sometimes a considerable increase in toxicity compared to that of the initial product. In particular, there can be a change in the routes or mechanisms of penetration into the organisms and in the effects at cellular or molecular level (e.g., emergence of mutagenic or carcinogenic effects). Among abiotic mechanisms of transformation can be distinguished photochemical, hydrolysis, oxidation, or reduction reactions:

1. Photolysis can occur by direct absorption of light, by sensitization via energy transfer, or by indirect processes. The aquatic environment does not constitute such a good matrix as the atmosphere for photochemical alterations. Light intensity is variably attenuated by natural waters, and photochemical rates are strongly depth-dependent in most freshwater systems.[32] Reactions in photodegradation are, in general, very complex, and studying them in aquatic environment is a particularly delicate task, as a large number of factors interact including spectral distribution and intensity of light, concentration of reactants, temperature, medium (solvent) in which reaction occurs, and pH.[33,34]

2. The effects of many pesticides are neutralized by the action of hydrolysis in a natural environment. Take, for instance, the chemical half-life (expressed in days) of some organophosphates in solution in water at 20°C: 690 for parathion, 175 for methylparathion, and only 1.1 for disulfoton.[2] Hydrolysis is very much influenced by temperature and only slightly less so by the pH. Some elements present in the environment, e.g., copper, also influence these chemical processes.

3. Oxidations in aquatic systems are much less common than in the atmosphere where the major oxidizing species are the HO radical and O_3 (ozone). Oxidation in water of both aromatic and aliphatic compounds produces alcohols, ketones, and hydroperoxides.[33]

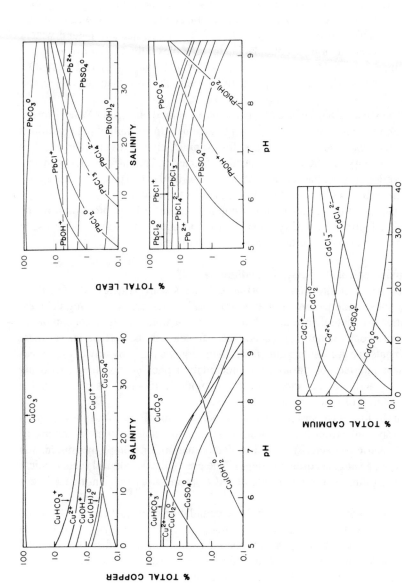

FIGURE 7. Variation of copper, lead, and cadmium chemical species over a range of salinity and pH (equilibrium models: salinity variation is at pH 8.0 and pH variation at 35.0 salinity). (From Kester, D. R., in *The Importance of Chemical "Speciation" in Environmental Processes*, Bernhard, M., Brinckman, F. E., and Sadler, P. J., Eds., Springer-Verlag, Berlin, 1986, 337; and Turner, D. R., Whitfield, M., and Dickson, A. G., *Geochim. Cosmochim. Acta*, 45, 855, 1981. With permission.)

4.   The reduction of organic compounds requires an environment with a low redox potential, and this can be found, for instance, in the deep layers of lake systems in the eutrophication phase. Many mechanisms have been described, many of them requiring catalysts, hydrogenation, halogenation, coupled oxidation-reduction, etc.[33]

Suspended matter in the water column is a "complex mixture of polyligand materials"[27] which acts as a trap for trace elements present in the medium. The complexation capacity of natural waters is based, for a given metal, on the chemical equilibrium between the free metal ions and the inorganic and the organic complexes. The particulates in natural waters consist predominantly of detrital organic colloidal matter (complex polymers known as humic and fulvic acids), inorganic solids such as metal oxides and hydroxides (SiO_2, Al_2O_3, $FeOOH$), carbonates, aluminosilicate clays, and living cells (bacteria and phytoplankton algae).[35]

Physical sorption on the external surface of a particulate is based on Van der Waal's forces of the relatively weak ion-dipole or dipole-dipole interactions. Under normal pH conditions of surface waters, silica, clay minerals, feldspars, for example, are negatively charged, thus provoking a stronger affinity for the cationic metallic species.[25] The different mechanisms of contaminant adsorption by solid particulate matter are described in Chapter 6. An important conceptual development has occurred in this field in the last decade.

In this way the suspended matter in the water column ensures a large proportion of the transport of contaminants in freshwater systems. It also contributes, by way of deposits, to the accumulation of these elements in the sediment compartment.

B. Principal Ecotoxicological Processes in Sediments

Sediments are composed of silts and clays (fine sediments with particles <50 μm) and sands and gravels (coarse sediments with grains >50 μm). They are, to varying degrees, rich in organic matter originating from autochthonous production and from allochthonous supplies. In addition to contaminant that reaches the sediment via the particle phase, there are metals that sediments can accumulate directly from the aquatic phase by way of exchanges at the water-sediment interface or through the precipitation phenomenon (e.g., reducing sulfide type) under the action of variations in certain abiotic factors such as pH, oxidation potential, or concentration of precipitating substances.[25]

Sediments can thus be likened to vast reservoirs for metal storing (Figure 8). From analyses of sediment cores, especially from lake environments, it is possible to combine dating of the deposits (granulometric characteristics, pollen determination, or isotope measurements) and metal concentration dosing, according to depth. The precivilizational level (geochemical background values) can thus be clearly differentiated from the story of man-induced contamination.

The importance of the sediment reservoir in the accumulation of metals can be illustrated by results obtained from a study of mercury distribution and flux mercury in the Ottawa River (Canada). In the 4.9-km stretch that was the study section, more than 96% of total mercury and 97% of methylmercury were localized in the bed sediments (see Chapter 9.2).

The main physical, chemical, and biological mechanisms that are responsible for the accumulation of trace metals in sediments are presented in Chapter 7. Let us simply underline here the importance of contaminant localization in the sediments and its evolution. Many sedimentological characteristics condition the modes in which metals are stored and the stability of their bonding with inorganic and organic particles; for example, grain size and abundance of humic material. However, during diagenesis, sedimentary constituents are modified, especially the organic part which is gradually degraded by the decomposing action of microorganisms. These transformations lead to physicochemical changes, and, thus, new equilibria need to be established between the dissolved species and the solid part. As an

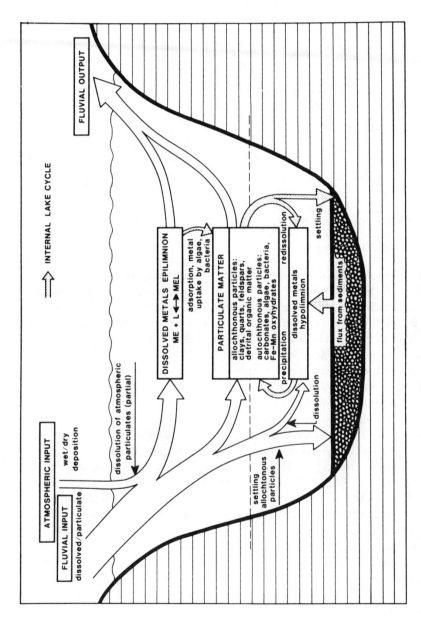

FIGURE 8. Processes affecting trace metals in lakes. (From Salomons, W. and Förstner, U., *Metals in the Hydrocycle*, Springer-Verlag, Berlin, 1984. With permission.)

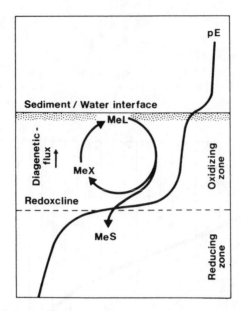

FIGURE 9.   Schematic presentation of the cycling, chelated heavy metals above the redoxcline. (From Hallbert, R. O., in *Metals in the Hydrocycle,* Salomons, W. and Förstner, U., Springer-Verlag, Berlin, 1984. With permission.)

example, let us mention the vital role of the oxidizing and reducing zones, which create a sharp redox gradient (Figure 9). Despite the diversity of the phenomenon, several common mechanisms can nevertheless be found: the near universality of bound oxygen as an available electron donor, the high degree of polyfunctionality, and the structural complexity and heterogeneity with respect to ligand-site microscopic physical and chemical environments.[35] Pore or interstitial water has a predominant place with regard to the transfer of trace metals to the water column and, in particular, to the intrasedimentary or benthic organisms (bioavailability) (see Chapter 7). However, "very little information is available on the chemical species composition of pore water."[21]

As with the case of metals in aquatic ecosystems, sediments are a favored place for storing most organic pollutants, either from the depositing of suspended matter or from the water-sediment interface. The reactions of sorption and fugacity (escaping tendency from water) are very closely dependent on the solubility and hydrophobicity of the substances. The formation of complexes with the different sedimentary constituents generally favors remanence of organic contaminants by a reduction of the rate of physicochemical transformations.[32] Thus, for organochlorine compounds, concentrations measured in the sediments are always very high, compared with other compartments of the biotopes.[36] Analyses carried out on the upper strata of sediments from Lake Geneva (Switzerland) showed, in some areas, high levels of contamination by PCBs.[37] The same is true for the Great Lakes (see Chapter 9.3). Most of the polycyclic aromatic hydrocarbons (PAH) in freshwater ecosystems are accumulated in sediments; the ratio between average concentrations detected in sediment and water, respectively, was usually greater than 1000.[38]

As well as having an important role in the storage of metals in freshwater ecosystems, sediments are also the seat of chemical transformations in these elements. Methylation reactions are most often quoted, especially for mercury. Since Jensen and Jernelov's work,[39] a great deal is now known about these mechanisms. They require the presence of a donor of the methyl radical, which can be nonenzymatic (methylcobalamine, acetic acid, propionic acid, tin and lead alkyls) or enzymatic (e.g., methionine-synthetase).[40] In the upper strata of sediments, in aerobic or anaerobic conditions, or in the water column, many bacterial populations are capable of biomethylating the mercury.[41] The efficiency of these transformations depends on their metabolic activity, which is modified by such factors as environment pH, the potential for oxidoreduction, and the presence of sulfides to immobilize the metal.[42] Contrary to these reactions, certain bacterial species are capable of degrading methylmercury ($CH_3Hg \rightarrow Hg_{II} \rightarrow Hg_o$), especially by means of an enzyme — organomercurial lyase — the functioning of which is still only poorly understood.[27] At the present time, the output from these biotransformations and the consequences in a natural environment of the production of methylmercury are not known. Yet this is one of the fundamental aspects of the ecotoxicology of this metal, in view of the toxicity of the methylated form for living beings including man and its importance in transfers along alimentary networks. Indeed, it represents more than 80 to 90% of accumulated mercury in terminal consumers. Other metals (arsenic, for example) are liable to be methylated in natural conditions, but unlike mercury, the organic form is much less toxic than the inorganic compounds.

While abiotic mechanisms ensure the transformation of organic contaminants in an aquatic environment, in many higher species biodegradation reactions depend on the activity of microorganisms (microbial degradation) and on that of enzymes — catalyzed conversions (biotransformation) — at cellular level. The conversion and degradation of natural organic compounds and xenobiotics by bacteria fall within the scope of the principal activity of such microorganisms — mineralization — which predominates in the nutrient cycle. Depending on the physicochemical conditions of the environment, especially oxygen concentration (aerobic or anaerobic conditions), the transformation of organic matter into unbound-oxidized states is more or less total; e.g., ammonia, nitrites, nitrates for organically bound nitrogen. Many laboratory investigations have been carried out on these mechanisms, and they have led to the developing of standardized methodologies, thus making it possible to know the chemical stability of certain pollutants when faced with the processes of biodegradation (see Chapter 2.4 in Volume II). Thus, the role of microorganisms in oil pollution has been the subject of a great deal of research, the qualitative and quantitative differences in the hydrocarbon content of petroleum influencing the susceptibility to degradation.[43] Today many questions still remain concerning, for example, the efficiency of the biodegradation of certain xenobiotics in anaerobic conditions; the importance of cometabolisms; the toxic effects of pollutants on the bacterial populations, modified by synergic or antagonistic interactions between the molecules and the factors predominating in the environment; and resistance mechanisms (incidence of plasmids). The difficulties involved in extrapolating from experimental studies of biodegradation processes to the reality of these phenomena in a natural environment should also be stressed. In aquatic biotopes, especially, there is great movement and mixing of bacterial populations, which are characterized by their very great ubiquitousness.[44] In the microlayer at the surface of the water column and at the water-sediment interface, the free-swimming and attached bacteria act jointly with abiotic mechanisms to transform organic pollutants under the influence of the factors in the environment (temperature, pH, oxygen, concentrations of organic and inorganic materials, etc.).

Last, let us look at the third essential role of sediments in relation to the contamination of freshwater systems by metals: the releasing phenomenon (flux from sediment). Because of the high levels that generally accumulate in the upper strata, this phenomenon can constitute an important source of pollution. Three main processes can be distinguished:

1. First is the mechanical stirring of the sediments that is natural in origin (currents, variations in flow and erosion, benthic and intrasedimentary organisms) or anthropic (dredging, navigation, construction work on the banks). When particles are once more in suspension in the water, with a new chemical context, this may lead to desorption reactions which increase metallic species concentration in the aquatic phase.

2.   Second are physicochemical modifications of the biotope occurring at the water-sediment interface and through the intermediary of interstitial water. Published results have mainly derived from experimental approaches and express the major controversy over the role of the various factors considered: pH, temperature, oxidoreduction potential, of the chlorine concentration, etc.[45]
3.   Third are the transfers involving intrasedimentary flora and fauna. In addition to the bacterial species already mentioned, many animal organisms live in sediments or at the water-sediment interface in the larval or adult phase.

These, then, are the principal mechanisms which distribute toxic products throughout the various compartments of freshwater biotopes and the main chemical transformations they undergo from abiotic and biotic factors. They enable us to understand the complexity of ecotoxicological phenomena in a natural environment. All these phenomena define the contamination conditions of living beings, either from pollutants present in the water or by the intermediary of consumed prey.

V. CONTAMINATION OF AQUATIC BIOCENOSES AND EFFECTS ON THE STRUCTURE AND FUNCTIONING OF ECOSYSTEMS

Two principal and closely interdependent components characterize the contamination of living beings: first, bioaccumulation, which is the result of all contaminant transfers between the surrounding environment and the organisms; second, the toxic effects, due to the action of contaminants on biological structures and on the different metabolic pathways.

To present a synthesis of the processes of bioaccumulation and of trophic transfer of the contaminants within freshwater ecosystems, the principal ecotoxicological mechanisms which concern the level "organism" are first considered. After this are discussed those mechanisms which characterize interspecific relations, especially the prey → predators transfers within the trophic chains and networks.

In the second part the effects of toxic products on the structure and functioning of aquatic ecosystems are examined.

A. Processes of Bioaccumulation and Trophic Transfer
1. Concept of Bioaccumulation

For an aquatic uni- or pluricellular organism, bioaccumulation represents the quantity of toxic product present in the individual at any given moment. It is the resultant of two antagonistic mechanisms:

1.   Biouptake is a set of adsorption and absorption processes of the exogenous products through the aqueous phase or through the intermediary of ingested food.
2.   Effluxes ensure substance transfers between an organism and the surrounding environment (processes of excretion, in the broadest sense). To this we should add the different endogenous reactions to biotransformation, especially toward organic contaminants. These two features — efflux and biotransformation — can be grouped within the category of decontamination mechanisms.

Bioaccumulation is, therefore, the difference between biouptake and decontamination, each being described in terms of flux (either continuous or discontinuous) between the two compartments; i.e., aquatic biotopes or consumed prey, on the one hand, and the organism on the other.

From an ecotoxicological point of view, we can quantify bioaccumulation using two criteria from the measurement of the element in the whole organism or in different tissue samples:

1. Content, or burden, represents the amount of contaminant in the whole organism or in the subcompartments. It is associated with the idea of transfer potential with regard to predators, for example. (See Chapters 1.1 and 2.2 in Volume II.)
2. Concentration corresponds to the quantity of contaminant per unit of live weight — μg/kg or ppb; mg/kg or ppm. This expresses indirectly the number of fixation sites occupied at organism, organ, or cellular level, and also reflects the toxicological risk associated with the presence of the toxic products.

These two criteria, concentration and content, are closely dependent on the increase in weight of the organism or the organ considered. Thus, the evolution of the concentration may show a decrease, while the content remains constant or even increases. This is the phenomenon of growth dilution, which can be seen particularly well when studying decontamination in fish with a rapid growth rate, like rainbow trout *Salmo gairdneri* (see Chapter 2.2 in Volume II).

Almost all results from research devoted to the study of bioaccumulation processes are expressed as concentrations. Very often, because of the size of individuals, measurements are applied to a limited number of tissue fragments, and it is therefore not possible to deduce bioaccumulation directly at the level of the organism as a whole. In fish caught in a natural environment, for example, the concentration of contaminant in the skeletal muscle is very often studied because of the alimentary importance of this tissue to man. Other organs, like liver, kidneys, and adipose tissue, are also often studied because of their ecotoxicological properties. For many authors,[4] the concentration criterion constitutes a good basis for the comparison of contamination levels between individuals or between species, within the same biotope or on a much wider scale. However, one must always be particularly careful to check the mode of expression of this criterion, especially with regard to the reference weight selected as a basis for calculation (fresh weight, dry weight, lipid or protein contents). Sometimes there are large differences within the same organism between the richness of the organs in water or in lipids, but this is also true between individuals, depending on their stage of development, their diet, and other factors.

2. Ecotoxicological Importance of Biological Barriers and Membrane Barriers

In any contamination process, whether by direct or trophic route, penetration of the toxic products into the organism depends on their crossing the biological barriers which separate the internal medium from the surrounding environment. The structural and functional characteristics of these barriers are extremely varied, depending on the species considered, the organ, or even the stage of development of the individual. They are also very closely dependent on the physicochemical properties of the biotope and its fluctuations. Three examples will illustrate this diversity:

1. Bacterial cell walls: by using various methods of investigation, such as the electron microscope and biochemical analysis, the very complex structure of these envelopes can be studied, especially for Gram-negative bacteria.[46] In fact, four structures are superimposed: the plasmic membrane, the peptidoglycane, the external membrane, and the peripheral exopolysaccharides. The periplasmic space, an aqueous compartment with no defined structure, separates the two intermediate surfaces (Figure 10). The chemical composition of these envelopes and the distribution of the constituent molecules undergo almost permanent variation. Thus, exopolysaccharides, the peripheral structure of the bacterial surface, are not present in all species, and their thickness is linked with the physicochemical characteristics of the environment and the physiological condition of the microorganisms. In some cases it constitutes a capsule with a thickness of several tens of microns, which can detach itself, partially or totally, thus contributing to an increase in the number of potential ligands for the contaminants present in the external environment.[47]

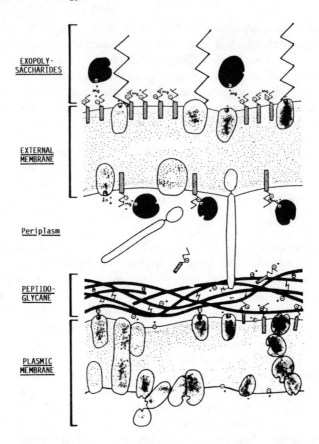

FIGURE 10.   Schematic presentation of the Gram-negative bacteria
cell walls. (From Bauda, P., Accumulation et Toxicité du Cadmium
chez les Bactéries Gram Négatives: Rôle des Enveloppes Bactériennes,
Doctorate thesis, Université de Metz, France, 1986.)

2.   The cell barriers of phytoplanktonic algae: these envelopes are characterized by their
     extreme diversity, from a normal plasmic membrane (as in *Dunaliella*) to a pectocel-
     lulosic wall, superimposed on the cellular membrane, composed mainly of proteins,
     lipids, and cellulose, and with an average thickness (in *Chlorella pyrenoidosa,* for
     example) of 210 Å for a cellular diameter of between 3 and 4 μm.[48]
3.   Epithelial barriers in fish: the penetration of toxic products by direct route (from
     contaminants present in the water) depends on crossing the branchial barrier or the
     cutaneous layers. The presence of scales and mucus considerably reduces the acces-
     sibility of the exogenous molecules to the epidermal layer. The lamellar structure of
     the respiratory epithelium ensures that there is a vast surface area for exchange with
     the external environment, representing two to ten times the total surface area of the
     fish.[49] As the water in the branchial cavity is being renewed continuously, this facilitates
     transfers through cellular monolayers. The distance separating the water from the blood
     is very small, about 1 to 6 μm.[50]

As far as the trophic route is concerned, absorption of the contaminants conveyed by
consumed prey occurs in the digestive tract, or more precisely, in the absorbent areas of
the intestine. As for the gills, the structure of this organ is particularly well adapted to its
exchange function, and the microvilli of the apical face of the enterocytes considerably
increase the interfaces with the intestinal lumen (Plate 1).* Our experimental approach to

* Plate 1 appears following page 74.

the bioaccumulation of mercury compounds with *Salmo gairdneri* showed the very strong specificity of the branchial and intestinal barriers with regard to the absorption of the metal (see Chapters 1.1 and 2.2 in Volume II).

We should remember that, whatever the organism or organ, each biological barrier has, at cell level, a sole component — the plasmic membrane — which controls, in the end, the exchanges between the external environment and the cytoplasm. It is an organized structure of about 80 Å in thickness and is composed of proteins and lipids, the lipids representing, on average, about two thirds of the membrane mass.[51] It also contains sugars covalently bonded to the proteins (glycoproteins) and to the lipids (glycolipids) (Figure 11). The principal characteristic of this membrane is the arrangement of phospholipids: "a two-dimensional solution of a mosaic of integral membrane proteins embedded in a fluid lipid bilayer with peripheral proteins bound loosely to either surface."[52] This arrangement makes horizontal and vertical molecular displacements possible, in relation to the layout of the membrane: lateral migrations of the phospholipids and proteins (average speed estimated at 10^{-8} cm/s for the lipids); flip-flop movement of the phospholipids from one monolayer onto the other; aggregation of certain lipids (phase separation).[53] The constituents of the membrane are generally arranged asymmetrically: glycoproteins and glycolipids on the external face, uneven distribution of the peripheral proteins and the phospholipids between the two monolayers.[54] All plasmic membranes are subject to the influence of a strong electric field, mainly due to the electrochemical gradient caused by the difference in concentration of ions on either side of the membrane. Thus, when in a stable state, a difference in potential of the order of a few tens of millivolts is established between the two compartments — extra- and intracellular — giving an electric field of 10^5 V/cm². The plasmic membrane defines the volume and form of the cell; it controls exchanges with the outside environment, being a selective permeability barrier.

From the ecotoxicological point of view, interactions between toxic products and biological membranes are based on two essential aspects: first, the crossing of these barriers, which is a precondition for the accessibility of contaminants into the cell or organism; second, the perturbations brought about at the level of structure and membrane functions that present a risk to the life of the cell.

Relationships between toxicants and the external face of the membranes have traditionally been described in terms of adsorption reactions. For metallic ions, for example, these reactions are due to electrostatic attractions, to ion exchanges, or to covalent bonding with the different membrane constituents. They can be analyzed, as can the solid particulate matter in the water column, by establishing adsorption isotherms (such as Freundlich's) for bacteria and photoplanktonic algae (see Chapter 6 in this volume and Chapters 2.3 and 2.4 in Volume II).

It is freely acknowledged in the literature that the crossing of the plasmic membrane is dependent on the physicochemical properties of the molecule considered, the essential criterion being liposolubility. It is important to stress, however, that the transport mechanisms are complex because of the extreme diversity of the three components in question: the membrane, the surrounding environment, and the toxic product being considered. The transport of elements can be passive or active, depending on whether the variation of free energy is positive or negative. Passive transports correspond to simple diffusion, due to a concentration gradient on either side of the membrane. They can be mediated, or facilitated, by (1) the presence of "pores", or aqueous canals, protein in origin, that cross the hydrophobic bilayer; or (2) certain macromolecular carriers, after complexation on the external face.

Active transports are necessarily accompanied by energy consumption, which derives either from hydrolysis of the ATP molecules (e.g., Na-K pumps), or from an ionic gradient (proton gradient in bacteria or in the absorbent cells of the root in plants). There are also exchanges by endocytosis, by which means solid or liquid elements are able to penetrate

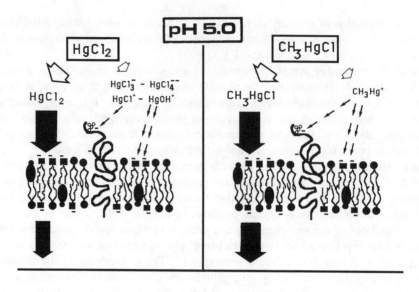

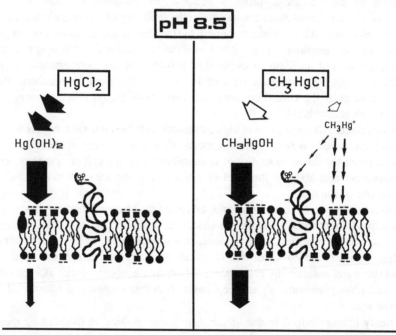

FIGURE 11.   Schematic presentation of the fixation and transport processes of two mercury compounds ($HgCl_2$ and CH_3HgCl) across model membranes, in relation to pH of the medium (5.0 and 8.5).

- phosphatidylserine (PS)
- phosphatidylcholine (PC)
- cholesterol
- protein
- polysaccharide
- negative charge
- mercury flux

(From Boudou, A., Georgescauld, D., and Desmazès, J. P., in *Aquatic Toxicology*, Nriagu, J. O., Ed., John Wiley & Sons, New York, 1983, chap. 4. With permission.)

the cytoplasm. This is the process that occurs when macromolecules are absorbed by pinocytosis in the second segment of the intestine of fish.

In the case of toxic products, active transport mechanisms are currently considered to be very infrequent. In certain species of bacteria — *Staphylococcus aureus* — cadmium appears to use the active transport system of manganese to reach the intracellular medium.[55]

Thus, crossing the plasmic membranes for most contaminants is based on passive transport systems, conditioned by the size of the molecules, their electric charge, and their solubility. From a chemical point of view, it is particularly important to consider the unstirred aqueous layers situated between the external environment and the cell membrane. They constitute a barrier to diffusion, and crossing it depends on hydrosolubility of the molecule (see Chapter 5). Thus, the resistance of this barrier (reciprocal of permeability) is based on solubility criteria which are the opposite of those of the membrane and its hydrophobic, phospholipid bilayer.[56]

It is always very difficult, however, to determine with any precision the quantities adsorbed. There are methods available such as cellular fractioning associated with separating techniques, microlocalization by means of histochemistry or with an electron microprobe (X-ray fluorescence), and the use of complexants in the environment. However, the follow-up to these methods and interpretation of the results pose fairly delicate problems, given the artifacts associated with the treatments that the biological material has to undergo and given the extreme lability of some bonds at the interfaces with the surrounding environment.

There is currently very little knowledge in ecotoxicology concerning the mechanisms ensuring the fixation of contaminants on membranes and the crossing of these barriers, despite the importance of these phenomena in relation to the processes of bioaccumulation and the toxicological effects to which they may lead. Let us take as an example the case of mercury compounds. For a long time, the thioloprive properties of this metal made it possible to analyze its fixation on membrane proteins (SH groups and S–S bridges) and show up many alterations to the enzymes or structure proteins.[57] On the other hand, very little work has been devoted to the role of the lipid component of membranes.[58] In the research program that we are developing on the processes of bioaccumulation and transfer of mercury compounds (see Chapter 1.1 in Volume II), we have undertaken to study the interactions between this heavy metal and the phospholipid bilayers of membranes. The basis of our work consists of taking into account the physicochemical factors of the medium (e.g., temperature, pH, chlorine concentrations) and some properties of the membrane (e.g., natural and synthetic phospholipids, presence or absence of electric charges on the polar heads, saturated or unsaturated aliphatic chains). By using several biophysical techniques applied to different lipid membrane models (multilamellar vesicles, liposomes, monolayers, and bimolecular lipid membranes [BLMs]), we were able to make, in particular, the following observations:

1. The accessibility of mercury to the hydrophobic intramembrane core (measuring the quenching of the fluorescence emission of pyrene) is very strongly influenced by the chemical form of the metal ($HgCl_2$ and CH_3HgCl or the different species associated with them), by the physicochemical conditions of the medium, and by the nature of the membrane phospholipids and the electric charges carried by their polar heads.[59]

2. We found a very important variation in the global transmembrane fluxes of mercury, according to the chemical form of the metal, the pH, and the chlorine concentration of the medium. These fluxes are due to the transport of neutral species, as the chlorine complexes — $HgCl_2$ and CH_3HgCl — are much more capable of crossing the hydrophobic barriers.[60] These results agree with those of Gutknecht[61] who showed that the average speed of diffusion of the species $HgCl_2$, through BLMs, is $1.3 \cdot 10^{-2}$ cm/s, which corresponds to a permeability 20 times greater than that of water and a million times greater than that of sodium, potassium, or chlorine ions. When the diffusion coefficients are calculated, these also reveal the significantly important role played by unstirred aqueous layers (Figure 11).

3.   We also found an effect of the mercury compounds on the dynamics of the phospholipids within the membrane models. Thanks to the technique of fluorescence polarization, we were able to demonstrate a very important action of inorganic mercury
($HgCl_2$) on the thermotropic properties of phosphatidylserine (PS) and phosphatidy-
lethanolamine (PE), two abundant phospholipids in natural membranes. The heavy
metal induces a strong rigidification of the bilayers with, in some experimental conditions, the complete disappearance of the lipid transition. A comparative analysis of
different phospholipids indicates a specific interaction of Hg (II) with headgroups
bearing a primary amine. The effects observed are very different from those of di- or
trivalent cations; they are modulated by the pH of the medium in relation to the chemical
speciation of mercury.[62]

3. Intracellular Storage and Biotransformation of Contaminants

Once they have crossed the plasmic membrane, toxicants reach the cytoplasmic compartment. This is a different physicochemical environment, but clearly (from the number
and diversity of potential ligands, for example) it is of the same complexity as the aquatic
biotopes already described. Distribution of contaminants throughout the intracellular compartments — hyaloplasm, organelles — is essentially based on shifts in the chemical equilibria
and, in certain cases, on biotransformation reactions. Originally brought to notice in mammals, biotransformation reactions are currently being studied in aquatic organisms.[63] They
seem to be integrated into the excretion function, inasmuch as they transform the lipid-
soluble compounds into more hydrosoluble products, thus making their elimination easier.
The majority of these reactions come under the heading of detoxification of pollutants, after
their chemical transformation. Some, however, produce metabolites which are sometimes
more toxic than the molecule absorbed (biotoxification). The enzymes catalyzing the biotransformation reactions are mainly localized in the soluble, mitochondrial, or microsomal
part of the liver. They are also present in the intestine, lungs or gills, and kidneys. These
enzymes are practically aspecific and become active only when the substratum is strongly
lipophilic. There are two main types of reactions: phase 1, consisting of nonsynthetic reactions
(hydrolysis, reduction, oxidation), and phase 2, corresponding to synthetic reactions (con-
jugation). Some examples of biotransformation reactions on pesticides by the hepatic enzymes from fish are shown in Figure 12. The microsomes, which derive from homogenization
of the smooth endoplasmic reticulum, contain highly active enzymes which are capable of
oxidizing a large number of xenobiotics. These enzymes — monooxygenases — are comprised of a group of hemoproteins called cytochrome P-450. Research carried out on fish
shows the influence of many abiotic and biotic factors on these mechanisms: e.g., temperature, photoperiod, season, diet, species and strains, age, sex.[63]

Concomitantly with these biotransformation processes, neutralization or sequestration
processes may also take place. These do not reduce the amounts bioaccumulated, but they
do withdraw certain elements from the metabolic circuits, either temporarily or definitively.
Several phenomena have been described, the most important being the fixation of metals
such as cadmium, mercury, copper, or zinc by metallothioneins, proteins rich in sulfhydryl
groups (see Chapters 2.2 and 2.3 in Volume II). We should also mention complexation
reactions leading, at cytoplasm level, to precipitations of insoluble salts (Figure 13) or
granular deposits in concentric layers (spherocrystals). This, however, is not a type of
decontamination, as the toxic product is still present in the organism and will be taken into
account when measurement is carried out, for example. However, the toxicological risks
are greatly reduced, both for the individual and for its predators.

4. Organotropism and Decontamination Mechanisms

In pluricellular organisms, especially the higher organisms, biouptake is not limited simply
to crossing the biological barriers between the external environment and the cytoplasm of

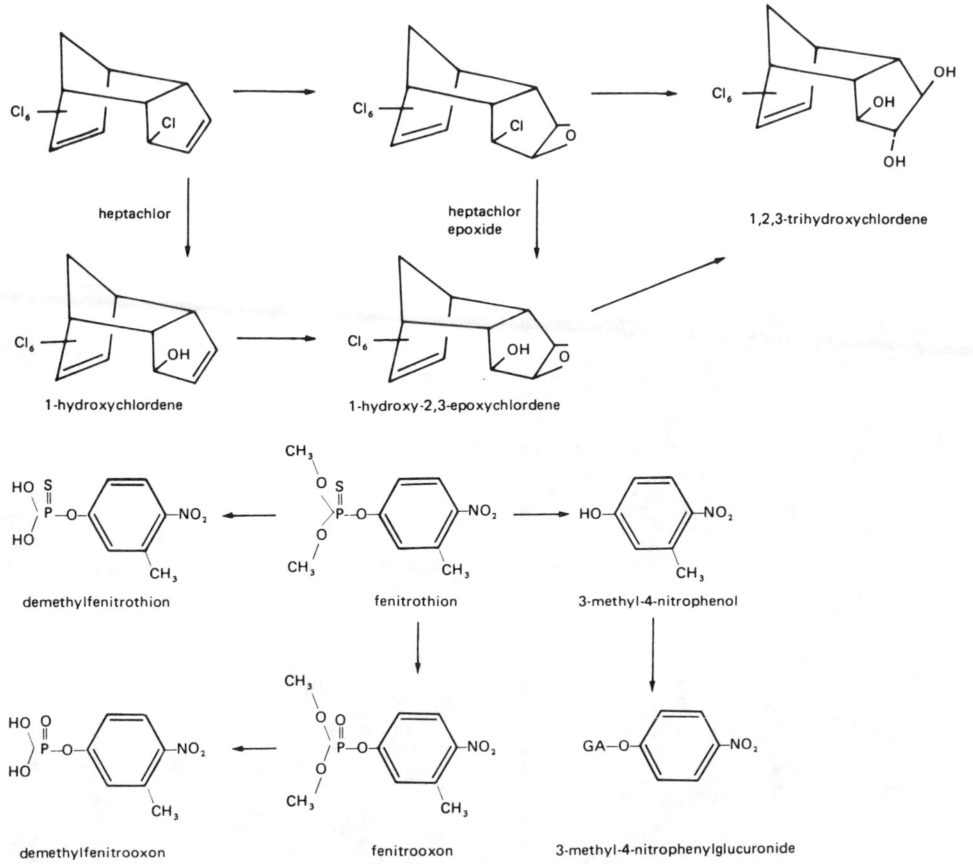

FIGURE 12. Biotransformation reactions of pesticides. (Above) Heptachlor (goldfish); (below) fenitrothion (rainbow trout). (From Kahn, M. A. Q., et al., in *Fundamentals of Aquatic Toxicology,* Rand, G. M. and Petrocelli, S. R., Eds., Hemisphere, Washington, D.C., 1985, 526. With permission.)

the epithelial cells. Indeed, many toxicants are able to leave their fixation sites in the biological barriers or in the cytoplasm (apparently without these being saturated) to pass into adjacent cells and reach the blood fairly quickly at the level of the capillary network. Although it is relatively easy to admit that processes similar to those already described form the basis of these intercellular exchanges, it is, however, a much more delicate problem to imagine what triggers these fluxes and to explain the rapidity of the phenomena for inorganic and organic pollutants: "Many of the biochemical processes involved in xenobiotic transport are poorly understood."[49]

In the overall mechanism of bioaccumulation, on the scale of the organism, blood plays two principal roles: first, the storage of a number of toxicants, at plasma level or in blood cells; second, the transportation of these products to the different tissues. Mercury compounds are once again a revealing example for the complexity and efficiency of the mechanisms. In fish blood inorganic compounds are localized in the serum fraction, fixed for the most part on the thiol groups of circulating proteins (albumins). Methylmercury, on the other hand, is stored almost exclusively in the erythrocytes on hemoglobin molecules. In both cases, fixation results from covalent bondings with (for the organic compound) intracellular localization. All studies carried out *in vitro* and *in vivo* indicate that the exchange capacities between the metal transported in the blood and the tissue compartments of the fish are very large and particularly fast.[65,66]

No typical scheme can be presented to show the distribution of a toxicant in the different

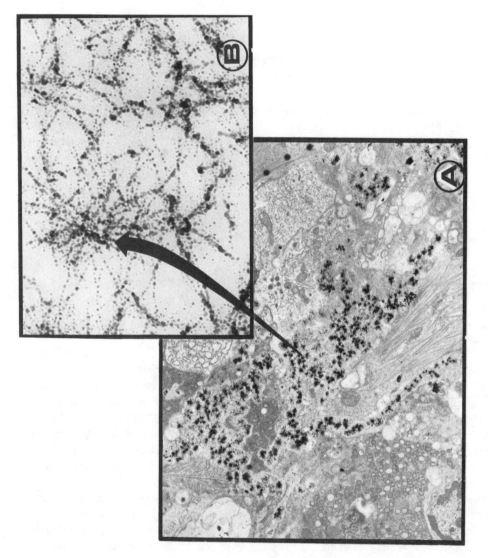

FIGURE 13.   Silver-sulfide precipitation in oyster cells after 4-week exposure (20 ppb Ag$^+$ in seawater). (A) Basement membranes of digestive cells. (B) silver precipitates at high magnification. (Magnification × 230,000). (Documents from Lab. d'Histophysiologie fondamentale et appliquée, Université de Paris VI, Dr. R. Martoja.[64])

organs of an individual. In fact, the process is the result of a very large number of parameters which all interact together, from the physicochemical properties of the contaminant, the characteristics of the biotope, and the modes of contamination (penetration routes, absorbed doses, etc.) to the anatomical and physiological particularities of the organism, the modes of transporting the element in the blood, and so on. In any given tissue, accumulation is a function of vascularization, of the properties of the biological barriers, of the abundance of fixation sites and their accessibility, and of the cell turnover, among other factors.

In the case of certain toxicants, a limited number of criteria have a dominating role with respect to the bioaccumulation processes. DDT, for example, a very liposoluble organochlorine insecticide, is mainly influenced in its localization in tissue by the fat content of the organs.

In fact, bioaccumulation of toxicants in organisms and their localization in the different organs or organelles, undergo almost constant evolution at cell level. This evolution is the result of the biouptake phenomena that have just been described and of decontamination processes that work toward reducing the amounts accumulated. On a larger scale, that of the whole organism, the elimination of toxicants into the external environment is generally based on excretion mechanisms. In aquatic vertebrates, the principal route is the production of urine. Glomerular filtration makes it possible for very small hydrosoluble molecules (molecular weight <60,000) to pass into the kidneys. This is also true for certain elements dissolved in the blood (e.g., metals). Other physiological mechanisms work toward eliminating toxicants: transfers at the epithelial structures, which ensure penetration of the contaminants into the organism, but also their elimination (gills, cutaneous barrier), accumulation in the gall bladder and discharge with bile into the small intestine, and indirect elimination of certain liposoluble chemicals through egg-laying or molting.

5. "Bioaccumulator" Organisms

Any aquatic organism living in a contaminated biotope is a potential target for accumulation. However, a very large number of parameters will condition its capacity to bioaccumulate. Let us look, for example, at its anatomical and physiological characteristics linked with its species, its stage of development, its sex, its diet; its lifestyle and ecological habitat (benthic, planktonic, intrasedimentary species); its position in the trophic networks, etc. Hence, faced with similar contamination conditions, certain species, known as biological "concentrators", show a very strong capacity for bioaccumulation. Such is the case for the following examples:

1. Unicellular organisms — bacteria or algae — living in the water column or at the sedimentary interface: thanks to their vast surface area for exchange with the environment, they are able to fix large quantities of toxic products by processes of adsorption or absorption. Expressed in the form of a bioconcentration factor (BF = concentration in the organism/concentration in the water), the bioaccumulation of organochlorine insecticides or heavy metals can be as high as 60,000 for certain bacteria or phytoplanktonic algae.[3,66]
2. Filtering organisms such as bivalve mollusks: their microphagic nutrition, together with their respiratory needs, necessitate the daily filtration of very large volumes of water (48 l/d for an oyster weighing 20 g).[3] This, therefore, favors the bioaccumulation of contaminants present in the environment.
3. Aquatic plants, especially bryophytes (*Fontinalis, Cinclitodus, Scapania*): these fixed species are perennial and characterized by a very strong capacity to accumulate trace metals (e.g., cadmium, chrome, zinc, lead) and organic micropollutants. This is due in particular to their richness in lipid components compared with other plants (1.4 to 4% fresh weight).[67] The rooted macrophytes which colonize lentic and shallow biotopes

should also be mentioned. Their root system is buried in the sediment, ensuring their anchorage and, for many species, drawing their mineral nutrients from the substratum. These species are likely to bioaccumulate the contaminants stored in sediments, thus representing a potential support for contaminant transfer into the benthic trophic networks.

4. Intrasedimentary species: considering the ecotoxicological properties of the sediment compartments — storage of contaminants, chemical transformations — organisms living permanently or temporarily in the substratum are particularly exposed to toxic products, either by direct transfer from interstitial water or during ingestion of the sediment. Let us mention, as an example, the mayfly — *Hexagenia limbata* and *H. rigida* — whose nymphal stage lasts about 1 year. The detritivorous larvae burrow into the substratum and feed by means of almost continuous ingestion of sediment. The estimated production of these two species in Lake Winnipeg (Canada) is about 100,000 tons/year; this represents the principal alimentary source for many carnivores, especially the fish fauna.[68]

The unicellular or multicellular organism represents the base unit in aquatic biocenoses, and we have already seen how it is integrated into complex structures, based essentially on prey-predator relationships: the food chains and networks. The transfer of contaminants by the trophic route, from autotrophic organisms to terminal consumers, confers a new dimension on contamination processes which cannot be fully understood except by means of an integrated ecotoxicological approach, such as has been described in the introduction to this chapter.

6. Transfer of Contaminants within Trophic Networks

At the base of trophic networks are autotrophic organisms, living in freshwater ecosystems — phytoplanktonic algae, macrophytes, certain species of bacteria — and generally speaking, they have a great capacity for bioaccumulation of most micropollutants present in biotopes. Thus, they constitute a significant potential source of contamination for primary consumers (zooplanktonic species, mollusks, certain herbivorous fish). The importance of the trophic route will be closely dependent on the levels of contamination provided by the foods consumed, but it will also depend, as was mentioned earlier, on the efficiency of the transport mechanisms through the intestinal barrier of the predators. A comparative study on trophic transfers of mercury compounds between *Chlorella vulgaris,* a unicellular phytoplanktonic algae, and *Daphnia magna,* a zooplanktonic microcrustacean, shows the importance of these phenomena. Indeed, with similar exposure conditions — $HgCl_2$ and CH_3HgCl, 1 µg Hg/l, 18°C, 24 h — the quantities of mercury fixed by the *Chlorella* are similar. However, the consumption by the *Daphnia* of these contaminated algae shows a transfer rate of nearly 100% for methylmercury and only 6% for the inorganic compound.[69,70]

The trophic transfers begun in this way at the start of alimentary networks can continue through herbivorous species and primary carnivores up to terminal consumers.

Several *in situ* ecotoxicological studies have shown a more or less progressive increase in the concentrations of certain contaminants, as one moves up through the trophic networks. From among these studies, let us mention the now famous case of Clear Lake in California. This ecosystem was treated with organochlorine insecticides (DDD), in order to eliminate the aquatic larvae of gnats — *Chaoborus astictopus* — as the adult stage used to swarm around the banks of the lake in summer. Because the pollutants cumulated during the transfer processes, they contaminated the higher levels of the trophic networks to a very great degree, which led to massive death rates among the populations of western grebes — *Aechmophorus occidentalis,* a fish-eating bird. The average concentration of DDD in their fat was more than 2500 mg/kg (ppm), which corresponds to a BF of around 150,000 (Figure 14).[71]

A similar phenomenon was the cause of the "Minamata illness" in Japan. Mercury from

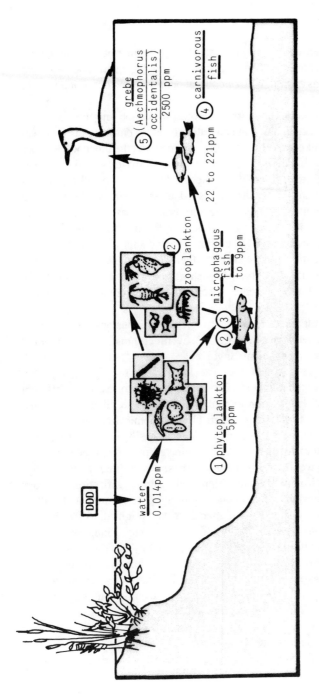

FIGURE 14. Schematic presentation of DDD transfer through the trophic networks of Clear Lake (California). (From Ramade, F., *Ecotoxicologie*, Masson, Paris, 1977. With permission.)

industrial waste had been discharged into the bay. The combination of the chemical transformations of the metal (biological methylation) and the transfers along the marine food
chains led to some extremely high levels of contamination in carnivorous fish. For certain
species, the estimated BF was greater than 500,000. Because their diet was not very varied,
the local population of fishermen absorbed large doses of mercury daily, leading to behavioral
problems, severe neurological attacks, many cases of teratogenesis, and several dozen, if
not hundreds, of deaths.[72,73]

Several other examples have been described in lake, estuary, or marine environments,
though with less dramatic consequences for animal species and for man. They involve
mercury, but also other metals such as cadmium or zinc, organochlorine and organophosphate
pesticides, certain radionuclides (e.g., ^{137}Ce).[74-79]

These mechanisms are known as bioamplifaction or biomagnification and are the result
of the ecotoxicological processes that contribute, at ecosystem level, to ensuring the transfers
of contaminants between the different compartments of the biotope and the biocenose. Many
articles have been written on biomagnification. They reveal much controversy on the subject,
as some authors contest the concept and others even reject it totally.[80] Several arguments
are used, especially the contradictions in results published after *in situ* studies: "Field
observations give conflicting accounts."[81] They show, in certain cases, that the highest
concentrations correspond to the species situated at the base of the trophic networks.[4,81,82]
We should also stress that several research projects that revealed this phenomenon were
based on protocols over which very little care was taken, either when samples were collected
or when contaminants were measured: "Data for organisms from lower trophic levels are
often based on analysis of the whole animal, whereas specific tissues are commonly analyzed
from larger animals."[4]

Although it would be very difficult for us to draw any precise conclusions regarding the
basis of biomagnification and its generalization, there are several comments that we would
make relating to the ecotoxicological mechanisms involved:

1. Only those contaminants with a chemical stability great enough to resist the different
    abiotic and biotic processes of degradation in natural conditions are likely to be trans-
    ferred along the trophic chains. Furthermore, their physicochemical properties should
    enable them to cross the intestinal barriers of organisms easily in order to ensure
    maximum efficiency for the transfers between contaminated prey and predators.
2. As a general rule, the average weight of organisms increases, sometimes considerably,
    when one passes from a given trophic level to a higher one. Biomagnification is always
    expressed in relation to the criterion "global concentration". The accumulated content
    ratio between predators and prey should be higher than the weight ratio for there to
    be an increase in concentrations between the trophic levels. Only a long-lasting con-
    tamination phase can ensure a sufficient supply of contaminant by the alimentary route.
    In the natural environment, the average life span of aquatic species, like the weight
    factor, increases as one moves up trophic networks: a few hours for the unicellular
    algae; a few days for the zooplanktonic microcrustaceans; several months for carniv-
    orous fish of the first order; several years for terminal consumers.
3. The efficiency of contaminant transfers is also dependent on the number of trophic
    levels and the diversity of consumed prey.
4. Last, we should remember that heterotrophic aquatic organisms can be contaminated
    by the trophic route and by the direct route. The respective importance of these two
    routes is extremely difficult to quantify in a natural environment. According to some
    authors, food is often the most important source of pollutants in marine or freshwater
    systems; for others, the contaminants present in the water column are much more
    easily accumulated by organisms.[83,84] Once again, there is no simple solution that can

be applied generally. A very large number of parameters interact, in relation to the characteristics of the contamination and of the ecosystem being considered. In both cases, the length-of-exposure factor plays an essential role.

Thus, the biomagnification of toxic products through alimentary transfers within trophic networks is a complex and progressive phenomenon. It requires the convergence of a set of favorable conditions, strongly linked with the time factor. A brief theoretical scenario, based on the evolution of the contamination of a continental aquatic ecosystem, can serve to illustrate these phenomena. During the first phase, which corresponds to the dispersal of the toxicant in the biotope, concentrations in the environment are relatively high and can bring about rapid contamination in the species representative of the first trophic levels of the biocenose. Samples and analyses made at that point would reveal an inverse biomagnification. Next, in the middle term, toxicant concentrations in the higher heterotrophic organisms increase progressively by means of cumulative transfers. It is during this second phase that contamination levels in the terminal consumers could reach maximal values, compared with the other links in the food chains. Furthermore, expressing the results with the aid of concentration factors will have the effect of amplifying the phenomena, as the concentrations measured in the water will have decreased (many pollutant complexation reactions, with the suspended matter and sediments). With a decrease in the exogenous supplies of contaminants in the biotope, the biomagnification previously observed will progressively stagnate and then drop away as the transfers by the direct and trophic routes have diminished and no longer compensate for the decontamination processes, especially in terminal consumers.

To sum up, we shall quote Moriarty's words:[4] "Biomagnification is too simple an idea for aquatic habitats."

Having considered the main ecotoxicological mechanisms ensuring the distribution of toxicants throughout the different compartments of aquatic biotopes and their accumulation by living beings, at organism level and within trophic networks, we shall now deal with the effects of contaminants on the structure and functioning of ecosystems.

B. Effects of Contaminants on the Structure and Functioning of Aquatic Ecosystems

The toxic capacity of contaminants with regard to living beings takes the form of structural damage and physiological or biochemical dysfunctions, some quite specific and some less so. It is closely dependent on the characteristics of the bioaccumulation, in particular the distribution of toxic products at cellular and molecular levels in the organs.

In this chapter, our intention is not to draw up an inventory of the many effects of contaminant action on aquatic organisms. Indeed, it can be assumed that these are as numerous and diverse as the levels and criteria for analysis that have been selected: e.g., morphological, physiological, biochemical, ethological. Moreover, several chapters in this book refer to this subject (Chapters 2.2 and 2.3 in Volume II, in particular) by describing the basic points of methodologies currently used in ecotoxicology with fish and phytoplanktonic algae as biological models.

Many authors have established direct links between the dose or concentration of a pollutant in the environment or food and the adverse effects on the organism (dose-response model). In fact, as we mentioned earlier, the accessibility of contaminants to the fixation sites within the organism is influenced by a very great number of parameters relating to the properties of the product studied, the physicochemical characteristics of the biotope and the biological barriers separating the individual from its environment. In this connection, the concentration criterion, at whole-organism level or principal organs, is a better indicator of the toxicological effects and of the risks incurred.

As was the case in the analysis of bioaccumulation and transfer processes, the organism represents the biological base level for understanding the toxic effects of contaminants.

According to Sheehan,[85] five main responses can be defined, depending on the levels of exposure and the severity of the damage:

1. Acute toxicity causing mortality
2. Chronically accumulating damage ultimately causing death
3. Sublethal impairment of various aspects of physiology and morphology
4. Sublethal behavioral effects
5. Measurable biochemical changes

When one is dealing with the population level, other criteria must be taken into account in order to understand the toxic effects of contaminants. Thus, among a group of individuals of the same species living in the same biotope, significant differences in toxic effect appear between individuals depending on, among other factors, age, development stage, sex, and morphological characteristics of organisms.

Contaminants are liable to affect the population size and dynamics, which are based on four principal measurements: birthrates, deathrates, gains from immigration, and losses from emigration.[4] They can thus cause a reduction or an increase in the natural fluctuation of numbers, in the biomass, in the sex ratio, and so on.

The effect of toxicants on the reproductive functions is particularly disturbing, as this has consequences for the future of the population. For those aquatic species that reproduce sexually, e.g., fish, there are several targets that are liable to be affected, independently or globally: gamete development, fertilization, embryo development, hatching, sexual maturation, courting and mating, etc.[85] (see Chapter 2.2 in Volume II).

In the paragraph relating to the ecological bases of ecotoxicology, we mentioned the importance of genetic factors with respect to the structure, functioning, and evolution of populations. There are many natural chemical substances or xenobiotics which are capable of interacting with the support molecule of genetic inheritance, deoxyribonucleic acid (DNA). Two basic processes can be distinguished, in relation to the mechanisms at work and their consequences for organisms and populations:

First, the genetic toxicity of contaminants affects the hereditary material and brings about cellular dysfunctions, which could lead to the death of individuals (for example, by inducing cancers). The effect of ionizing radiation on DNA has been recognized for a long time now. It is seen especially in the appearance of mutations, the fate of which depends on whether they are dominant, neutral, or recessive, and also on their relative fitness.[4] A heavy metal such as cadmium is able to interact by complexation with nucleic acids, with the purine or pyrimidine bases, or with the phosphate groups; it can also disturb their metabolism by inhibiting the enzymes involved.[86] In the same way, PAH, through their metabolites produced by liver microsomal transformations (diol-epoxides), are able to bond covalently with DNA and, thus, have a very strong mutagenic and carcinogenic capacity.[38] Genetic toxicity can become evident at gene (gene mutations) or at chromosome level (chromosome breaks, aneu- and polyploidy). These disruptions will either be restricted to the cell or to the organism (metabolic dysfunctions, cancers, etc.) or, in certain cases, they will be transmitted to descendants through the gametes. We must stress, however, that enzymatic systems are capable of repairing DNA when one or sometimes even both of its polynucleotide chains are injured (endonucleases, exonucleases, polymerases, etc.).

Second, the action of contaminants exerts a selection pressure on the genetic polymorphism of populations. In contrast with the natural mechanisms of evolution from which the extreme diversification of species within the biosphere originates, pollution is characterized by the speed with which it can bring about changes: "The critical difference between evolutionary change and that wrought by pollution is the speed."[4] Faced with the contamination of their biotope, the individuals of one species may react differently and set up a variety of adaptive reactions. The evolution of a population in an ecosystem, especially the percentage of mortality, depends on the impact of the toxicants.

Certain organisms are naturally endowed with defense mechanisms and confront the appearance of contamination as a new selection factor (genetic preadaption theory). As they are insensitive to this stress, they form the basis for resistant populations. They represent a change in the gene pool which will be transmitted by heredity to future generations. Such ability to resist pollutants is currently the subject of a great deal of research work, especially on insects and their reactions to treatment with insecticides.[87] While it is a well-established fact that very many insect species which devastate crops have become resistant to one or more insecticides through elimination of individuals without resistant genotypes,[88] certain phenomena may not necessarily be due to genetic mechanisms.[4]

A study of the genetic, biochemical, and physiological mechanisms of resistance shows the complexity of the modifications developed, and these mechanisms can be found at every stage of contamination of the organism: penetration, detoxification rate (amplification of the genes and increased production of biotransformation enzymes), excretion, alteration of action sites (e.g., structural modification of the active site of the acetylcholinesterase and resistance to carbamates or organophosphates).[89] In bacteria, studies into resistance to cadmium have shown five main mechanisms: increased fixation of the metal on cell walls due to exopolymers; synthesis of metallothioneins and intracellular sequestration of the metal; formation of insoluble complexes (cadmium sulfides); decrease in bioaccumulation and absorption capacities; methylation of cadmium (volatile form).[46]

In the aquatic environment, relatively little is known about resistance in nontarget species, except those with known economic value such as fish.[90] From an ecotoxicological point of view, these resistance phenomena can contribute, in certain cases, to furthering the bioaccumulation processes of trace metals or persistent xenobiotics, thus increasing the risks attached to their cumulative transfer along trophic networks.

The effects of contaminants on the biocenose are centered on disruptions to interspecific relations, especially predator-prey interactions. Any source of pollution that affects the biomass and productivity of one of these trophic levels will necessarily lead to imbalances, sometimes catastrophic, within the populations of the higher levels. Let us quote, for example, the very sharp increase in the numbers of amphipods (*Gammarus lacustris* and *Hyalella azteca*) on the banks of two Canadian lakes after the destruction of the fish population by rotenone;[91] or the phytoplankton blooms after treatment of artificial ponds by organophosphate pesticides which led to the death of most of the zooplanktonic herbivores.[92]

As in extreme environments — ecosystems at altitude, deserts, etc. — pollution of aquatic environments causes, at biocenose level, a decrease in species richness with, in some cases, proliferation in certain species, thus ensuring the maintenance of, or even an increase in, global biomasses (Figure 15). As Sheehan[93] points out, "such effects cannot be totally attributed to direct toxic mortality, but may be due also to induced reductions in the abilities of organisms to function successfully in competitive and trophic interactions, or may be the result of increased emigration or reduced immigration." A diminution such as this in biocenotic diversity is always a sign of weakening in the ecosystem. Its ability to adapt to new stresses, whether natural or anthropic, is reduced (see Chapter 8).

The effects of contamination on the structure of biocenoses is seen in repercussions, appreciable in varying degrees, on the fundamental functional characteristics of ecosystems: primary productivity, energy transfers, decomposition of organic matter, flow of matter within the cycle, control and internal feedback reactions. So far, our knowledge of these topics is very limited, and this is a reflection of the hypercomplexity of ecotoxicological mechanisms and the difficulties involved in setting up investigation methods. It is usually necessary to wait months or decades to be able to reveal significant symptoms of a dysfunctioning, compared with variations caused by the natural fluctuations of ecological factors (Figure 16).

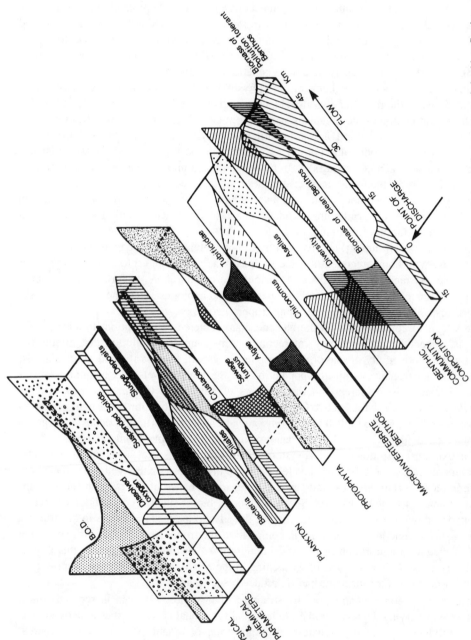

FIGURE 15.   Spatial variation of physical, chemical, and biological consequences of the continuous discharge of a severe organic load into flowing water, (From Bartsch, A. F., in *Biological Indicators of Freshwater Pollution and Environmental Management*, Hellawell, J. M., Ed., Elsevier, London, 1986. With permission.)

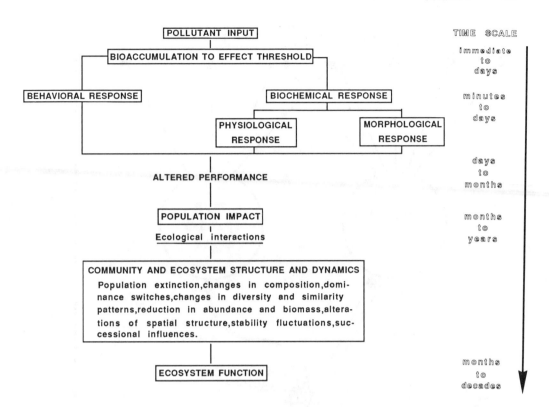

FIGURE 16.   A conceptual chronology of induced effects following exposure to toxic pollutants, emphasizing changes in community and ecosystem structure and dynamics. (From Sheehan, P. J., in *Effects of Pollutants at the Ecosystem Level*, SCOPE 22, Sheehan, P. J., Miller, D. R., Butler, G. C., and Bourdeau, P., Eds., John Wiley & Sons, Chichester, 1984, 51. With permission.)

Among the most important questions with which ecotoxicology is today confronted are those concerned with the capacities of an ecosystem to adapt and rehabilitate itself when faced with different sorts of contamination. Compared with normal history, which leads to an optimal stage (climax), man's action causes disturbances of varying degrees of severity which contribute to modifying the evolution of natural systems and to gradually increasing the divergence between the theoretical and real states[94] (Figure 17). The ecotoxicological approach to these phenomena requires setting up a diagnosis of the extent of this divergence and establishing methodologies to evaluate the potential adaptability of ecosystems. A whole set of concepts can be linked with this analysis: inertia (ecological buffering capacity), resilience (disruption that a system can absorb and then return to a stable state), hysteresis (degree to which restoration path is an exact reversal of degradation path), malleability (degree to which stable state established after disturbance differs from the original steady state).[93]

Among studies recently carried out at ecosystem level, those dealing with the Thames estuary have established close links between the pollution zones and their evolution over the last few decades, the physicochemical factors of the biotope, and the dynamics of the macrofauna populations and fish communities.

VI. CONCLUSION

In an ecotoxicological approach to the contamination of natural systems, one is confronted with the extreme complexity of the mechanisms involved. These mechanisms represent not only the result of action by the abiotic, biotic, and contamination factors, but also of their

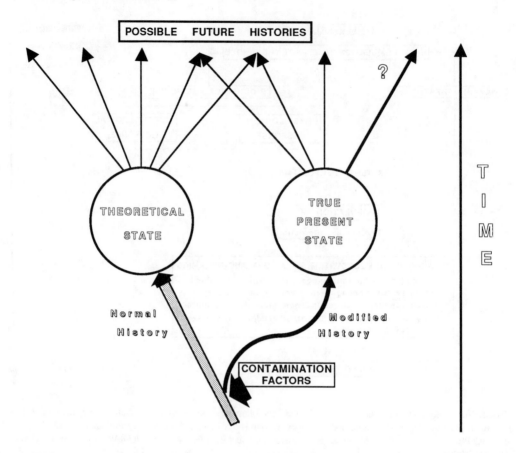

FIGURE 17.   Theory of adaptative strategies: evaluation patterns for an ecosystem. (Adapted from Blandin, P., *Bull. Ecol.*, 17, 215, 1986.)

interactions. What is more, these factors are constantly varying, according to both time and space.

In this chapter, a synthesis has been developed that covers the principal processes that enable toxicants to enter freshwater ecosystems, the distribution of these toxicants in the different compartments of the biotope (aquatic phase, suspended matter, sediments), their accumulation in living beings, and their transfer within the trophic networks. All these processes are at the origin of the dysfunctions which affect ecosystems, dysfunctions of varying degrees of severity and duration, depending on the level of contamination, the scale of the effects, and capacity of the systems concerned to adapt or recuperate.

The role of the chemical fate of the toxicants in biotopes and biocenoses is paramount, both for metal pollutants and organic products: chemical speciation, complexation reactions with the suspended matter and the sediments, chemical transformations, etc. On it depend the bioavailability of contaminants and their biouptake by living beings. The majority of ecotoxicological processes are associated with reactions occurring at the interfaces between the aquatic phase, the atmosphere, suspended matter, sediments, and organisms (biological barriers and cell membranes).

Because of the wide-ranging aims of ecotoxicology, this discipline involves a vast area of research. Indeed, the mechanisms and effects observed on the scale of ecosystem or even ecosphere are the combined result of an almost infinite number of processes which can be observed at lower levels, from the atomic and molecular structure up through the different biological integration levels.

The current extent of knowledge is an indicator of the progress made in this field over the last decade, but it also reveals the great heterogeneity of available information and the gaps in our knowledge as to the causes and consequences of the dysfunctions in contaminated systems.

An assessment of the present situation shows that two steps are necessary: first, we must continue with work already begun; second, we must define new research objectives, either in specialized fields or based on a multidisciplinary approach. In any case, all increase in our knowledge is directly dependent on our devising and setting up appropriate methods to produce new results capable of making new contributions to the discipline or to reinforce those concepts already established.

REFERENCES

1. **Truhaut, R.**, Ecotoxicology: objectives, principles and perspectives, *Ecotoxicol. Environ. Saf.*, 1, 151, 1977.
2. **Butler, G. C., Ed.**, *Principles of Ecotoxicology*, John Wiley & Sons, Chichester, 1978.
3. **Ramade, F.**, *Ecotoxicologie*, Masson, Paris, 1977.
4. **Moriarty, F.**, *Ecotoxicology: The Study of Pollutants in Ecosystems*, Academic Press, London, 1983.
5. **Butler, G. C.**, Developments in ecotoxicology, *Ecol. Bull.*, 36, 9, 1984.
6. **Rand, G. M. and Petrocelli, S. R., Eds.**, *Fundamentals of Aquatic Toxicology*, Hemisphere, Washington, D.C., 1985.
7. **Duvigneaud, P.**, *La Synthèse Écologique*, Doin, Paris, 1974.
8. **Cole, L. C.**, The ecosphere, *Sci. Am.*, 198, 83, 1953.
9. **Barbault, R.**, *Ecologie Générale*, Masson, Paris, 1983.
10. **Blandin, P. and Lamotte, R.**, Ecologie des systèmes et aménagements: fondements théoriques et principes méthodologiques, in *Fondements Rationnels de l'Aménagement d'un Territoire*, Lamotte, M., Ed., Masson, Paris, 1985, 139.
11. **Dobzhansky, T.**, *Génétique du Processus Évolutif*, Flammarion, Paris, 1977.
12. **Blandin, P.**, Evolution des écosystèmes et spéciation: le rôle des cycles climatiques, *Bull. Ecol.*, 18, 59, 1987.
13. **Odum, E. P.**, *Fundamentals of Ecology*, W. B. Saunders, Philadelphia, 1959.
14. **Hellawell, J. M.**, *Biological Indicators of Freshwater Pollution and Environmental Management*, Elsevier, London, 1986.
15. **Miller, D. R.**, Distinguishing ecotoxic effects, in *Effects of Pollutants at the Ecosystem Level*, Sheehan, P. J., Miller, D. R., Butler, G. C., and Bourdeau, P., Eds., John Wiley & Sons, Chichester, 1984, 15.
16. **Ramade, F.**, *Les Catastrophes Écologiques*, McGraw-Hill, Paris, 1987.
17. **Turco, R. P., Toon, O. B., Ackerman, T. P., and Sagan, C.**, Nuclear winter: global consequences of multiple nuclear explosions, *Science*, 222, 1283, 1983.
18. **Fritsch, P. and De Saint Blanquat, G.**, Nitrates, nitrites et nitrosamines, in *Toxicologie et Sécurité des Aliments*, Derache, R., Ed., Lavoisier, Paris, 1986, 281.
19. **Miller, D. R.**, Chemicals in the environment, in *Effects of Pollutants at the Ecosystem Level*, Sheehan, P. J., Miller, D. R., Butler, G. C., and Bourdeau, P., Eds., John Wiley & Sons, Chichester, 1984, 7.
20. **Sievering, H., Dave, M., Dolske, D. A., and McCoy, P.**, Transport and deposition of trace metals over southern Lake Michigan, in *Atmospheric Pollutants in Natural Waters*, Eisenreich, S. J., Ed., Ann Arbor Science Publishers, Ann Arbor, MI, 1981.
21. **Salomons, W. and Baccini, P.**, Chemical species and metal transport in lakes, in *The Importance of Chemical "Speciation" in Environmental Processes*, Bernhard, M., Brinckman, F. E., and Sadler, P. J., Eds., Springer-Verlag, Berlin, 1986, 193.
22. **Jamet, P.**, Dispersion des pesticides dans l'environnement, in *Les Produits Antiparasitaires à Usage Agricole: Conditions d'Utilisation et Toxicologie*, Fournier, E. and Bonderf, J., Eds., Lavoisier, Paris, 1983, 189.
23. **Andrews, M. J.**, Thames estuary: pollution and recovery, in *Effects of Pollutants at the Ecosystem Level*, Sheehan, P. J., Miller, D. R., Butler, G. C., and Bourdeau, P., Eds., John Wiley & Sons, Chichester, 1984, 195.
24. **Luckey, T. D. and Venugopal, B.**, *Metal Toxicity in Mammals. I. Physiology and Chemical Basis of Metal Toxicity*, Plenum Press, New York, 1977.
25. **Salomons, W. and Förstner, U.**, *Metals in the Hydrocycle*, Springer-Verlag, Berlin, 1984.

26. **Maurin, C.**, Accidental oil spills: biological and ecological consequences of accidents in French waters on commercially exploitable living marine resources, in *Effects of Pollutants at the Ecosystem Level*, Sheehan, P. J., Miller, D. R., Butler, G. C., and Bourdeau, P., Eds., John Wiley & Sons, Chichester, 1984, 311.

27. **Bernhard, M., Brinckman, F. E., and Sadler, P. J., Eds.,** *The Importance of Chemical "Speciation" in Environmental Processes*, Springer-Verlag, Berlin, 1986.

28. **Kester, D. R.**, Equilibrium models in seawater: applications and limitations, in *The Importance of Chemical "Speciation" in Environmental Processes*, Bernhard, M., Brinckman, F. E., and Sadler, P. J., Eds., Springer-Verlag, Berlin, 1986, 337.

29. **Turner, D. R., Whitfield, M., and Dickson, A. G.**, The equilibrium speciation of dissolved components in freshwater and seawater at 25°C and 1 atm pressure, *Geochim. Cosmochim. Acta*, 45, 855, 1981.

30. **Chau, Y. K. and Wong, P. T. S.**, Direct speciation analysis of molecular and ionic organometals, in *Trace Element Speciation in Surface Waters and Its Ecological Implications*, Leppard, G. C., Ed., Plenum Press, New York, 1983, 87.

31. **Nimmo, D. R.**, Pesticides, in *Fundamentals of Aquatic Toxicology*, Rand, G. M. and Petrocelli, S. R., Eds., Hemisphere, Washington, D.C., 1985, 335.

32. **Baughman, G. L. and Lassiter, R. R.**, Prediction of environmental pollutant concentration, in *Estimating the Hazard of Chemical Substances to Aquatic Life*, Cairns, J., Dickson, K. L. and Maki, A. W., Eds., American Society for Testing and Materials, Philadelphia, 1978, 35.

33. **Stern, A. M. and Walker, C. R.**, Hazard assessment of toxic substances: environmental fate testing of organic chemicals and ecological effects testing, in *Estimating the Hazard of Chemical Substances to Aquatic Life*, Cairns, J., Dickson, K. L., and Maki, A. W., Eds., American Society for Testing and Materials, Philadelphia, 1978, 81.

34. **Tissot, A., Boule, P., and Lemaire, J.**, Photochemistry and environment: the photohydrolysis of chlorobenzene, *Chemosphere*, 13, 381, 1984.

35. **Leckie, J. O.**, Adsorption and transformation of trace element species at sediment/water interfaces, in *The Importance of Chemical "Speciation" in Environmental Processes*, Bernhard, M., Brinckman, F. E., and Sadler, P. J., Eds., Springer-Verlag, Berlin, 1986, 237.

36. **Wanchope, R. D.**, The pesticide content of surface water draining from agricultural fields: a review, *J. Environ. Qual.*, 7, 459, 1978.

37. **Thomas, R. L., Vernet, J. P., and Franck, R.**, DDT, PCBs and HCB in the sediments of Lake Geneva and upper Rhône River, *Environ. Geol.*, 5, 103, 1984.

38. **Neff, J. M.**, Polycyclic aromatic hydrocarbons, in *Fundamentals of Aquatic Toxicology*, Rand, G. M. and Petrocelli, S. R., Eds., Hemisphere, Washington, D.C., 1985, 416.

39. **Jensen, S. and Jernelov, A.**, Biological methylation of mercury in aquatic organisms, *Nature (London)*, 223, 753, 1969.

40. **Reisinger, K., Stoeppler, M., and Nurnberg, H. W.**, On the biological methylation of lead, mercury and arsenic in the environment, in *Heavy Metals in the Environment*, CEP Consultants, Edinburgh, 1981, 649.

41. **Craig, P. J.,** *Organometallic Compounds in the Environment*, Longman, London, 1985.

42. **Pritchard, P. H. and Bourquin, A. W.**, Microbial toxicity studies, in *Fundamentals of Aquatic Toxicology*, Rand, G. M. and Petrocelli, S. R., Eds., Hemisphere, Washington, D.C., 1985, 177.

43. **Colwell, R. R.**, Toxic effects of pollutants on microorganisms, in *Principles of Ecotoxicology*, Butler, G. C., Ed., John Wiley & Sons, Chichester, 1978, 275.

44. **Jackson, T. A.**, Effects of hydroelectric development on microbial methylation and demethylation of mercury in riverine lakes of northern Manitoba (Canada), in *Heavy Metals in the Environment*, CEP Consultants, Edinburgh, 1985, 282.

45. **Jackson, T. A., Kipphut, G., Hesslein, R. H., and Schindler, D. W.,** Experimental study of trace metal chemistry in soft-water lakes at different pH levels, *Can. J. Fish. Aquat. Sci.*, 30, 387, 1980.

46. **Bauda, P.**, Accumulation et Toxicité du Cadmium chez les Bactéries Gram Négatives: Rôle des Enveloppes Bactériennes, Doctorate thesis, Université de Metz, France, 1986.

47. **Brown, M. J. and Lester, J. N.**, Role of bacterial extracellular polymers in metal uptake in pure bacterial culture and activated sludge. I. Effects of metal concentrations, *Water Res.*, 16, 1549, 1982.

48. **Northcote, D. M. and Goulding, K. J.**, The chemical composition and structure of the cell wall of *Chlorella pyrenoidosa*, *Biochemistry*, 70, 391, 1958.

49. **Spacie, A. and Hamelink, J. L.**, Bioaccumulation, in *Fundamentals of Aquatic Toxicology*, Rand, G. M. and Petrocelli, S. R., Eds., Hemisphere, Washington, D.C., 1985, chap. 17.

50. **Hughes, G. H. and Perry, S. F.**, Morphometric study of trout gills: a light microscopic method suitable for the evaluation of pollutant action, *J. Exp. Biol.*, 64, 447, 1976.

51. **Berkaloff, A., Bourguet, J., Favard, P., and Lacroix, J. C.,** *Biologie et Physiologie Cellulaires. I. Membrane Plasmique*, Herman, Paris, 1977.

52. **Singer, S. J. and Nicholson, G. L.,** The fluid mosaic model of the structure of cell membranes, *Science,* 175, 720, 1972.
53. **Wallack, D. F. H.,** *Plasma Membrane and Disease,* Academic Press, London, 1979.
54. **Kimelberg, H. K.,** The influence of membrane fluidity on the activity of membrane-bound enzymes, in *Dynamic Aspects of Cell Surface Organization,* Poste, G. and Nicholson, G. L., Eds., Elsevier/North-Holland, Amsterdam, 1977, 205.
55. **Tynecka, Z., Gos, Z., and Zajac, J.,** Reduced cadmium transport determined by a resistance plasmid in *Staphylococcus aureus, J. Bacteriol.,* 147, 305, 1981.
56. **Saarikoski, J., Lindstrom, R., Tyynela, M., and Viluksela, M.,** Factors affecting the absorption of phenolic and carboxylic acids in the guppy — *Poecilia reticulata, Ecotoxicol. Environ. Saf.,* 11, 158, 1986.
57. **Rothstein, A.,** Mercurials and red cell membranes, in *The Function of Red Blood Cells: Erythrocyte Pathobiology,* Alan R. Liss, New York, 1981, 105.
58. **Boudou, A., Georgescauld, D., and Desmazès, J. P.,** Ecotoxicological role of the membrane barriers in transport and bioaccumulation of mercury compounds, in *Aquatic Toxicology,* Nriagu, J. O., Ed., John Wiley & Sons, New York, 1983, chap. 4.
59. **Boudou, A., Desmazès, J. P., and Georgescauld, D.,** Fluorescence quenching study of mercury compounds and liposomes interactions: effect of charged lipids and pH, *Ecotoxicol. Environ. Saf.,* 6, 379, 1982.
60. **Bienvenue, E., Boudou, A., Desmazès, J. P., Gavach, C., Georgescauld, D., Sandeaux, J. R., and Seta, P.,** Transfer of mercury compounds across biomolecular lipid membranes: effect of lipid composition, pH and chloride concentration, *Chem. Biol. Interact.,* 48, 91, 1984.
61. **Gutknecht, J.,** Inorganic mercury transport through lipid bilayer membranes, *J. Membr. Biol.,* 61, 61, 1981.
62. **Delnomdedieu, M., Boudou, A., Desmazès, J. P., Faucon, J., and Georgescauld, D.,** Mercury compounds interactions with phospholipid bilayers: a fluorescence polarization study, in *Molecular Mechanism of Metal Toxicity and Carcinogenicity,* First Int. Meet., Urbino, Italy, 1988.
63. **Lech, J. J. and Vodicnik, M. J.,** Biotransformation, in *Fundamentals of Aquatic Toxicology,* Rand, G. M. and Petrocelli, S. R., Eds., Hemisphere, Washington, D.C., 1985, 526.
64. **Martoja, R., Ballan-Dufrançais, C., Jeantet, A. Y., Gouzerh, P., Amiard, J. C., Amiard-Triquet, C., Berthet, B., and Band, J. P.,** Effets chimiques et cytologiques de la contamination expérimentale de l'huître *Crassostrea gigas* par l'argent administré sous forme dissoute et par voie alimentaire, *Can. J. Fish. Aquat. Sci.,*45, 1827, 1988.
65. **Giblin, F. J. and Massaro, E. J.,** Pharmacodynamics of methylmercury in the rainbow trout: tissue uptake, distribution and excretion, *Toxicol. Appl. Pharmacol.,* 24, 81, 1973.
66. **Boudou, A. and Ribeyre, F.,** Contamination of aquatic biocenoses by mercury compounds: an experimental ecotoxicological approach, in *Aquatic Toxicology,* Nriagu, J. O., Ed., John Wiley & Sons, New York, 1983, chap. 3.
67. **Mouvet, C.,** The use of aquatic bryophytes to monitor heavy metals pollution of freshwaters, as illustrated by case studies, in *Proc. Int. Cong. Soc. Int. Limnol.,* Lyon, 1983.
68. **Flannagan, J. F. and Cobb, D. G.,** Production of Hexagenia in Lake Winnipeg (Canada), in *Proc. 4th Int. Conf. Ephemeroptera,* Landa, V., Ed., CSAV, 1984, 307.
69. **Ribeyre, F. and Boudou, A.,** Experimental trophic chain contamination with methylmercury: importance of the system "producer-primary consumer", *Environ. Pollut. (A),* 24, 193, 1980.
70. **Boudou, A. and Ribeyre, F.,** Comparative study of the trophic transfer of two mercury compounds — HgCl$_2$ and CH$_3$HgCl — between *Chlorella vulgaris* and *Daphnia magna.* Influence of temperature, *Bull. Environ. Contam. Toxicol.,* 27, 624, 1981.
71. **Hunt, E. G. and Bischoff, A. I.,** Inimical effects on wildlife of periodic DDD applications to Clear Lake, *Calif. Fish Game,* 46, 91, 1960.
72. **Takisawa, Y.,** Epidemiology of mercury poisoning, in *The Biogeochemistry of Mercury in the Environment,* Nriagu, J. O., Ed., Elsevier/North-Holland, Amsterdam, 1979, 325.
73. **Ui, J.,** Mercury pollution of sea and freshwater: its accumulation in water biomass, *Rev. Int. Oceanogr. Med.,* 22, 79, 1971.
74. **Ratkowsky, D. A., Dix, T. G., and Wilson, K. C.,** Mercury in fish in the Derwent estuary (Tasmania) and its relation to the position of the fish in the food chain, *Aust. J. Mar. Freshwater Res.,* 26, 223, 1975.
75. **Cox, J. A., Carnahau, J., Dinunzio, J., McCoy, J., and Meister, J.,** Source of mercury in fish in new impoundments, *Bull. Environ. Contam. Toxicol.,* 23, 779, 1979.
76. **Potter, L., Kidd, D., and Standiford, D.,** Mercury levels in Lake Powell: bioamplification of mercury in manmade reservoirs, *Environ. Sci. Technol.,* 8, 41, 1974.
77. **Tanabe, S., Mori, T., Tatsukawa, R., and Miyazaki, N.,** Global pollution of marine mammals by PCBs, DDT and HCHs, *Chemosphere,* 12, 1269, 1983.

78. **Amiard-Triquet, C. and Amiard, J. C.,** *Radioécologie des Milieux Aquatiques,* Masson, Paris, 1980.
79. **Dallinger, R. and Kautzky, H.,** The passage of Cu, Zn, Cd and Pb along a short food chain into the fish *Salmo gairdneri,* in *Heavy Metals in the Environment,* CEP Consultants, Edinburgh, 1985, 694.
80. **Hamdy, M. K. and Prabhu, N. V.,** Behavior of mercury in bio-systems: biotransference of mercury through food chains, *Bull. Environ. Contam. Toxicol.,* 21, 170, 1979.
81. **Phillips, D. J. H.,** *Quantitative Aquatic Biological Indicators: Their Use to Monitor Trace Metal and Organochlorine Pollution,* Applied Science, London, 1980.
82. **Fowler, S. W. and Elder, D. L.,** PCB and DDT residues in a Mediterranean pelagic food chain, *Bull. Environ. Contam. Toxicol.,* 19, 244, 1978.
83. **Bryan, G. W.,** Bioaccumulation of marine pollutants, *Philos. Trans. R. Soc. London,* 286, 483, 1979.
84. **Prosi, F.,** Heavy metals in aquatic organisms, in *Metal Pollution in the Aquatic Environment,* Förstner, U. and Wittmann, G. T. W., Eds., Springer-Verlag, Berlin, 1979, 271.
85. **Sheehan, P. J.,** Effects on individuals and populations, in *Effects of Pollutants at the Ecosystem Level,* Sheehan, P. J., Miller, D. R., Butler, G. C., and Bourdeau, P., Eds., John Wiley & Sons, Chichester, 1984, 23.
86. **Jacobson, K. B. and Turner, J. E.,** The interaction of cadmium and certain other metal ions with proteins and nucleic acids, *Toxicology,* 16, 1, 1980.
87. **Delorme, R.,** Les phénomènes de résistance: mécanismes biochimiques et physiologiques, in *Insectes, Insecticides et Santé,* Acta, Paris, 1985, 257.
88. **Wood, R. J.,** Insecticide resistance: genes and mechanisms, in *Genetic Consequences of Man-made Change,* Bishop, J. A. and Cook, L. M., Eds., Academic Press, London, 1981, 53.
89. **Hama, H.,** Resistance to insecticides due to reduced sensitivity of acetylcholinesterase, in *Pest Resistance to Pesticides,* Georghiou, G. P. and Saito, T., Eds., Plenum Press, New York, 1983, 299.
90. **Heckman, C. W.,** Reactions of aquatic ecosystems to pesticides, in *Aquatic Toxicology,* Nriagu, J. O., Ed., John Wiley & Sons, New York, 1983, 355.
91. **Anderson, R. S.,** Effects of rotenone on zooplankton communities and a study of their recovery patterns in two mountain lakes in Alberta, *J. Fish. Res. Board Can.,* 27, 1335, 1970.
92. **Hughes, D. N., Boyer, M. G., Papst, M. H., Fowle, C. D., Rees, G. A. V., and Baulu, P.,** Persistence of three organophosphorous insecticides in artificial ponds and some biological implications, *Arch. Environ. Contam. Toxicol.,* 9, 269, 1980.
93. **Sheehan, P. J.,** Effects on community and ecosystem structure and dynamics, in *Effects of Pollutants at the Ecosystem Level,* SCOPE 22, Sheehan, P. J., Miller, D. R., Butler, G. C., and Bourdeau, P., Eds., John Wiley & Sons, Chichester, 1984, 51.
94. **Blandin, P.,** Bioindicateurs et diagnostic des systèmes écologiques, *Bull. Ecol.,* 17, 215, 1986.

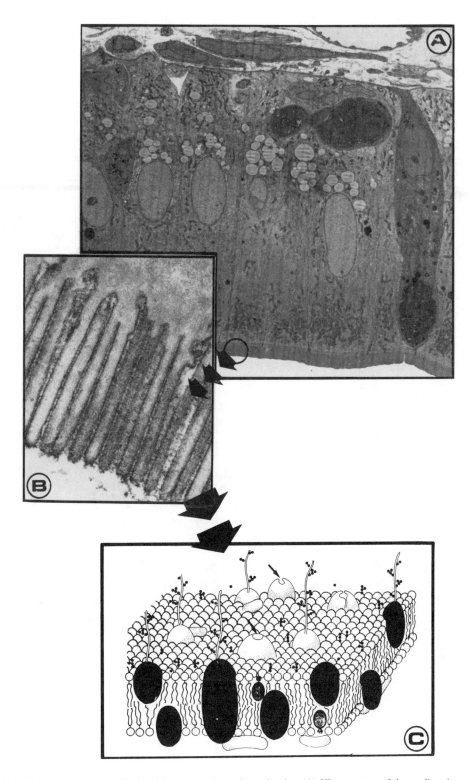

PLATE 1. The concept of biological barrier and membrane barrier. (A) Ultrastructure of the medium intestine enterocytes of *Salmo gairdneri*. (Magnification × 1000); (B) ultrastructure of the enterocyte microvilli. (Magnification × 30,000); (C) schematic illustration of the plasmic membrane. (A and B from Bergot, INRA, St. Pée-sur-Nivelle, France — personal documents; C from Berkaloff, A., Bourguet, J., Favard, P., and Lacroix, J. C., *Biologie et Physiologie Cellulaires. I. Membrane Plasmique,* Herman, Paris, 1977. With permission.)

Analytical Chemistry and Chemical Assessment of Contaminants in Freshwater Ecosystems

Chapter 4

PROBLEMS AND ANALYTICAL METHODS FOR THE DETERMINATION OF TRACE METALS AND METALLOIDS IN POLLUTED AND NONPOLLUTED FRESHWATER ECOSYSTEMS

M. Stoeppler

TABLE OF CONTENTS

I. INTRODUCTION

Trace metals and metalloids in freshwater ecosystems such as lakes, rivers, and estuaries occur either dissolved or bound to particulate matter.[1-8] Total amounts and concentration ratios (dissolved/particulate bound) depend on the aquatic chemistry of the elements considered, on man-made and geological impacts, and on general properties of the respective aquatic ecosystem.[4-6]

This chapter deals with the present state of the art, reliable approaches for the quantitation of dissolved and particulate-bound stages from sampling to final determination for 13 important metals and metalloids in fresh water. These are aluminum, arsenic, cadmium, cobalt, chromium, copper, mercury, nickel, lead, selenium, tin, thallium, and zinc. Aluminum nowadays is of particular environmental significance due to its mobilization, e.g., in lakes as a result of acid rain impact.[9] There are many recent original and review papers that provide basic information on the occurrence of these elements, typical concentration levels in freshwater systems, and their environmental chemistry.[1,2,9-25]

It has to be stressed, however, that only papers have been considered that are based mainly on studies performed during approximately the last decade with proven and sensitive methods applying open-ocean, high-quality working procedures and standards for the collection and analysis of freshwater samples. This is of paramount importance for reliable information about the levels of dissolved elements, which are commonly much — sometimes many times — lower than formerly supposed and typically range from a few nanograms per liter to micrograms per liter. This is true even for waters with considerable man-made inputs of trace metals. For example, Nürnberg and co-workers already reported a decade ago levels in some West German aquatic ecosystems considered polluted areas — Lake Constance, Jade Bay, and Weser — on the order of less than ten up to a few hundred nanograms per liter of dissolved cadmium, lead, and even copper.[26] It should be mentioned, however, that high percentages of, e.g., lead[4] and mercury[27] are bound to particulate matter, while, e.g., cadmium, cobalt, and copper (at least in rivers) occur in quite remarkable amounts in dissolved form.[4,7,8]

II. SAMPLING, SAMPLE PRESERVATION, AND STORAGE

Collection of water samples, their preservation, and intermediate- or long-term storage are probably the most crucial parts of the whole analytical procedure. Doubtless, the most critical step, however, is collection of water samples. It should be performed with utmost care and expertise. Probably inadequate performance of sampling was one of the most frequent reasons for erroneous results in the past.[4] Thus, it appears to be of eminent importance to draw the reader's attention to reliable approaches that, if properly performed, will avoid systematic errors in the presampling, collection, and preparation of samples for trace metal analysis.

In general, sampling procedures for rivers, lakes, and estuaries should be organized to use, preferably, already available information from recent literature dealing with collection and short-term preservation of seawater samples. However, an extremely important prerequisite is that the laboratory staff be already experienced in low-level analytical determinations or at least be able to gain the required expertise by training programs and collaboration with laboratories already experienced in the field. Also, current quality control assessment (discussed below) is absolutely necessary.

All this is very important if one considers how expensive this type of analysis might be and that sometimes rigid — and legal — measures have to be taken if results of monitoring appear to show considerable pollution or, in current programs, an increase of pollution. If the reported data are not reliable enough from this point of view, one can imagine the waste of manpower and technical investment in those regrettably still-frequent cases.

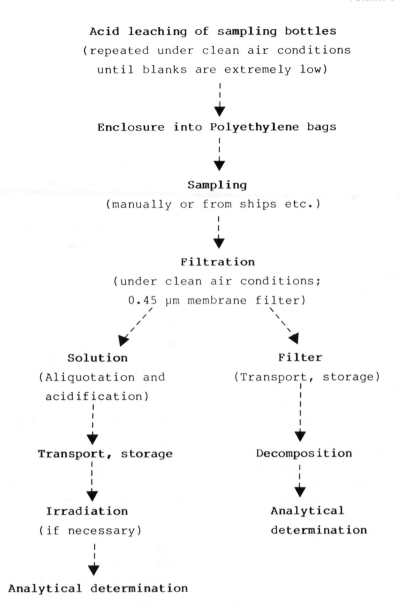

FIGURE 1. Scheme for preparative steps, sampling, and determination for trace and ultratrace contents of metals and metalloids in fresh water.[31]

Current analytical studies in various water samples, which very often have relatively low levels of trace elements compared to other environmental materials, thus require, in addition to the basic needs for an experienced staff, appropriately equipped laboratories with an instrumentation that allows the reliable quantification of even the lowest levels present in the investigated samples.

If all these prerequisites are met, one can at least expect that the technical part of the task can be performed appropriately.

All factors that might limit accuracy and measures taken from sampling (including filtration) to sample storage (including dealing with preparation and proper use of all sampling gear for, e.g., subsequent voltammetric determination of Zn, Cd, Pb, Bi, Cu, Co, Ni, and Hg) are shown schematically in Figure 1 and are described in detail in several papers by

Mart.[28-31] Similar approaches have to be taken for other analytical methods, too. These papers are based on long-term experience with ultratrace metal analysis down to a few nanograms per liter in aqueous media and, thus, provide an excellent introduction and guidance for practical work. For all elements discussed here, with the exception of mercury (for which glass or quartz bottles are required to avoid contamination by metallic mercury from outside and diffusion of metallic mercury after reduction from inside),[31,32] polyethylene bottles should be used for collection and storage. If total elemental determination is required for freshwater samples, an acidification to pH 2 directly after collection is necessary to avoid wall adsorption.[32,33] Even for speciation studies acidification has been reported to be successful for total selenium, selenium(IV), and selenium(VI) species at microgram-per-liter levels.[34]

Since losses of trace metals in fresh water due to wall adsorption are a serious error source, pretreatment with Al^{3+} or shock freezing at liquid-nitrogen temperature immediately after sampling has been proposed recently.[35] Only the latter, of course, is applicable if aluminum has to be determined as well.

The British Standing Committee of Analysts[33] has given a detailed description of appropriate sample containers and storage conditions for waters and effluences, including the elements treated here (Al, As, Cd, Cr, Co, Cu, Pb, Hg, Ni, and Zn). For the determination of total metals it is recommended to add either 20 ml of 5 *M* HCl per liter of sample or 2 to 10 ml concentrated HNO_3 per liter of sample and to store at 4°C in a refrigerator, "but be guided by the conditions for the most sensitive metal likely to be present." If necessary, two or more samples should be taken where preservation techniques or bottle material (Hg) are incompatible. Also, influence of air should be considered.[33]

Recent investigations performed on the influence of the time of acidification after sample collection for lead in drinking water showed that acidification of water samples after collection could be delayed up to 14 d without any adverse effect on lead content data.[36] This can give some guidance for freshwater samples as well, in that acidification immediately after sampling might not be necessary, although it has to be investigated in each case. Finally, it should be mentioned that analysis of water samples always requires careful planning in order to obtain the required information which strongly determines sampling concepts that also have to consider the particular properties of freshwater systems for a reliable sampling strategy (including sampling position, time and frequency of sampling, discrete and composite samples, and sample volume). This is also discussed in detail in the already-mentioned booklet[33] which is recommended as a very valuable source of information. A second, improved and enlarged edition with new data on stabilizers is in preparation at present.[37]

III. SAMPLE PREPARATION

Preparation of freshwater samples for subsequent analytical determinations strongly depends on the elements to be determined and the methods applied. Here, only preparation for total elemental determinations will be discussed.

While acidification is sufficient for filtrates and most elements above determination limit, if graphite furnace atomic absorption spectrometry (GFAAS) and total reflection X-ray fluorescence (TXRF) (the latter only if the concentration of interfering matter is low) have to be applied, voltammetry for those samples often requires a particular pretreatment. That usually consists of UV irradiation to decompose dissolved organic matter (DOM) which might interfere with voltammetric analysis due to complexation and electrode reactions. Samples from rivers and estuaries can be particularly affected by DOM. Since the risk of contamination during UV irradiation can be significant due to corroding connections and solders of UV lamps, specially designed irradiation devices have been developed to avoid contamination. Systems that fulfill these needs, those made mainly from very pure quartz, are described elsewhere in detail.[31,38] The latter system was primarily designed as a first-step module coupled to a microprocessor-controlled voltammetric analyzer,[39] but can also be used, of course, independently and in connection with other analytical methods. This system is based upon photodigestion by UV irradiation under addition of ultrapure 30% hydrogen peroxide.[38]

For multielement determinations in fresh waters by, e.g., inductively coupled plasma atomic emission spectroscopy (ICP-AES) and in some cases also for GFAAS and TXRF, a chemical preconcentration/separation step is necessary. This is to be achieved either by use of chelating resins (e.g., Chelex-100) followed by an appropriate elution step or solvent extraction with various chelates (critically reviewed in general[40] and recently reviewed for seawater analysis[41]), also applicable in principle as already mentioned for the tasks discussed here.

Because the content of DOM might interfere in these steps, too, a photodigestion prior to preconcentration steps similar to that described for subsequent voltammetric analysis is sometimes also advantageous for these methods.[38]

Particular problems might be encountered with mercury. Total mercury determinations by, e.g., cold vapor atomic absorption spectrometry (AAS) — at present the approach by far predominantly applied and also very sensitive — require in most cases (if there are methylmercury contents in filtrates and if unfiltered samples have to be analyzed) an effective pretreatment. UV irradiation,[33,42] frequently supported by addition of acids using partly and completely closed quartz tubes,[43] has proved to be reliable even at the very low mercury content of nanogram-per-liter levels of fresh waters.

With the exception of mercury (because of its losses during transport and storage of filters), the digestion of unfiltered parallel samples in comparison to filtrates is recommended more. For all other elements the digestion of 0.45-μm filters is common for the determination of metal contents in particulate matter. Due to the generally elevated contents of trace elements in particulate matter from fresh waters, the contamination is not severe and can be controlled without exceptional efforts. Not only for voltammetry, but also for the application of other methods, low temperature ashing in an oxygen plasma is a promising approach. Trace metal losses, except for mercury (see remarks above) and possibly for arsenic, are practically negligible if this method is applied, as the ashing temperature is below 150°C and the decomposition is complete. After decomposition (approximately 3 h) the residue can be easily dissolved by appropriate acids (HNO_3, HCl)[31] and the resulting analyte solution used for determination. However, other digestion procedures such as many different modes of wet digestion[44] and pressurized digestion in polytetrafluoroethylene vessels[45] are suited as well for methods other than voltammetry.

IV. ANALYTICAL METHODS

A. General Overview

Besides voltammetric methods and a few very sensitive modes of AAS (graphite tube furnace, hydride, and cold vapor techniques), other — mainly multielement — methods are in increasing use for the determination of trace and ultratrace elemental levels in fresh water.

Table 1 summarizes the most important analytical approaches and detection limits at present, based not only on 3 σ of noise or blank values from recent literature, but also from the author's experience where voltammetry and AAS are concerned.[46] However, the table requires some comments and additional information.

For the estimation of the detection limits of ICP-AES (probably still valid for most routine instruments), two recent compilations have been used.[47,48] However, progress in this area is fast, and it is claimed by some manufacturers that even in routine work and with the sequential mode detection limits can be achieved now that are on average approximately a factor of 3 to 5 lower than given in Table 1. See, for example, Reference 49.

Table 1

**TYPICAL RELATIVE DETECTION LIMITS BY VARIOUS METHODS
FOR 13 ELEMENTS OF IMPORTANCE IN AQUATIC SYSTEMS**

Element	ICP-AES	FAAS	GFAAS	HYAAS/ CVAAS	DPSV ADPV	TXRF	INAA	Nuclide
Al	30	30	0.08		≤0.15[a]	—	4	^{28}Al
As	50	30	0.8	≤0.03	≤0.2	0.2	0.05	^{76}As
Cd	2.5	3	0.0035		≤0.0002	0.4	1.5	^{111m}Cd
Co	6.0	15	0.17		≤0.005	0.1	0.03	^{60m}Co
Cr	6.0	4.5	0.08		0.02[b]	0.4	20	^{51}Cr
Cu	5.5	1.5	0.08		0.002	0.1	0.1	^{64}Cu
Hg	25	300	≤10[c]	≤0.001	0.005	0.2	0.03	^{197}Hg
Ni	10	3	0.2[d]		0.001	0.1	15	^{65}Ni
Pb	40	15	0.08		0.001	0.2	3000	^{207m}Pb
Se	75	150	0.7	≤0.03	0.1	0.2	1	^{77m}Se
Sn	30	30	0.6[e]	0.8	0.1[f]	0.5	2	^{123m}Sn
Tl	40	30	0.5		0.015[g]	0.2		—
Zn	2	1.5	0.003		0.02	0.1	2.5	^{69m}Zn

Note: Detection limit values are in μg/l. Aqueous solutions are noninterfering. Detection limit is based
on 3 6 of noise or blank values. For detailed comment on the values listed in this table, see Section
IV.A. FAAS, flame atomic absorption spectrometry. HYAAS/CVAAS, hydride atomic absorption
spectrometry/cold vapor atomic absorption spectrometry.

[a] See Reference 55.
[b] See Reference 56.
[c] See Reference 50.
[d] Wall atomization.
[e] See Reference 51.
[f] See Reference 76.
[g] See Reference 77.

As far as GFAAS is concerned, in former compilations the given detection limits were
based on 100-μl injections into the tube, but recently on more realistic 50-μl injections.[46]
However, the introduction of platform techniques, certainly an improvement in method
reliability, reduces the volume pipetted onto the platform to around 30 μl. Thus, the detection
limits given for GFAAS in Table 1 are calculated now for 30-μl injections onto the L'vov
platform. On the other hand, many modern sample injection systems in GFAAS do allow
multiple injections that are applicable for many freshwater types. In this case, the attainable
detection limits are lower in comparison to those listed in the table.

The detection limits given for the hydride method are based on a realistic sample volume
of 20 ml, attainable with most commercially available systems. If, however, preconcentration
techniques (not yet commercially available) are used, absolute detection limits around 20
pg can be attained for, e.g., selenium.[32] For mercury with the cold vapor method, a preconcentration on mercury absorbers (nets, wire, and wool) made from precious metals
(mainly gold) is possible so that under proper working conditions and with volumes (≥100
ml, detection limits ≤0.001 μg/l can also be attained[42] with commercial instruments.[53] The
values for differential pulse stripping voltammetry (DPSV) and adsorption differential pulse
voltammetry (ADPV) are taken mainly from work carried out by the late Professor H.W.
Nürnberg and co-workers at the author's institute.[54] These values usually refer to a PAR
174 instrument with hanging mercury drop, mercury film, and gold electrodes, respectively,
and an average cell volume of 20 ml.

The values for TXRF are estimated from research work done at GKSS, Geesthacht, West

Germany and based on a drop of a noninterfering sample of 50-μl initial volume.[57] As for GFAAS, the use of higher volumes and preconcentration techniques is possible if lower detection limits are required.[58] Given values for instrumental neutron activation analysis (INAA) refer to a sample size of 0.5 ml and a thermal flux of 10^{13} n/cm/s and a defined detection system.[59] Since for fresh water in many cases higher volumes can be taken if the water is evaporated or preconcentrated[60] and for some elements radiochemical procedures are applied, frequently significantly lower contents are detectable by this method. In Sections IV.B and C some detailed information is given on the most promising methods at present of voltammetry, atomic absorption, atomic emission, and on some selected applications in the field of water analysis. The remaining methods, TXRF, neutron activation analysis and (not included in Table 1) isotope dilution mass spectrometry are less frequently applied, but important additional independent methods are discussed in Section IV.D.

B. Voltammetry

Electroanalytical methods are of particular significance for the determination of numerous trace metals and metalloids in many different materials. If, with respect to the required detection power, advanced techniques are applied, it is possible to attain for various metals and metalloids (Cd, Co, Cu, Hg, Ni, Pb; see Table 1) extraordinarily low detection limits linked with a precision $\leqslant 5\%$ even at nanogram-per-liter levels. Oligoelement determination from the same analyte solution is usually performed by several versions of stripping voltammetry (DPSV), square wave voltammetry (SWV), and adsorption voltammetry (AV), employing the hanging mercury drop electrode (HMDE), mercury film electrode (MFE), and solid (e.g., gold) electrodes as working electrodes.[54-56, 61] A significant broadening of voltammetric techniques is offered by AV in that it has an analytical potential for metals that do not form stable amalgams. The principle consists in the formation of trace metal chelates with a suitable organic ligand added to the analyte. For example, for nickel and cobalt, dimethylglyoxime is frequently used,[63] for aluminum either 1,2-dihydroxyanthraquinone-3-sulfonic acid[64] or solochrome violet RS,[55] for chromium(III) diethylene triaminepentaacetic acid.[56] The formed chelates are adsorbed at the surface of the HMDE at a potential that is in the range of the zero charge potential. In this manner preconcentration is achieved. The adsorption time is adjusted to such small intervals, depending on the concentration of the respective trace metals in the analyte, that complete coverage of the electrode surface cannot take place. Subsequently, the electrode potential is scanned toward the reduction potential for the determination step. This technique has also been applied successfully for amalgam-forming elements.[61] ADPV in connection with the MFE provides no particular advantage, compared to the more easily manipulated HMDE.[61] Particularly high precision and speed can be attained if the SWV mode is combined with AV.[65]

Owing to their foundation on Faraday's Law (1 mol of a substance undergoing an electrode process is equivalent to the very high electrical charge of n $\times$ 96,500 C with n usually between 1 and 3) these voltammetric approaches provide a very high detection power, supported by the possible *in situ* enrichment on the electrodes. This is linked with good to excellent precision and accuracy if interferences are kept to a minimum by proper sample preparation. Moreover, these methods are based on a detailed and comprehensive understanding of their physicochemical properties, i.e., the kinetics and thermodynamics of electrode processes. This offers, after a proper training in electrochemistry and voltammetry, a broad potential for trace analysis in many materials.[55,61,62]

Thus, for the elements listed in Table 1, voltammetric approaches compete favorably with the other analytical methods if detection power and reliability are concerned. For the simultaneous determination of cadmium, lead, copper, selenium, zinc, nickel, and cobalt in seawater, in wet deposition, and in drinking water — due to the matrix (seawater) and extremely low contents (all these matrices) — and also to some extent for determination in fresh water, voltammetry can often be considered as the most promising method at present as compared to GFAAS. The latter has for some elements comparable detection power, but is only a single-element method and needs for many elements and matrices separation and/or chemical preconcentration steps that are not required for voltammetry. It has to be mentioned, however, that even in aqueous matrices, particularly in fresh water, UV irradiation (see Section II) is necessary prior to voltammetric determination in order to destroy all organic matter present completely. This is not needed if GFAAS is applied.

Another advantage of voltammetric techniques is that instrumentation and running costs are less expensive than for most other analytical techniques.[46,61] An earlier disadvantage of voltammetry, the time needed for a single determination, i.e., the manpower required in routine analysis, is becoming more and more balanced by the commercial availability of at least partly automated and computer-controlled systems that nevertheless are offered at quite moderate costs in comparison to instrumentation for other methods. These instruments, of course, allow the application of all relevant polarographic and voltammetric modes. The compact and rugged design of this instrumentation makes it additionally a very favorable tool for field studies, e.g., in mobile laboratories.[66]

A limitation of voltammetry certainly is that only a definite, though important, number of metals (including all metals of ecotoxic influence) are accessible to this analytical approach.[67,68] As is usually the case for most other methods, the determination of trace metals in solid materials (i.e., in particulate matter from rivers and lakes after filtration) requires a decomposition prior to the determination step. For all voltammetric modes a complete decomposition of any interfering organic matter is mandatory. The reason is that a large number of organic and surface-active compounds often inhibit or even suppress electrode reactions. This sensitivity to a wide variety of different compounds, including organometallic, makes voltammetry an effective tool for speciation studies of trace metals in natural waters.[6]

As far as the metals listed in Table 1 are concerned, the following references provide detailed methodological information on the voltammetric determination of aluminum,[55,64] arsenic,[61,68-72] cadmium,[39,54,61,65,68,71,72] cobalt,[61,63,65,71] chromium,[56,61,73] copper,[39,61,65,68,71,74] mercury,[61,68] nickel,[61,63,65,71] lead,[39,61,65,68,71] selenium,[61,71,75] tin,[76,77] thallium,[68,78-81] and zinc.[39,61,68,71]

C. Atomic Absorption and Emission Spectrometry

The theoretical background, practical aspects, and the state of the art have been recently, critically, and extensively reviewed in general,[67,81-84] for AAS[86-92] as well as for atomic emission spectroscopy (AES).[93-96] Relevant detection limits for the elements to be discussed in this chapter are listed in Table 1 (cf. Section IV.A). Because of this ample recent literature only a few highlights and recommendations for practical applications in water analysis will be given.

The application of AAS underwent a tremendous increase from the introduction of the method in 1955 until today, with the potential to determine more than 60 elements including all ecotoxic metals with the different modes of flame, graphite furnace, hydride, and cold vapor AAS. Despite the fact that AAS still poses many interference problems (especially with the graphite furnace) considerable progress has been achieved in the last couple of years. First of all, owing to contributions of numerous workers, some insight into chemical, physicochemical, and physical interferences has been gained.[97,98] Based on L'vov's ideas on absolute AAS methods,[99] considerable improvements have been made. These combine matrix modification, background correction, more sophisticated graphite furnaces including L'vov platforms, versatile temperature programming, and sample injection, with alternate gas possibilities and very capable computerization in general. Thus, the present state of the art is characterized by an increased availability of better-controlled techniques such as the stabilized temperature platform furnace (STPF) concept in conjunction with further progress in matrix modification.[101,102] With such an important gain in reliability for many elements and matrices, an approach to standardless GFAAS appears now to be within reach, and with the introduction of the "characteristic mass" (pg/0.0044 absorbance · second) a way has been shown how this might be controlled in practical analysis.[103-106] The STPF and addition of a substance (matrix or element modifier) to the sample that stabilizes the analyte and permits charring at higher temperatures leads to a removal of potentially interfering matter.[107] This has been applied to the determination of 12 trace elements (Al, As, Be, Cd, Co, Cr, Cu, Mn, Ni, Pb, Se, and V) in natural waters with detection limits below 1 μg/l with a precision and accuracy ≤20%.[108] Standardization can be simply performed against aqueous working curves. Compared to former approaches based mainly on preconcentration techniques due to matrix influences, this was, of course, an important improvement. Meanwhile, further progress has been achieved with matrix modifiers. It has been shown that a mixture of palladium and magnesium nitrates are a very powerful modifier for many elements such as As, Bi, In, Pb, Sb, Se, Sn, Te, and Tl.[109-111] Recent studies have shown that palladium sometimes (also in combination with NH_4NO_3) can be used as a modifier for Cu, Ag, Au, Zn, Cd, Hg, Ga, Ge, and P[112] which makes it very promising for most metals and metalloids treated in this chapter.

Even with multiple injections, possible in some cases, GFAAS has slightly to significantly poorer detection power for a number of elements (e.g., As, Cd, Co, Cr, Cu, Hg, Ni, Pb, Se, Sn, and Tl) in solution compared to voltammetry. Thus, for some fresh waters with trace metal contents at the nanogram-per-liter level, direct GFAAS can only provide upper limits. However, this is not very often the case, and there are preconcentration techniques proved to be reliable even for extremely low metal contents in seawater.[113-120] This, however, needs more skills and contamination precautions than the direct approach or use of voltammetry. The significantly improved performance of the graphite furnace technique as well as the potential for direct analyses without any pretreatment makes this approach comparatively quick and well suited for many important duties in water control and surveillance at somewhat elevated levels. However, for the analysis of particular matter after an appropriate pretreatment or even with direct solid sampling,[92,121] there are more advantages than limitations of the method compared to voltammetry.

For the metals listed in Table 1 the quoted recent compilations and original papers provide detailed methodological information on the GFAAS determination of aluminum,[46,87,89,90,108,122] arsenic,[46,87,89,90,108,110,115] cadmium,[46,87,89,90,108,112-114,117,123] cobalt,[46,87,90,108] chromium,[46,87,89,90,108] copper,[46,87-90,108,113,114] lead,[46,87-90,108,110,114] selenium,[46,87,89,90,108,110,118] tin,[23,51,77,87,89,90,110,124,125] thallium,[46,81,89,90,109,110,120] and zinc.[46,87,89,90,112-114]

In the last couple of years there has been significant progress in the determination of arsenic, selenium, and tin as volatile hydrides and mercury as Hg°, the latter after preconcentration on noble metal absorbers. For this purpose various more or less sophisticated commercial systems are available[126] that offer, due to the use of higher sample volumes, very low relative detection limits (see Table 1). However, if directly applied for water samples, there are significant differences in signal form and signal appearance for different valency states (Se, As) and organometallic species (Hg) that can be favorably used for speciation. Methylmercury, for example, is not reduced to metallic mercury if stannous chloride solutions are used, while sodiumtetrahydroborate quantitatively reduces methylmercury to elemental mercury, but introduces some contamination problems at lower levels.

For the analysis of total elemental levels in water, this has to be considered and measures taken for pretreatment of samples and maintaining experimental conditions so that the total elemental content can be estimated reliably. This usually is performed by UV irradiation or wet digestion for mercury compounds, and peak area evaluation and proper adjustment of working conditions to avoid interferences during the hydride evolution step or in the quartz absorption tube, for hydride-forming elements. This is discussed and working conditions given for total and reactive forms of elements in, e.g., some recent papers and reviews for mercury,[42,43,53,87,88,90,127-131] arsenic,[10,87,88,90,132,133] selenium,[21,52,87,90,119,134] and tin.[22,23,90,124,135]

Analysis of higher elemental levels in fresh waters can be performed with flame AAS as well.[82,84,87,90] With its somewhat superior (for some — mainly refractory — elements) detection power and its multielement potential, ICP-AES successfully competes with flame AAS.[49,82,84,89,94,95] The method is additionally favored by the now commercially introduced low-cost ICP-AES systems, on the one hand, and a further improved detection power even for routine work in recently introduced instruments, on the other.[49] For lower elemental levels various preconcentration techniques can be used that range from ion exchange and solvent extraction[40,136] to simple evaporation.[137] However, if less powerful methods requiring concentration levels ≥ 100 are used, contamination problems always have to be faced which make those methods less suitable for routine determination of metals that are only present at very low levels. For ICP-AES problems may also arise if waters with higher salt contents have to be analyzed.[138]

D. Other Methods

A comparatively new analytical approach for simultaneous multielement determinations of metals in aqueous samples is a particular version of X-ray fluorescence — TXRF. It uses the total reflection of the primary X-ray beam at the plane and smooth surface of a reflector. This significantly reduces the radiation-induced background, and, hence, also detection limits (see Table 1), if the sample is an extremely thin layer on the reflector.[139,140] This technique has gained importance within the last couple of years. It has been successfully applied in a number of research studies not only for rain, fresh water and seawater, but also for other matrices as well, after an appropriate sample preparation.[141] Most favorable conditions exist for rainwater, but also fresh water can be analyzed after direct measurement or a relatively simple pretreatment. This is performed using either a freeze-drying or reverse-phase technique down to 5 to 20 µg/l for many trace metals that do not require extreme preconcentration factors.[58] A commercial system is available; however, it is at relatively high cost compared to other competing approaches, and only a few instruments are in routine use worldwide at present. However, in the near future this method might be of increasing importance for many duties in research and surveillance and as an independent analytical approach in quality assessment tasks.

INAA and radiochemical neutron activation analysis (RCNAA), considered classical multielement trace analytical methods,[59,142] were for many years nearly the only ultratrace methods with detection potential down to the picogram level. Thus, they have also been extensively used for aqueous samples after various preconcentration techniques.[40,60] Progress in trace and ultratrace chemistry would not have been possible without these methods as basis for new developments and as indispensable aids in many areas if general problems of accuracy are concerned. RCNAA, especially, still plays an important role for the accurate determination of a number of selected elements for quality control, which also is the case for trace elements in freshwater samples.

Another method with multielement potential and high inherent accuracy, although comparatively slow-sample throughput, is isotope dilution mass spectrometry (IDMS).[143,144] This method is indispensable for, e.g., certification of reference materials.[145] It has already successfully demonstrated its remarkable potential for reliable analysis of trace and ultratrace levels of metals such as lead, cadmium, thallium, iron, zinc, copper, nickel, and uranium in water samples.[146,147]

Ion chromatographic techniques are relatively inexpensive and can be used for separation and determination of cations and anions.[148] With these it is possible to determine a number of metals after a rather simple transformation into metal chelates. This is especially valuable in aqueous samples and has been demonstrated recently for, e.g., V, Cr, Co, Ni, Cu, Zn, Se, Mo, Cd, Tl, Hg, Pb, and Bi after preconcentration from water with an on-line system using [18]C-reversed-phase liquid chromatography and UV detection[149] and for other slightly different systems as well.[150,151] Limits of detection of these techniques, regarded as a complement to atomic spectroscopy,[150] are at the absolute nanogram level. Thus, it is possible with a preconcentration of some 20 to 100 ml to quantitate contents in the upper nanogram-per-liter range.

At present, probably the most promising technique for multielement determination is the combination of a plasma for the production of ions and the introduction of them into a mass spectrometer (ICP-MS).[152,153] More than 90% of the elements of the periodic table can be determined in principle by this method with typical detection limits between 1 and 0.01 μg/l in an aqueous analyte solution and in an impressively wide dynamic range. At present two different systems with prices around $200,000 (U.S.) are commercially available. It has to be mentioned, however, that the performance of present instrumentation is far from being perfect and that, compared to other analytical methods, some limitations still exist. For example, for higher salt contents in analyte solutions, extremely high dilutions are needed, and accuracy as well as precision appear to be somewhat worse than for other currently applied (with respect to detection power) competing methods.[153] This, however, is less pronounced for the analysis of trace elements in water. First experiences with aqueous matrices are quite good and show that freshwater analysis might become an important application area of this method,[154,155] particularly if isotope dilution is applied when utmost accuracy is required.[156]

V. QUALITY ASSESSMENT

Especially for fresh waters and, most importantly, for rivers and lakes used as reservoirs for potable water, objective data on water quality are required for assessments of the present situation and of long-term trends in water quality that are as accurate as possible. For rivers, reliable data on metal levels of dissolved and particulate phases are equally important, in order to estimate the amount of materials discharged by them to the sea. Thus, current analytical quality control for data produced by the laboratories involved in this type of work (which might also be purely scientific) is of paramount importance. In this context, efforts carried out in the U.K., that attempt to ensure that results of adequate accuracy are made,[156] shall be discussed here. They consist, for each determinant, of sequential completion of a number of individual but closely linked stages.[157]

1. Establishment of a working group to plan and coordinate all activities
2. Definition of determinants and required accuracy to ensure a clear specification of the analytical requirements
3. Selection of appropriate methods
4. Detailed and unambiguous description of methods to ensure that these are properly applied
5. Estimation of within-laboratory precision
6. Accuracy check of standard solutions for elimination of this source of bias
7. Setup of quality control charts to maintain a continuous check of precision in each laboratory
8. Performance of tests at distinct intervals to maintain a continuing check to ensure that each laboratory achieves adequately small bias

Based on this or similar schemes numerous interlaboratory comparisons have been performed for the determination of trace elements in fresh water. The results of studies carried out from 1976 to 1985, however, show that the general situation — despite some quite acceptable results — still cannot be regarded as satisfactory.[158-166] It has to be mentioned that these intercomparisons do not include a check for errors during sample collection. Thus, even unsatisfactory results might underestimate discrepancies in data from real samples that also include sample collection. This might be attributed to a number of reasons, e.g., less satisfactory laboratory skills, still not completely mastered sample preparation techniques, methodological bias or use of less-suited ones, and a lack in availability and use of appropriate reference and control materials.

From the author's view and experience, besides the already mentioned training of staff to achieve and maintain proper skills, the current application of independent analytical methods[67,77,115,155,167,168] and the use of appropriate certified reference materials[167] is highly recommended. Regrettably, for fresh water only five reference materials are available at present to the author's knowledge (see Table 2). For particulates, sediment reference materials could be used where appropriate. Also, in-house control materials made from excess amounts of real samples are useful for current quality assessment of analytical procedures and methodological development. Control of losses, sometimes also of contamination, is favorably possible with a proper use of radiotracers,[32,169] i.e., an adjustment of the chemical form of the radiotracers (isoformation) to that of the inactive elements in the sample.

VI. CONCLUSIONS AND PROSPECTS

There is significant progress in sampling and analysis of seawater under clean conditions in conjunction with new or modified and more accurate methods for direct analysis. Because of this progress, the accumulation of more accurate data on trace metal analysis in fresh waters can be expected in the near future. In addition, working laboratories could be set up, if careful training of staff and appropriate equipment can be combined. This has to be linked with improved internal and external quality control measures as well as with preparation and use of a broader selection of reference and certified reference materials.

As far as methods are concerned, further progress can be expected for AV (more elements) and for multielement methods. The latter are urgently needed for the current assessment of the pollution situation in rivers and lakes and the relationship between various elements of ecological and ecotoxicological importance. Some progress has been achieved with multielement GFAAS. Recently, an oligoelement Zeeman-GFAAS instrument was commercially introduced.[170] For years, multielement atomic absorption with a continuum source, although not commercially available, has shown a detection power close to that of monoelement GFAAS and a significantly wider dynamic range.[171-173]

Also, simultaneous multielement potential provides a new analytical principle, furnace atomization nonthermal emission spectrometry (FANES) with impressive detection limits already reported for many elements.[173] With graphite tubes as atomization cells and laser excitation atomic fluorescence spectrometry, there is a remarkable potential for even lower detection limits.[174] Laser techniques certainly will play a very important role in future analytical instrumentation. Promising multielement possibilities can be also expected from flow injection analysis, coupled with various determination systems.[175]

Finally, it should be mentioned that even classical furnace AAS and ICP-AES might provide a significant potential for improvements as will certainly be the case for recently introduced TXRF and ICP-MS instrumentation. Thus, the trend toward more detection power, multielement potential, improvement in reliability, precision, and accuracy, and refinements in computerization will certainly continue in all analytical branches.

Table 2
PRESENTLY AVAILABLE REFERENCE MATERIALS FOR TRACE METAL DETERMINATIONS IN FRESH WATER (AQUEOUS PHASE) IN µg/l

Material	Element	Remarks
IAEA W-4 Synthetic fresh water, containing also principal constituents of fresh water	Al, As, B, Ba, Be, Cd, Co, Cr, Cu, Fe, Mn, Ni, Pb, Sr, U, V, Zn	Simulating natural levels, 5 ml of concentrated reference material and blank solution (each in quartz ampule) to be diluted to 2l; noncertified information; values for Hg, Mo, Se available
NBS 1643a Trace elements in water	Fe(88), Pb(27), Mn(31), Hg(0.2), Ni(55), Se(11), Ag(2,8), Sr(239), V(53), Zn(68), As(76), Ba(46), Be(19), Cd(10), Cr(17), Co(19), Cu(18)	950-ml bottle
NBS 1641b Mercury in water	1520.0	6 × 20-ml ampules
NBS 1642b Mercury in water	1.49	950-ml bottle
NRCC riverine water reference material SLR S-1	Al(23.5), V(0.66), Cr(0.36), Mn(1.77), Cu(3.58), Zn(1.34), Sr(136), Mo(0.78), Sb(0.63), Ba(22.2), Pb(0.106), U(0.28)	Filtered water, acidified to pH 1.6 (ultrapure nitric acid) 2 l polyethylene bottles, containing also principal constituents of fresh water[156]

Note: These reference materials can be obtained from the following:

IAEA: International Atomic Energy Agency
Analytical Quality Control Services
Laboratory Seibersdorf
P.O.Box 100
A-1400 Vienna
Austria

NRCC: National Research Council of Canada
Division of Chemistry
Ottawa, K1A OR6
Canada

NIST: Office of Standard Reference Materials
Room B311, Chemistry Building
National Institute for Standards and Technology
Gaithersburg, MD 20899

River sediment reference materials to check analytical procedures for particulate matter can be obtained from the above manufacturers, from BCR, and from NIES as well:

BCR: Community Bureau of Reference (BCR)
Commission of the European Communities
200 Rue de la Loi
B-1049 Brussels
Belgium

NIES: National Institute for Environmental Studies
Japan Environment Agency
P.O. Yatabe
Tsukuba Ibaraki 300-21
Japan

REFERENCES

1. **Bowen, H. J. M.,** *Environmental Chemistry of the Elements,* Academic Press, London, 1979.
2. **Förstner, U. and Wittmann, G. T. W.,** *Metal Pollution in the Aquatic Environment,* 2nd ed., Springer-Verlag, Berlin, 1981.
3. **Martin, J.-M. and Meybeck, M.,** Elemental mass balance of material carried by major world rivers, *Mar. Chem.,* 7, 173, 1979.
4. **Mart, L., Nürnberg, H. W., and Rützel, H.,** Levels of heavy metals in the tidal Elbe and its estuary and the heavy metal input into the sea, *Sci. Total Environ.,* 44, 25, 1985.
5. **Sigg, L., Sturm, M., and Stumm, W.,** Schwermetalle im Bodensee; Mechanismen der Konzentrations-regulierung, *Naturwissenschaften,* 69, 546, 1982.
6. **Malle, K. G. and Müller, G.,** Metallgehalt und Schwebstoffgehalt im Rhein, *Z. Wasser Abwasser Forsch.,* 15, 11, 1982.
7. **Breder, R.,** Die Belastung des Rheins mit toxischen Metallen, Doctoral thesis, University of Bonn, Bonn, West Germany, 1981.
8. **Breder, R.,** Cadmium in European inland waters, in *Cadmium,* Vol. 2, Stoeppler, M. and Piscator, M., Eds., Springer-Verlag, Berlin, 1988, 159.
9. **Borg, H.,** Trace metals and water chemistry of forest lakes in northern Sweden, *Water Res.,* 21, 65, 1987.
10. **Savory, J. and Wills, M. R.,** Arsen, in *Metalle in der Umwelt,* Merian, E., Ed., Verlag Chemie, Weinheim, West Germany, 1984, 3.
11. **Stoeppler, M.,** Cadmium, in *Metalle in der Umwelt,* Merian, E., Ed., Verlag Chemie, Weinheim, West Germany, 1984, 7.
12. **Elinder, C.-G.,** Cadmium: uses, occurrence, and intake, in *Cadmium and Health: A Toxicological and Epidemiological Appraisal,* Vol. 1, Friberg, L., Elinder, C.-G., Kjellström, T., and Nordberg, G. F., Eds., CRC Press, Boca Raton, FL, 1985, chap. 3.
13. **Mislin, H. and Ravera, O.,** *Cadmium in the Environment,* Birkhäuser, Basel, 1986.
14. **Scheinberg, H.,** Kupfer, in *Metalle in der Umwelt,* Merian, E., Ed., Verlag Chemie, Weinheim, West Germany, 1984, 12.
15. **May, K. and Stoeppler, M.,** Studies on the biogeochemical cycle of mercury. I. Mercury in sea and inland water and food products, in *Proc. Int. Conf. Heavy Metals in the Environment,* Vol. 1, Müller, G., Ed., CEP Consultants, Edinburgh, 1983, 241.
16. **Figueres, G., Martin, J. M., Meybeck, M., and Seyler, P.,** A comparative study of mercury contamination in the Tagus estuary (Portugal) and major French estuaries (Gironde, Loire, Rhone), *Estuarine, Coastal Shelf Sci.,* 20, 183, 1985.
17. **Nojiri, Y., Otsuki, A., and Fuwa, K.,** Determination of subnanogram-per-liter levels of mercury in lake water with atmospheric pressure helium microwave induced plasma emission spectrometry, *Anal. Chem.,* 58, 544, 1986.
18. **Oskarsson, A.,** Nickel, in *Metalle in der Umwelt,* Merian, E., Ed., Verlag Chemie, Weinheim, West Germany, 1984, 16.
19. **Ewers, U. and Schlipköter, H. W.,** Blei, in *Metalle in der Umwelt,* Merian, E., Ed., Verlag Chemie, Weinheim, West Germany, 1984, 6.
20. **Einbrodt, H. J. and Michels, S.,** Selen, in *Metalle in der Umwelt,* Merian, E., Ed., Verlag Chemie, Weinheim, West Germany, 1984, 19.
21. **Robberecht, H. and Van Grieken, R.,** Selenium in environmental waters: determination, speciation and concentration levels, *Talanta,* 29, 823, 1982.
22. **Braman, R. S. and Tompkins, M. A.,** Separation and determination of nanogram amounts of inorganic tin and methyltin compounds in the environment, *Anal. Chem.,* 51, 12, 1979.
23. **Weber, G.,** The importance of tin in the environment and its determination at trace levels, *Fresenius Z. Anal. Chem.,* 321, 217, 1985.
24. **Kemper, F. H. and Bertram, H.,** Thallium, in *Metalle in der Umwelt,* Merian, E., Ed., Verlag Chemie, Weinheim, West Germany, 1984, 22.
25. **Henkin, R. H.,** Zink, in *Metalle in der Umwelt,* Merian, E., Ed., Verlag Chemie, Weinheim, West Germany, 1984, 25.
26. **Valenta, P., Mart, L., Nürnberg, H. W., and Stoeppler, M.,** Voltammetrische simultane Spurenanalyse toxischer Metalle in Meerwasser, Binnengewässern, Trink- und Brauchwasser, *Vom Wasser,* 48, 89, 1977.
27. **Airey, D. and Jones, P. D.,** Mercury in the River Mersey, its estuary and tributaries during 1983 and 1984, *Water Res.,* 16, 565, 1982.
28. **Mart, L.,** Ermittlung und Vergleich des Pegels toxischer Spurenmetalle in nordatlantischen und mediterranen Küstengewässern, Doctoral thesis, Rheinisch-Westfälische Technische Hochschule, Aachen, West Germany, 1979.

29. **Mart, L.,** Prevention of contamination and other accuracy risks in voltammetric trace metal analysis of natural waters. I. Preparatory steps, filtration and storage of water samples, *Fresenius Z. Anal. Chem.,* 296, 350, 1979.

30. **Mart, L.,** Prevention of contamination and other accuracy risks in voltammetric trace metal analysis of natural waters. II. Collection of surface water samples, *Fresenius Z. Anal. Chem.,* 299, 97, 1979.

31. **Mart, L.,** Minimization of accuracy risks in voltammetric ultratrace determination of heavy metals in natural waters, *Talanta,* 29, 1035, 1982.

32. **May, K., Reisinger, K., Flucht, R., and Stoeppler, M.,** Radiochemical studies on the behaviour of mercury chloride and methylmercury chloride in freshwater and seawater, *Vom Wasser,* 55, 63, 1980.

33. Standing Committee of Analysts, General principles of sampling and accuracy of results, in *Methods for the Examination of Waters and Associated Materials,* Her Majesty's Stationery Office, London, 1980.

34. **Cheam, V. and Agemian, H.,** Preservation and stability of inorganic selenium compounds at ppb levels in water samples, *Anal. Chim. Acta,* 113, 237, 1980.

35. **Scheuermann, H. and Hartkamp, H.,** Stabilization of water samples for the analysis of traces of heavy metals, *Fresenius Z. Anal. Chem.,* 315, 430, 1983.

36. **Miller, R. G., Doerger, J. U., Kopfler, F. L., Stober, J., and Roberson, P.,** Influence of the time of acidification after sample collection on the preservation of drinking water for lead determination, *Anal. Chem.,* 57, 1020, 1985.

37. **Pittwell, L. R.,** Secretary, Standing Committee of Analysts, private communication, 1987.

38. **Dorten, W., Valenta, P., and Nürnberg, H. W.,** A new photodigestion device to decompose organic matter in water, *Fresenius Z. Anal. Chem.,* 317, 264, 1984.

39. **Valenta, P., Sipos, L., Kramer, I., Krumpen, P., and Rützel, H.,** An automatic voltammetric analyzer for the simultaneous determination of toxic trace metals in water, *Fresenius Z. Anal. Chem.,* 312, 101, 1982.

40. **Bächmann, K.,** Multielement concentration for trace elements, *CRC Crit. Rev. Anal. Chem.,* 12, 1, 1981.

41. **Ashton, A. and Chan, R.,** Monitoring of microgram per litre concentrations of trace metals in sea water: the choice of methodology for sampling and analysis, *Analyst (London),* 112, 841, 1987.

42. **Ahmed, R. and Stoeppler, M.,** Decomposition and stability studies of methylmercury in water using cold vapour atomic absorption spectrometry, *Analyst (London),* 111, 1371, 1986.

43. **May, K. and Stoeppler, M.,** Pretreatment studies with biological and environmental materials. IV. Complete wet digestion in partly and completely closed quartz vessels for subsequent trace and ultratrace mercury determinations, *Fresenius Z. Anal. Chem.,* 317, 248, 1984.

44. **Bock, R. A.,** *A Handbook of Decomposition Methods in Analytical Chemistry,* Marr, I. L., Ed., International Textbook Company, Glasgow, 1978.

45. **Jackwerth, E. and Gomišček, S.,** Acid pressure decomposition in trace element analysis, *Pure Appl. Chem.,* 56, 479, 1984.

46. **Stoeppler, M.,** Analytical aspects of the determination and characterization of metallic pollutants, in *Pollutants and Their Ecotoxicological Significance,* Nürnberg, H. W., Ed., John Wiley & Sons, New York, 1985, 317.

47. **Winge, R. K., Peterson, V. J., and Fassel, V. A.,** Inductively coupled plasma-atomic emission spectroscopy: prominent lines, *Appl. Spectrosc.,* 33, 206, 1979.

48. **Boumans, P. J. M.,** *Line coincidence tables for ICP-AES,* Pergamon Press, Elmsford, NY, 1980.

49. Product information: Jobin-Yvon ICP-Spectrometer JY 38 Plus, Instruments S.A. GmbH, May 1987.

50. **Welz, B.,** Hg determination with GFAAS and palladium modifier, personal communication.

51. **Pinel, R., Benabdallah, M. Z., Astruc, C., and Astruc, M.,** Determination of inorganic tin and organotin compounds by graphite-furnace atomic absorption spectrometry with a new matrix modifier, *Anal. Chim. Acta,* 181, 187, 1986.

52. **Alt, F., Messerschmidt, J., and Tölg, G.,** A contribution towards the improvement of Se-determination in the pg-region by hydride AAS, *Fresenius Z. Anal. Chem.,* 327, 233, 1987.

53. **Welz, B. and Melcher, M.,** Picotrace determination of mercury using the amalgamation technique, *At. Spectrosc.,* 5, 37, 1984.

54. **Nürnberg, H. W.,** The voltammetric approach in trace metal chemistry of natural waters and atmospheric precipitation, *Anal. Chim. Acta,* 164, 1, 1984.

55. **Wang, J., Farias, P. A. M., and Mahmoud, J. S.,** Stripping voltammetry of aluminium based on adsorptive accumulation of its solochrome violet RS complex at the static mercury drop electrode, *Anal. Chim. Acta,* 172, 57, 1985.

56. **Golimowski, J., Valenta, P., and Nürnberg, H. W.,** Trace determination of chromium in various water types by adsorption differential pulse voltammetry, *Fresenius Z. Anal. Chem.,* 322, 315, 1985.

57. **Michaelis, A.,** private communication.

58. **Prange, A. and Kramer, K.,** Aufbereitung von wäßrigen Proben zur Bestimmung von Schwermetallen mit Hilfe der TRFA, in *Totalreflexions-Röntgenfluoreszenzanalyse,* 86/E/61, Michaelis, W. and Prange, A., Eds., GKSS, Geesthacht, West Germany, 1987, 48.

59. **Guinn, V. P. and Hoste, J.,** Neutron activation analysis, in *Elemental Analysis of Biological Materials,* Tech. Rep. Ser. International Atomic Energy Agency, Vienna, 1980, 105.

60. **van der Sloot, H. A.,** *Neutron Activation Analysis of Trace Elements in Water Samples after Preconcentration on Activated Carbon,* ECN-1, Netherlands Energy Research Foundation, Petten, 1976.

61. **Nürnberg, H. W.,** Potential of voltammetry in environmental oligoelement analysis of trace metals, in *Instrumentelle Multielementanalyse,* Sansoni, B., Ed., VCH Publishers, Weinheim, West Germany, 1985, 415.

62. **Nürnberg, H. W.,** Voltammetric trace analysis in ecological chemistry of toxic metals, *Pure Appl. Chem.,* 54, 853, 1982.

63. **Pihlar, B., Valenta, P., and Nürnberg, H. W.,** New high-performance analytical procedure for the voltammetric determination of nickel in routine analysis of waters, biological materials and food, *Fresenius Z. Anal. Chem.,* 307, 337, 1981.

64. **van den Berg, C. M. G., Murphy, K., and Riley, J. P.,** The determination of aluminium in seawater and fresh water by cathodic stripping voltammetry, *Anal. Chim. Acta,* 188, 177, 1986.

65. **Ostapczuk, P., Valenta, P., and Nürnberg, H. W.,** Square wave voltammetry: a rapid and reliable determination method for Zn, Cd, Pb, Cu, Ni and Co in biological and environmental samples, *J. Electroanal. Chem.,* 214, 51, 1986.

66. **Stoeppler, M. and Backhaus, F.,** Contributions to environmental research and surveillance. I. Design, construction and operation of a mobile trace analytical laboratory, *Ber. Kernforschungsanlage Juelich,* Jül, 1571, 1979.

67. **Stoeppler, M. and Nürnberg, H. W.,** Analytik von Metallen und ihren Verbindungen, in *Metalle in der Umwelt,* Merian, E., Ed., Verlag Chemie, Weinheim, West Germany, 1984, 4a.

68. **Wang, J.,** *Stripping Analysis, Principles, Instrumentation and Applications,* VCH Publishers, Deerfield Beach, FL, 1985.

69. **Bodewig, F. G., Valenta, P., and Nürnberg, H. W.,** Trace determination of As(III) and As(V) in natural waters by differential pulse anodic stripping voltammetry, *Fresenius Z. Anal. Chem.,* 311, 187, 1982.

70. **Leung, P. C., Subramanian, K. S., and Méranger, J. C.,** Determination of arsenic in polluted waters by differential pulse anodic-stripping voltammetry, *Talanta,* 29, 515, 1982.

71. **Adeloju, S. B., Bond, A. M., and Briggs, M. H.,** Multielement determination in biological materials by differential pulse voltammetry, *Anal. Chem.,* 57, 1386, 1985.

72. **Adeloju, S. B. and Brown, K. A.,** Determination of ultra-trace amounts of cadmium in natural waters by the combination of a solvent extraction procedure and anodic stripping voltammetry, *Analyst,* 112, 221, 1987.

73. **Golimowski, J., Sigg, L., Valenta, P., and Nürnberg, H. W.,** Chromium levels in European inland waters, in *Proc. Int. Conf. Heavy Metals in the Environment,* Vol. 1, Lekkas, T. D., Ed., CEC Consultants, Edinburgh, 1985, 107.

74. **Batley, G. E.,** Interferences in the determination of copper in natural waters by anodic stripping voltammetry, *Anal. Chim. Acta,* 189, 371, 1986.

75. **Batley, G. E.,** Differential-pulse polarographic determination of selenium species in contaminated waters, *Anal. Chim. Acta,* 187, 109, 1986.

76. **Weber, G.,** Determination of tin in the ng g^{-1} range by differential pulse polarography, *Anal. Chim. Acta,* 186, 49, 1986.

77. **Weber, G.,** Determination of traces of tin in aqueous matrices by voltammetry and atomic absorption spectrometry, *Fresenius Z. Anal. Chem.,* 322, 311, 1985.

78. **Bonelli, J. E., Taylor, H. E., and Skogerboe, R. K.,** A direct differential pulse anodic stripping voltammetric method for the determination of thallium in natural waters, *Anal. Chim. Acta,* 118, 243, 1980.

79. **Gemmer-Čolos, Kiehnast, I., Trenner, J., and Neeb, R.,** Inverse-voltammetric determination of thallium, *Fresenius Z. Anal. Chem.,* 306, 144, 1981.

80. **Riley, J. P. and Siddiqui, S. A.,** The determination of thallium in sediments and natural waters, *Anal. Chim. Acta,* 181, 117, 1986.

81. **Sager, M.,** *Spurenanalytik des Thalliums,* Georg Thieme Verlag, Stuttgart, 1986.

82. **Parsons, M. L., Major, S., and Forster, A. R.,** Trace element determination by atomic spectroscopic methods: state of the art, *Appl. Spectrosc.,* 37, 411, 1983.

83. **Greenfield, S., Hieftje, G. M., Omenetto, N., Scheeline, A., and Slavin, W.,** Twenty-five years of analytical atomic spectroscopy, *Anal. Chim. Acta,* 180, 69, 1986.

84. **Slavin, W.,** Flames, furnaces, plasmas, how do we choose?, *Anal. Chem.,* 58, 589A, 1986.

85. **Brokaert, J. A. C. and Tölg, G.,** Recent developments in atomic spectrometry methods for elemental trace determinations, *Fresenius Z. Anal. Chem.,* 326, 495, 1987.

86. **Slavin, W.,** Atomic absorption spectroscopy, the present and the future, *Anal. Chem.,* 54, 685A, 1982.

87. **Cantle, J. E., Ed.,** *Atomic absorption spectrometry,* Elsevier, Amsterdam, 1982.

88. **Stoeppler, M.,** Atomic absorption spectrometry — a valuable tool for trace and ultratrace determination of metals and metalloids in biological materials, *Spectrochim. Acta,* 38B, 1559, 1983.

89. **Slavin, W.,** *Graphite furnace AAS — A Source Book,* Perkin Elmer, Ridgefield, 1984.

90. **Welz, P.,** *Atomic absorption spectrometry,* 2nd ed., VCH Publishers, Weinheim, West Germany, 1985.

91. Special issue: Colloq. on "Present status and perspectives of the analysis of solid samples by AAS", *Fresenius Z. Anal. Chem.,* 322, 653, 1985.

92. Special issue: 2nd Int. Colloq. Solid Sampling with Atomic Spectroscopic Methods, *Fresenius Z. Anal. Chem.,* 328, 315, 1987.

93. **Brokaert, J. A. C.,** Spectral interferences and matrix effects in optical emission spectroscopy, in *Instrumentelle Multielementanalyse,* Sansoni, B., Ed., VCH Publishers, Weinheim, West Germany, 1985, 337.

94. **Fassel, V. A.,** Analytical inductively coupled plasma spectroscopies — past, present and future, *Fresenius Z. Anal. Chem.,* 324, 511, 1986.

95. **Boumans, P. W. J. M., Ed.,** *Inductively Coupled Plasma Emission Spectroscopy,* Vols. 1 and 2, John Wiley & Sons, New York, 1987.

96. **Ramsey, M. H., Thompson, M., and Banergee, E. K.,** Realistic assessment of analytical data quality from inductively plasma atomic emission spectrometry, *Anal. Proc.,* 24, 260, 1987.

97. Special issue: Post Colloquium Spectroscopicum Internationale, "Selected topics from graphite furnace and hydride-generation AAS", *Fresenius Z. Anal. Chem.,* 323, 673, 1986.

98. **Sturgeon, R.,** Graphite furnace atomic absorption spectrometry: fact and fiction, *Fresenius Z. Anal. Chem.,* 324, 807, 1986.

99. **L'vov, B. V.,** Electrothermal atomization — the way toward absolute methods of AAS, *Spectrochim. Acta,* 33B, 153, 1978.

100. **Beaty, M. and Barnett, W.,** Techniques for analyzing difficult samples with the HGA graphite furnace, *At. Spectrosc.,* 1, 72, 1980.

101. **Slavin, W., Manning, D. C., and Carnrick, G. R.,** The stabilized temperature platform furnace, *At. Spectrosc.,* 2, 137, 1981.

102. **Kaiser, M. L., Koirtyohann, S. R., and Hinderberger, E. J.,** Reduction of matrix interferences in furnace AA with the L'vov platform, *Spectrochim. Acta,* 36B, 773, 1981.

103. **Herber, R. F. M.,** Some instrumental improvements in electrothermal atomization AAS, *Spectrochim. Acta,* 39B, 271, 1984.

104. **Slavin, W., Carnrick, G. R., Manning, D. C., and Pruszkowska, E.,** Recent experiences with the stabilized temperature platform furnace and Zeeman background correction, *At. Spectrosc.,* 4, 69, 1983.

105. **Slavin, W. and Carnrick, G. R.,** The possibility of standardless furnace atomic absorption spectroscopy, *Spectrochim. Acta,* 39B, 271, 1984.

106. **Slavin, W. and Carnrick, G. R.,** A survey of applications of the stabilized temperature platform furnace and Zeeman correction, *At. Spectrosc.,* 6, 157, 1985.

107. **Ediger, R. D.,** Atomic absorption analysis with the graphite furnace using matrix modification, *At. Absorpt. Newsl.,* 14, 127, 1975.

108. **Manning, D. C. and Slavin, W.,** The determination of trace elements in natural waters using the stabilized temperature platform furnace, *Appl. Spectrosc.,* 37, 1, 1983.

109. **Xiao-quan, S., Zhe-ming, N., and Li, Z.,** Application of matrix modification in determination of thallium in waste water by graphite furnace atomic-absorption spectrometry, *Talanta,* 31, 150, 1984.

110. **Schlemmer, G. and Welz, B.,** Palladium and magnesium nitrates, a more universal modifier for graphite furnace atomic absorption spectrometry, *Spectrochim. Acta,* 41B, 1157, 1986.

111. **Voth-Beach, L. M. and Shrader, D. E.,** Investigations of a reduced palladium chemical modifier for graphite furnace atomic absorption spectrometry, *J. At. Absorp. Spectrom.,* 2, 45, 1987.

112. **Schlemmer, G.,** personal communication.

113. **Bruland, K. W., Franks, R. P., Knauer, G. A., and Martin, J. H.,** Sampling and analytical methods for the determination of copper, cadmium, zinc, and nickel at the nanogram per liter level in seawater, *Anal. Chim. Acta,* 105, 233, 1979.

114. **Danielsson, L.-G., Magnusson, B., Westerlund, S., and Zhang, K.,** Trace metal determination in estuarine waters by electrothermal atomic absorption spectrometry after extraction of dithiocarbamate complexes into Freon, *Anal. Chim. Acta,* 144, 183, 1982.

115. **Subramanian, K. S., Leung, P. C., and Méranger, J. C.,** Determination of arsenic (III, V, total) in polluted waters by graphite furnace atomic absorption spectrometry and anodic stripping voltammetry, *Int. J. Envion. Anal. Chem.,* 11, 121, 1982.

116. **Willie, S. N., Sturgeon, R. E., and Berman, S. S.,** Determination of total chromium in seawater by graphite furnace atomic absorption spectrometry, *Anal. Chem.,* 55, 981, 1983.

117. **Statham, P. J.,** The determination of dissolved manganese and cadmium in sea water at low n mol l^{-1} concentrations by chelation and extraction followed by electrothermal atomic absorption spectrometry, *Anal. Chim. Acta,* 169, 149, 1985.

118. **Sturgeon, R. E., Willie, S. N., and Berman, S. S.,** Preconcentration of selenium and antimony from seawater for determination by graphite furnace atomic absorption spectrometry, *Anal. Chem.,* 57, 6, 1985.

119. **Willie, S. N., Sturgeon, R. E., and Berman, S. S.,** Hydride generation atomic absorption determination and selenium in marine sediments, tissues and seawater with in situ concentration in a graphite furnace, *Anal. Chem.,* 58, 1140, 1986.

120. **Suzuki, H. and Ohta, K.,** Determination of thallium by electrothermal atomic absorption spectrometry with a metal atomizer, *Fresenius Z. Anal. Chem.,* 322, 480, 1985.

121. **Langmyhr, F. J. and Wibetoe G.,** Direct analysis of solids by atomic absorption spectrophotometry, *Prog. Anal. At. Spectrosc.,* 8, 193, 1985.

122. **Craney, C. L., Swartout, K., Smith, F. W., and West, D. C.,** Improvement of trace aluminium determination by electrothermal atomic absorption spectrophotometry using phosphoric acid, *Anal. Chem.,* 58, 656, 1986.

123. **Lum, K. R. and Callaghan, M.,** Direct determination of cadmium in natural waters by electrothermal atomic absorption spectrometry without matrix modification, *Anal. Chim. Acta,* 187, 157, 1986.

124. **Andreae, M. O. and Byrd, J. T.,** Determination of tin and methyltin species by hydride generation and detection with graphite-furnace atomic absorption or flame emission spectrometry, *Anal. Chim. Acta,* 156, 147, 1984.

125. **Pruszkowska, E., Manning, D. C., Carnrick, G. R., and Slavin, W.,** Experimental conditions for the determination of tin with the stabilized temperature platform furnace and Zeeman background correction, *At. Spectrosc.,* 4, 87, 1983.

126. **Stoeppler, M.,** Commercially available instrumentation for atomic absorption spectrometry, in *Nachr. Chem. Tech. Lab.,* 33/4, M3, 1985.

127. **Oda, C. E. and Ingle, J. D., Jr.,** Continuous flow cold vapor atomic absorption determination of mercury, *Anal. Chem.,* 53, 2030, 1981.

128. **Freimann, P. and Schmidt, D.,** Determination of mercury in seawater by cold vapor atomic absorption spectrophotometry, *Fresenius Z. Anal. Chem.,* 313, 200, 1982.

129. **Bloom, N. S. and Crecelius, E. A.,** Determination of mercury in seawater of sub-nanogram per liter levels, *Mar. Chem.,* 14, 49, 1983.

130. **Temmerman, E., Dumarey, R., and Dams, R.,** Optimization and evaluation of reduction-aeration amalgamation for the analysis of mercury in drinking-water by cold vapor atomic absorption spectrometry (CVAAS), *Anal. Lett.,* 18 (A2), 203, 1985.

131. **May, K. and Stoeppler, M.,** Modified cold vapor AAS method for extremely low mercury concentrations, in preparation.

132. **Welz, B. and Schubert-Jacobs, M.,** Investigations on atomization and mechanisms in hydride-generation atomic absorption spectrometry, *Fresenius Z. Anal. Chem.,* 324, 832, 1986.

133. **Stoeppler, M., Burow, M., Backhaus, F., Schramm, W., and Nürnberg, H. W.,** Arsenic in seawater and brown algae of the Baltic and the North Sea, *Mar. Chem.,* 18, 637, 1986.

134. **Verlinden, M., Deelstra, H., and Adrieaenssens, E.,** The determination of selenium by atomic absorption spectrometry: a review, *Talanta,* 28, 637, 1981.

135. **Braman, R. S. and Tompkins, M. A.** Separation and determination of nanogram amounts of inorganic tin and methyltin compounds in the environment, *Anal. Chem.,* 51, 12, 1979.

136. **Nojiri, Y., Kawai, T., Otsuki, A., and Fuwa, K.,** Simultaneous multielement determination of trace metals in lake waters by ICP emission spectrometry with preconcentration and their background levels in Japan, *Water Res.,* 4, 503, 1985.

137. **Thompson, M., Ramsay, M. H., and Pahlavanpour, B.,** Water analysis by inductively coupled plasma atomic emission spectrometry after a rapid pre-concentration, *Analyst,* 107, 1330, 1982.

138. **Speer, R., Hoffmann, P., and Lieser, K. H.,** Problems in measuring water samples of different salt content by ICP-AES, *Fresenius Z. Anal. Chem.,* 325, 558, 1986.

139. **Wobrauschek, P. and Aiginger, H.,** Total-reflection X-ray fluorescence spectrometric determination of elements in nanogram amounts, *Anal. Chem.,* 47, 852, 1975.

140. **Knoth, J. and Schwenke, H.,** An X-ray fluorescence spectrometer with totally reflecting sample support for trace analysis at the ppb level, *Fresenius Z. Anal. Chem.,* 291, 200, 1979.

141. **Michaelis, W., Prange, A., and Knoth, J.,** Recent applications of total reflection X-ray fluorescence analysis in multielement analysis, in *Instrumentelle Multielement-Analyse,* Sansoni, B., Ed., VCH Publishers, Weinheim, West Germany, 1985, 264.

142. **Erdtmann, G. and Petri, H.,** Nuclear activation analysis: fundamentals and techniques, in *Treatise on Analytical Chemistry,* Vol. 14, 2nd ed., Elving, P. J., Ed., John Wiley & Sons, New York, 1986, 419.

143. **Heumann, K. G.,** High accuracy in the element analysis by mass spectrometry, *Fresenius Z. Anal. Chem.,* 324, 601, 1986.

144. **Heumann, K. G.,** Isotope dilution mass spectrometry of inorganic and organic substances, *Fresenius Z. Anal. Chem.,* 325, 661, 1986.

145. Meeting Report, Interlaboratory lead analysis of standardized samples of seawater, *Mar. Chem.,* 2, 69, 1974.

146. **Mykytiuk, A. P., Russel, D. S., and Sturgeon, R. E.,** Simultaneous determination of iron, cadmium, zinc, copper, nickel, lead and uranium in seawater by stable isotope dilution spark source mass spectrometry, *Anal. Chem.,* 52, 1281, 1980.

147. **Trettenbach, J. and Heumann, K. G.,** Determination of lead, cadmium and thallium in water samples at ppb and ppt levels by isotope dilution mass spectrometry, *Fresenius Z. Anal. Chem.,* 322, 306, 1985.

148. **Jones, V. K. and Tarter, J. G.,** Simultaneous analysis of anions and cations in water samples using ion chromatography, *Int. Lab.,* 1519, 36, 1985.

149. **Munder, A. and Ballschmiter, K.,** Chromatography of metal chelates. XI. Trace Analysis of cadmium, cobalt, copper, mercury and nickel in water using bis(ethoxyethyl)dithiocarbamate as reagent for RP C_{18}-HPLC and photometric detection, *Fresenius Z. Anal. Chem.,* 323, 869, 1986.

150. **Rubin, R. B. and Herberling, S. S.,** Metal determinations by ion chromatography: a complement to atomic spectroscopy, *Int. Lab.,* 1719, 54, 1987.

151. **King, J. N. and Fritz, J. S.,** Determination of Co, Cu, Hg and Ni as bis(2-hydroxyethyl)dithiocarbamate complexes by high performance liquid chromatography, *Anal. Chem.,* 59, 703, 1987.

152. **Houk, R. S.,** Mass spectrometry of inductively coupled plasmas, *Anal. Chem.,* 58, 97A, 1986.

153. **Selby, M. and Hieftje, G. M.,** Inductively coupled plasma-mass spectrometry: a status report, *Int. Lab.,* 17/8, 28, 1987.

154. **Dietz, F.,** Experiences with ICP mass spectrometry in water analysis, *Fresenius Z. Anal. Chem.,* 324, 222, 1986.

155. **Sansoni, B., Brunner, W., Wolff, G., Ruppert, H., and Dittrich, R.,** Comparative instrumental multi-element analysis. I. Comparison of ICP-MS with ICP-AES, ICP-AFS and AAS for the analysis of natural waters from a granite region, *Fresenius Z. Anal. Chem.,* 331, 154, 1988.

156. **Beauchemin, D., McLaren, J. W., Mykytiuk, A. P., and Berman, S. S.,** Determination of trace metals in a river water reference material by inductively coupled plasma mass spectrometry, *Anal. Chem.,* 59, 778, 1987.

157. **Wilson, A. L.,** Approach for achieving comparable analytical results from a number of laboratories, *Analyst,* 104, 273, 1978.

158. **Carron, J. M. and Aspila, K. I.,** Interlaboratory quality control studies Nos. 12 and 13: aluminium, cadmium, chromium, cobalt, copper, lead, iron, manganese, nickel, and zinc, Rep. Ser. No. 44, Inland Waters Directorate, Ontario Region, Water Quality Branch, Burlington, Ontario, 1976.

159. **Sugawara, K.,** Interlaboratory comparison of the determination of mercury and cadmium in sea and fresh waters, *Deep Sea Res.,* 25, 223, 1978.

160. **Dybczynski, R., Tugsavul, A., and Suschny, O.,** Problems of accuracy and precision in the determination of trace elements in water as shown by recent International Atomic Energy Agency intercomparison tests, *Analyst,* 103, 733, 1978.

161. **Franz, J. and Grubert, G.,** Results of an intercomparison analysis of thallium in surface water, *Z. Wasser Abwasser Forsch.,* 13, 138, 1980.

162. **Knöchel, A. and Petersen, W.,** Results of an interlaboratory test for heavy metals in Elbe water, *Fresenius Z. Anal. Chem.,* 314, 105, 1983.

163. The Severn Estuary Chemists' Sub-Committee, Results of an inter-laboratory analytical quality control programme for non-saline waters, *Analyst,* 109, 3, 1984.

164. The Committee for Analytical Quality Control (Harmonised Monitoring), Accuracy of determination of cadmium, copper, lead, nickel and zinc in river waters, Tech. Rep. TR 213, WRC Environment, Medmenham, U. K., 1984.

165. The Committee for Analytical Quality Control (Harmonized Monitoring), Accuracy of determination of trace concentrations of dissolved cadmium in river waters: analytical quality control in the harmonised monitoring scheme, *Analyst,* 110, 103, 1985.

166. The Committee for Analytical Quality Control (Harmonized Monitoring), Accuracy of determination of trace concentrations of dissolved cadmium in river waters: analytical quality control in the harmonised monitoring scheme, *Analyst,* 110, 247, 1985.

167. **Taylor, J. K.,** Principles of quality assurance of chemical measurments, NBSIR 85-3105, National Bureau of Standards, U.S. Department of Commerce, Washington, D. C., 1985.

168. **Bruland, K. W., Coale, K. H., and Mart, L.,** Analysis of seawater for dissolved cadmium, copper and lead: an intercomparison of voltammetric and atomic absorption methods, *Mar. Chem.,* 17, 285, 1985.

169. **Krivan, V.,** Application of radiotracers to methodological studies in trace element analysis, in *Treatise on Analytical Chemistry,* Vol. 14, 2nd ed., John Wiley & Sons, New York, 1986, 339.

170. Simultaneous Multi-Element Analysis Atomic Absorption Spectrophotometer, Z-9000, Hitachi, Tokyo, Japan, 1987.

171. **Marshall, J., Ottaway, B. J., Ottaway, J. M., and Littlejohn, D.,** Continuum-source atomic absorption spectrometry — new lamps for old?, *Anal. Chim. Acta,* 180, 357, 1986.

172. **Harnly, J. M.,** Multielement atomic absorption with a continuum source, *Anal. Chem.,* 58, 933A, 1986.

173. **deGalan, L.,** New directions in optical atomic spectrometry, *Anal. Chem.,* 58, 697A, 1986.

174. **Preli, F. R., Jr., Dougherty, J. P., and Michel, R. G.,** Laser excited atomic fluorescence spectrometry with a laboratory constructed tube electrothermal atomizer, *Anal. Chem.,* 59, 1784, 1987.

175. **Ružička, J.,** Flow injection analysis — a survey of its potential for spectroscopy, *Fresenius Z. Anal. Chem.,* 324, 745, 1986.

Chapter 5

CHEMICAL SPECIATION OF TRACE METALS

M. Astruc

TABLE OF CONTENTS

I. INTRODUCTION

One of the most crucial advances in the field of aquatic chemistry in these last years has been the development of new concepts about the speciation of trace elements in the aquatic environment. This became possible — even though the basic ideas have long been known — only when the development of methods for trace and ultratrace analysis were such that the reliable determination of total concentrations of elements became common. Then it appeared clear that the knowledge of the total concentration of trace elements was definitely insufficient to understand their behavior in water bodies, their biogeochemical cycles, and their ecotoxicity.[1-3]

II. AQUATIC CHEMISTRY OF TRACE ELEMENTS

The aquatic chemistry of trace elements is ruled by a complex competition of the following elementary processes:

1. Homogeneous chemical reactions (e.g., acid-base, complex formation with inorganic ligands such as CO_3^{2-}, OH^-, Cl^-, S^{2-}, or organic matter such as humic substances or biological ligands, oxidoreduction, chemical precipitation)
2. Surface reactions involving colloids, particulates, and sediment

The influence of biota is still a very cloudy subject. Excretion in water of powerful ligands by a large variety of living species has been demonstrated several times in the laboratory.[4,5] In natural conditions, however, this is not so clear, due to both dilution and the extreme complexity of the problem.[6,7] For some elements biologically mediated methylation (biomethylation) has been documented and may play a dominant role in the overall biogeochemical cycles of these elements and perhaps some others. The simultaneous possible occurrence of all these processes in a defined water makes the speciation problem a very difficult one, more so as the concentrations of interest may be very low.

Aquatic toxicology and, moreover, ecotoxicology belong to the specialties where the necessity of knowledge as detailed as possible of metal speciation must be emphasized: different chemical species of a single element often have deeply differing bioavailabilities and effects on target organisms.

III. CHEMICAL SPECIATION AND UPTAKE MECHANISMS

Let us describe schematically the chemist's point of view on aquatic toxicity of metals. Let the element considered, M, be present in the bulk water as species M_1 (only for sake of simplicity). The first process to consider is mass transfer of M from the bulk of the solution toward the cell membrane (Figure 1).

The bulk solution is homogenized by convection processes down to a boundary layer whose thickness δ is determined by the hydrodynamics of the system considered. In this layer mass transfer is ruled essentially by diffusion; electric effects cannot be completely ruled out, but may be thought of as usually negligible.

Species M_1 entering the diffusion layer at δ may finally exert toxic effects following a variety of eventual pathways (Figure 2). A similar, more simple model has already been proposed.[8]

Pathways I and II involve deleterious interactions of M with extracellular material or biomembrane components; in I the predominant aqueous species M_1 is directly active, whereas in pathway II, M_1 must first be converted into species M^{n+} before toxic effect occurs.

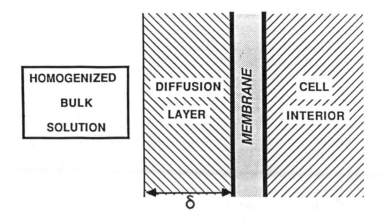

FIGURE 1.   Mass transfer processes towards a cell membrane.

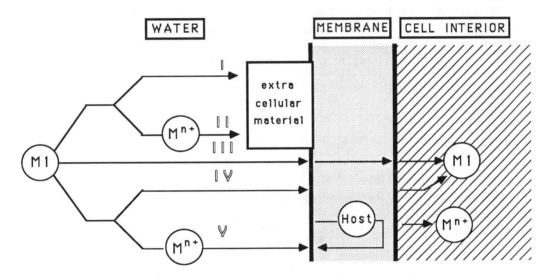

FIGURE 2.   Schematic illustration of possible ways of interaction between a dissolved species M_1 and a cell.

In pathways III to V the element M enters the cell through the biomembrane and acts in the cytosol. Pathway III describes the situation where the liposolubility of M_1 is such that direct diffusion through the membrane is possible. On the contrary, in IV and V hydrophilic species M_1 or M^{n+} can only cross the membrane by some host-mediated transport process.

Very little is actually known about the kinetics involved in these processes. However, an approximate comparative examination is possible. Let us suppose that the toxic incorporation processes themselves are fast enough not to be rate determining. Under this hypothesis the slower processes I to V may be compared as they control the overall process.

Process I should be very fast, being dependent only on mass transfer by diffusion through the boundary layer, whereas process II may eventually be slower, depending on the chemical kinetics involved in the reaction

$$M_1 \rightarrow M^{n+}$$

In process III the liposoluble species M_1 diffuses free through, successively, the diffusion layer and the cell membrane. High toxicity is predictable.[2,3] Host-mediated diffusion through the membrane (processes IV and V) is necessarily slower. The supplementary chemical step in process V may slow down again the overall process.

Aquatic chemistry has demonstrated that very often M_1 is some complex ML of the cation M with the organic or inorganic ligand L.

Toxic effects exerted through processes II and V are essentially dependent on the "lability" of species ML, i.e., its kinetic ability to dissociate. Most, but not all, inorganic species are very labile, but many organic ligands give complexes of low or even negligible lability.

The preceding discussion is concerned with dissolved species. There is no serious reason why colloidal species, such as humic complexes, for example, should not be considered in a similar manner, keeping in mind, however, their low mobility in water (diffusion coefficients lower by about two orders of magnitude).

Organometallic compounds involving some covalent metal-carbon bond are usually considered separately owing to their particular chemistry. Their hydrophobicity qualifies them at first sight for process III in Figure 2. However, one may wonder if the lack of consideration of other pathways is not merely a consequence of the almost complete ignorance of their aquatic chemistry. This field remains widely open to further investigations.

These considerations have prompted analytical chemists to undertake enormous efforts to create new methods able to differentiate, identify, and determine metal species down to the ultratrace level, and aquatic chemists to improve the fundamental knowledge of the thermodynamic properties of chemical species and the kinetics of their reactions.

IV. DISSOCIATION KINETICS AND LABILITY

The formation of the complex species ML is usually schematized by (assume $p > n$)

$$M^{p+} + L^{n-} = \overset{k_f}{\underset{k_b}{\rightleftharpoons}} ML^{(p-n)+}$$

process characterized by the equilibrium formation constant

$$K = \frac{(ML^{(p-n)+})}{(M^{p+})(L^{n-})}$$

Let k_f and k_b be, respectively, the forward and backward rate constants. The reaction rates

$$v_f = k_f(M^{p+})(L^{n-})$$

$$v_b = k_b(ML^{(p-n)+})$$

being equal in the equilibrium situation, then

$$K = \frac{k_f}{k_b}$$

A somewhat more detailed study of the complex formation[7,9] leads to the consideration of the following succession of elementary processes:

$$[M(H_2O)_q]^{p+} + L^{n-} \overset{K_{os}}{\rightleftharpoons} [M(H_2O)_q \dots L]^{(p-n)+}$$

$$k_b \uparrow \downarrow k_f \qquad\qquad \uparrow \downarrow f_{-w}$$

$$[M(H_2O)_{q-1} L]^{(p-n)+} \overset{K_{oi}}{\rightleftharpoons} [M(H_2O)_{q-1} \dots L]^{(p-n)+}$$

The first step is the outer sphere association of the ligand to the aqueous ion, a fast process as little energy is required. The value of K_{os} depends essentially on the electric charges of the reactants. An anionic ligand such L^{n-} will associate more rapidly with $M(H_2O)_q^{p+}$ than with a neutral ligand and yet more rapidly than with a positively charged ligand. Then a water molecule is eliminated from the inner coordination sphere. This water loss is usually the rate-determining step. Finally, the ligand rapidly enters the inner sphere of coordination. Further elimination of water molecules may be necessary if L is a pluridentate ligand; these complementary processes are fast.

The overall forward rate constant of complex formation, k_f, is thus determined essentially by k_{-w}, characteristic of the metal ion involved, not of the ligand. The dissociation rate constant of the complex, k_b, is defined by the ratio

$$k_b = \frac{k_f}{K}$$

For a given metal ion the dissociation rate constants of complexes with various similar ligands will, thus, vary essentially with K.

A complex will be said "labile" if the dissociation rate, v_b, in the experimental conditions used is higher than the rate of consumption of M. In toxicological studies involving uptake of toxicant through a pathway such as V, species M_1 will be labile only if v_b is higher than the rate of host-mediated transport through the biological membrane.

Physicochemical determination of the lability of a complex species is most often made by electroanalytical chemistry,[10-12] usually anodic stripping voltammetry (ASV). In ASV a labile complex may dissociate completely in the diffusion layer; hence, the metal ion is reduced and amalgamated at the mercury electrode. Similarity is obvious but is not an identity: ASV-lability is different from biolability as the time scales differ.

The practical upper limitation of k_f is about 10^9/mol/s. So even very stable complexes (i.e., $K \geqslant 10^7$/mol) may be labile if the water-loss value is high enough. Let us consider the dissociation of a hypothetical complex CdL of cadmium ion with a neutral ligand L. Let the equilibrium formation constant be 10^8/mol. The water loss value of cadmium is $k_{-w} = 9 \times 10^7$/s, and the outer sphere association constant K_{os} is approximately equal to 0.1.

Thus, the formation rate constant is $k_f = K_{os} k_{-w} = 9 \times 10^6$/mol/s, and the dissociation rate constant $k_b = k_f K^{-1} = 9 \times 10^{-2}$/s. Consider the ideally simple situation of a pure solution of CdL at concentration $C^o = 10^{-5}$ M. The equilibrium concentration of Cd^{2+} is 3.1×10^{-7} M that we will consider as negligible. Let us introduce now a reactant that consumes immediately and completely Cd^{2+}.

Dissociation of CdL occurs at a rate defined by:

$$\frac{d(CdL)}{dt} = -k_b(CdL)$$

At time t the concentration of undissociated complex is, thus, $(CdL) = C^o \exp[-k_b t]$. The time necessary to dissociate 50% of CdL is

$$\tau_{1/2} = 0.69 \, k_b^{-1} = 7.7 \text{ s}$$

This reaction is rather slow and the hypothetical complex CdL would appear labile or not, depending on the time scale of the experiment. For example, such a complex would be nonlabile in polarography[8] or quasilabile in ASV.

If the equilibrium constant were $K = 10^6/M$, then $\tau_{1/2}$ would be 100 times shorter. The complex CdL would then be quasi-labile in polarography, labile in ASV.

V. CHEMICAL MODELING

Chemical modeling of waters has been the object of many studies. Over 50 computer programs are available; some of them are very powerful. MINEQL,[13] for example, allows calculation of the equilibrium state of a system gathering up to 20 trace elements and 30 ligands. Several limitations to the application of these models to actual situations do exist. If they are well designed to deal with homogeneous solution chemistry, they hardly accommodate heterogeneous processes linked to the presence of suspended solids or colloids.

Recent theoretical efforts have already at least partly overcome these criticisms, but other limitations still remain: (1) these calculations deal with the equilibrium state, rather unlikely in actual situations, and (2) modeling needs a complete set of stoichiometric and thermodynamic data. In natural waters inorganic solutes and the equilibrium constants of their reactions are generally known well enough. However, organic compounds are only very badly defined, even after very sophisticated studies, and their complexes with trace elements are at best known on a semiquantitative basis.

As a consequence, chemical modeling gives very detailed information in special laboratory situations where the chemical composition of the medium may be perfectly defined. When dealing with actual environmental samples, this information must be handled with care, especially if the organic content is high, but it is nevertheless often very valuable combined with experimental observations.

Some general conclusions have been obtained from these equilibrium calculations for purely inorganic waters. In oxic fresh waters the metal species most commonly encountered are the free hydrated cation $M(H_2O)_q^{p+}$ and weak complexes of high to medium lability with hydroxyl or carbonate ions. In seawater the concentration of chloride ions is so high that chloride complexes are often predominant; their lability is high.

VI. SPECIATION OF TRACE ELEMENTS

Widely differing chemical species of trace elements occur in the aquatic environment. All are able to form inorganic complexes and combine with naturally occurring organic ligands. A few of them have been demonstrated to occur as well as organometallic compounds involving metal-carbon covalent bonds. These organometallics may be industrial products or produced by natural biotic or abiotic methylation.

Cadmium and tin will be taken here as examples of these respective classes.

A. Speciation of Cadmium
1. Suspended Solids

Experimental speciation of trace elements in water begins with the evaluation of the particulate fraction. This fraction may be more or less important, depending on the element and, moreover, on the kind of water considered. In the open ocean the level of particulate concentration is usually so low that its contribution to the speciation of metals is negligible.[14] In contrast, fresh waters, coastal seawaters, and overall estuarine waters contain high amounts of suspended solids able to play a dominant role in the speciation phenomena.

Laboratory studies of the interactions of dissolved metals with carefully defined suspensions have demonstrated that adsorption of cadmium is rather weak as compared to those of other elements such as iron, aluminum, cobalt, manganese, nickel, or even copper.[15]

Cadmium distribution between the dissolved and particulate species depends on several parameters other than the amount of suspended solids. The presence of organic complexing agents may reduce cadmium adsorption on particulates,[16] and concentrated inorganic ligands may act in the same way: chloride ions, for example, that produce rather stable complexes with cadmium may seriously influence its distribution between the dissolved and absorbed states.[17]

This distribution is also dependent on the experimental method used to evaluate it, such as the filtration process. In the Mississippi and Minnesota rivers most metals were demonstrated as associated with the 1000 to 10,000-D fraction by ultrafiltration.[18] This would have been qualified as "dissolved" were a classical filtration on a 0.45-μm membrane used.

Acidification should also result in a dissolution of cadmium from particulates. pH modification may be due not only to human impact, but also to seasonal changes such as algal blooms.

In estuaries the behavior of cadmium is complicated as several significant changes of the composition of water occur simultaneously. The salinity gradient promotes cadmium desorption from suspended solids, but flocculation of colloids of iron and manganese hydroxides enhances adsorption possibilities as it increases, sometimes enormously, the amount of suspended solids.

2. Dissolved Species

In unpolluted fresh waters dissolved inorganic cadmium species may be predicted by thermodynamic equilibrium calculations, but very little experimental information is available due to the lack of suitable speciation techniques. On this theoretical basis the aqueous cadmium ion ("free" cadmium) predominates, but several inorganic complexes with chloride, bicarbonate, or hydroxyl ligands have nonnegligible contributions, depending on pH and salinity.[19] In seawater the large concentration of chloride induces a predominance of chloride complexes, $CdCl^+$ and $CdCl_2$.

Dissolved organic matter may seriously interfere in metal speciation. Several categories of organic compounds naturally present in the aquatic environment have been identified[20] and their interactions with cadmium studied.

Amino acids are well-known ligands of transition metals. However, as they occur in natural water at submicromolar concentrations, they cannot compete with inorganic ligands, since their complexes with cadmium are too weak.[21] A special mention must be made of some amino acids such as cysteine where sulfur atoms may enhance the stability of cadmium complexes enough to coexist with inorganic species.

Humic substances are a major component of dissolved organic matter. In fresh waters and coastal seawaters, they derive mainly from the allochthonous degradation of vegetation and have a high degree of aromaticity.[22] In the open ocean autochthonous molecules, probably produced by condensation of smaller molecules, predominate; these humic compounds have a much lower degree of aromaticity. As a consequence of this wide diversity of chemicals gathered under a single name, very diverse conclusions have been published on their interactions with cadmium. In the open ocean it has been shown[23] that complexation of cadmium by humic substances is unimportant. In waters with higher contents of dissolved organic matter (coastal waters, estuaries, fresh waters) the stability of humic-cadmium complexes is sufficient to promote the existence of these species. In some cases it has even been reported[4] that cadmium complexes were more stable than their copper equivalents.

Excretion of strong or very strong organic ligands by aquatic organisms has been recently demonstrated several times. Microorganisms are concerned (plankton, fungi, bacteria) and also macrospecies such as algae[7] or snails.[6] Very few data on the complexation of cadmium by these compounds exist.

3. Lability of Cadmium Species

The rate of dissociation of the complex species CdL is defined by:

$$v_b = \frac{d(CdL)}{dt} = -k_b(CdL) = \frac{-k_f}{K}(CdL)$$

In a series of complexes of cadmium with similar ligands, v_b will thus be inversely proportional to K values.

Due to the water-exchange rate constant of the aqueous cadmium ion (9×10^{-7}/mol/s) and the relatively low stability of cadmium complexes, the lability of cadmium complexes is rather high. Inorganic cadmium complexes are ASV-labile. Organic complexes may be of lower lability; humic and fulvic[8] complexes have been demonstrated as ASV-quasi-labile. In natural waters over 70% of cadmium is ASV-labile.[24]

Experimental data correlating speciation of cadmium and toxicity are very scarce.[25,26] Complexation has been demonstrated as affecting toxicity with one unicellular alga.[27]

B. Speciation of Tin

The general picture is completely different. The solubility of inorganic tin species in natural waters is extremely low, as is their toxicity. It is only during the last 5 to 10 years that environmental significance of organotin compounds has been emphasized. Most of the research effort up to now has been devoted to the development of convenient analytical methods for the speciation and determination of tin species at very low significant levels.[28] Most of these compounds have an important solubility in lipids.

1. Dissolved Species

Organotin compounds are usually tetracoordinated derivatives of the general formula R_nSnX_{4-n} where R is some alkyl or aryl group and X some monovalent anion. In waters the exchange of X seems to be easy enough to enable one to consider only the species $R_nSn^{(4-n)+}$. This may be an oversimplification justified only by ignorance.

Many organotin compounds are produced by the industry. The main ones are compounds di- and trisubstituted by butyl (essentially), phenyl, or cyclohexyl radicals.[29] Due to an important biocidal use in water (antifouling paints), tributyltin Bu_3Sn^+ and its degradation products are most likely to occur in the aquatic environment.

Degradation of Bu_3Sn^+ is known to occur as follows:

$$Bu_3Sn^+ \rightarrow Bu_2Sn^{2+} \rightarrow BuSn^{3+} \rightarrow Sn^{IV}$$

Accordingly, all these species have been evidenced in many locations.

Natural methylation of inorganic tin has been demonstrated as possible in several conditions. $MeSn^{3+}$, Me_2Sn^{2+}, and Me_3Sn^+ have been found to occur in the ppt range in waters. A very limited number of determinations suggest the possible existence in the aquatic environment of other species such as the tetrasubstituted derivatives: butyl-methyl compounds or hydridated organotins.

All these compounds have widely different toxicities so that a detailed knowledge of the speciation of tin is necessary to predict ecotoxicological effects.

The physicochemical stability of these numerous species is generally high enough to open experimental speciation possibilities. All the speciation methods developed up to now combine a chromatographic separation (gas chromatography after derivatization or high performance liquid chromatography) of the species followed by a specific detection (atomic absorption or flame photometry or even mass spectrometry). A very high sensitivity is required as, even in the most heavily polluted environments tested thus far, total dissolved tin is in the low-ppb range. Moreover, the most toxic compounds, such as Bu_3Sn^+, seem to exert deleterious effects on sensitive species as filter feeders down to the 10- to 50-ppt range.[30]

Available speciation methods for dissolved species are now in operation in several laboratories in the world, and information is being gathered on the aquatic chemistry of organotin compounds. However, very little is already well established apart from the identity of the main organotin cations. Nothing is known of their possible complexes with either organic or inorganic ligands.

2. Suspended Solids

Adsorption of organotin compounds on particulate matter is very likely prone to play an important role. However, it is only very recently that precise information began to appear.[31]

VII. CONCLUSION

Chemical speciation of trace metals plays a very important role in aquatic toxicology. An enormous research effort has been devoted by analytical chemists to the development of convenient speciation methods during the last decade.

A precise definition of the physicochemical species involved and of their lability is now a prerequisite to any toxicological study.[32]

REFERENCES

1. **Florence, T. M. and Batley, G. E.,** Chemical speciation in natural waters, *CRC Crit. Rev. Anal. Chem.,* 9, 219, 1980.
2. **Florence, T. M.,** Trace element speciation and aquatic toxicology, *Trends Anal. Chem.,* 2, 162, 1983.
3. **Astruc, M. and Pinel, R.,** La spéciation des éléments en traces dans les systèmes aquatiques, *Actual. Chim.,* May 29, 1985.
4. **Giesy, J. P.,** Control of trace metal equilibria, in *Trace Elements Speciation in Surface Waters and its Ecological Implication,* NATO Conf. Ser., Leppard, G. G., Ed., Plenum Press, New York, 1983, 195.
5. **Leppard, G. G. and Burnison, B. K.,** *Trace Elements Speciation in Surface Waters and its Ecological Implication,* NATO Conf. Ser., Leppard, G. G., Ed., Plenum Press, New York, 1983, 105.
6. **Tournié, T. and El Mednaoui, H.,** Metal complexing agents released into the marine environment by the deposit feeder Hydrobia ulvae (Gastropoda: Prosobranchia): characterization and regulation processes, *Mar. Ecol. Prog. Ser.,* 34, 251, 1986.
7. **Morel, F. M. M.,** *Principles of Aquatic Chemistry,* John Wiley & Sons, New York, 1983.
8. **Florence, T. M.,** Electrochemical approaches to trace element speciation in waters — a review, *Analyst,* 111, 489, 1986.
9. **Pakenkopf, G. K.,** *Introduction to Natural Water Chemistry,* Marcel Dekker, New York, 1978.
10. **Van Leeuwen, H. P.,** Kinetic classification of metal complexes in electroanalytical speciation, *J. Electroanal. Chem.,* 99, 93, 1979.
11. **Davisson, W.,** Defining the electroanalytically measured species in a natural water sample, *J. Electroanal. Chem.,* 87, 395, 1978.
12. **Turner, D. R. and Whitfield, M.,** The reversible electrodeposition of trace metal ions from multiligand systems. II. Calculations on the electrochemical availability of lead at trace levels in seawater, *J. Electroanal. Chem.,* 103, 61, 1979.
13. **Westall, J. C., Zachary, J. L., and Morel, F. M. M.,** MINEQL: a computer program for the calculation of chemical equilibrium composition of aqueous systems, Tech. Note No. 18., Ralph M. Parsons Laboratory for Water Resources and Environmental Engineering, Department of Civil Engineering, Massachussetts Institute of Technology, Cambridge, 1976.
14. **Mart, L. and Nürnberg, H. W.,** The distribution of cadmium in the sea, in *Cadmium in the Environment,* Vol. 50, Mislin, H. and Ravera, O., Eds., Birkhaüser Verlag, Basel, 1986.
15. **Yeats, P. A. and Bewers, J. M.,** Discharge of metal from the St. Lawrence River, *Can. J. Earth Sci.,* 19, 982, 1982.
16. **Salomons, W. and Van Pagee, H.,** Prediction of NTA-levels in river systems and their effect on metal concentrations, in *Proc. Int. Conf. Heavy Metals in the Environment,* CEP Consultants, Edinburgh, 1981.

17. **Salomons, W. and Kerdijk, H. N.,** Cadmium in fresh and estuarine water, in *Cadmium in the Environment,* Vol. 50, Mislin, H. and Ravera, O., Eds., Birkhaüser Verlag, Basel, 1986.
18. **Alberts, J. J, Giesy, J. P., and Evans, D. W.,** Distribution of dissolved organic carbon and metal-binding capacity among ultrafiltrable fractions isolated from selected surface waters of the South Western United States, *Environ. Geol. Water Sci.,* 6, 91, 1984.
19. **Moore, J. W. and Ramamoorthy, S.,** *Heavy Metals in Natural Waters,* Springer-Verlag, New York, 1984.
20. **Astruc, M.,** Evaluation of methods for the speciation of cadmium, in *Cadmium in the Environment,* Vol. 50, Mislin, H. and Ravera, O., Eds., Birkhaüser Verlag, Basel, 1986.
21. **Valenta, P., Simoes, Gonçalves, M. L. S., and Sugawara, M.,** Voltammetric studies on the speciation of cadmium and zinc by aminoacids in sea water, in *Complexation of Trace Metals in Natural Waters,* Kramer, C. J. M. and Duinker, J. C., Eds., Junk, The Hague, 1984.
22. **Crawford, R. L.,** *Lignin: Biodegradation and Transformation,* John Wiley & Sons, New York, 1981.
23. **Raspor, B., Nünberg, H. W., Valenta, P., and Branica, M.,** Studies in sea water and lake water on interactions of trace metals with humic substances isolated from marine and estuarine sediments, *Mar. Chem.,* 15, 217, 1984.
24. **Florence, T. M.,** Trace metal species in freshwaters, *Water Res.,* 11, 681, 1977.
25. **Ravera, O.,** Cadmium in freshwater ecosystems, in *Cadmium in the Environment,* Vol. 75, Mislin, H. and Ravera, O., Eds., Birkhaüser Verlag, Basel, 1986.
26. **Smies, M.,** Biological aspects of trace element speciation in the aquatic environment, in *Trace Element Speciation in Surface Waters and Ecological Implication,* NATO Conf. Ser., Leppard, G. G., Ed., Plenum Press, New York, 1983, 177.
27. **Premazzi, G., Bertone, R., Freddi, A., and Ravera, O.,** Combined effects of heavy metals and chelating substances on Selenastrum cultures, in Proc. Seminar on Ecological tests relevant to the implementation of proposed regulations concerning environmental chemicals: evaluation and research needs, Berlin, 1977, 169.
28. **Thompson, J. A. J., Sheffer, M. G., Pierce, R. C., Chan, Y. K., Cooney, J. J., Cullen, W. R., and Maguire, R. J.,** Organotin compounds in the aquatic environment: scientific criteria for assessing their effects on environmental quality, Rep. No. NRCC 22494, Environmental Secretariat, Ottawa, 1985.
29. **Astruc, M. and Pinel, R.,** L'étain dans l'environnement, Rep. CNRS-PIREN, 1982.
30. **Alzieu, C. and Heral, M.,** Ecotoxicological effects of organotin compounds on oyster culture, in *Ecotoxicological Testing for the Marine Environment,* Vol. 2, Persoone, G., Jaspers, E., and Claus, C., Eds., Institute for Marine Scientific Research, State University of Ghent, Bredene, 1984, 187.
31. **Randall, L., Han, J. S., and Weber, J. H.,** Determination of inorganic tin, methyltin and butyltin compounds in sediments, *Environ. Technol. Lett.,* 7, 571, 1986.
32. **Samoiloff, M. R. and Wells, P. G.,** Future trends in marine ecotoxicology in *Ecotoxicological Testing for the Marine Environment,* Vol. 1, Persoone, G., Jaspers, E., and Claus, C., Eds., Institute for Marine Scientific Research, State University of Ghent, Bredene, 1984, 733.

Chapter 6

ADSORPTION OF TRACE INORGANIC AND ORGANIC CONTAMINANTS BY SOLID PARTICULATE MATTER

A. C. M. Bourg

TABLE OF CONTENTS

I. INTRODUCTION

The toxicity and the rate of transport of contaminants is highly dependent on their physical form. Indeed, the dissolved fraction of pollutants is usually more harmful to the biota than its counterpart, which is fixed on or trapped within natural solids. Also, dissolved species are often transported much more rapidly than suspended solids.

Even though many, important questions remain to be answered, it is now sufficiently established that adsorption/desorption phenomena at solid-solution interfaces are significant controls on the fate of contaminants in the hydrosphere.[1,2] Suspension of solids in turbulent aquatic systems, downward motion of particles in the water columns of lakes and oceans, and water movement through solids in aquifers are all susceptible to regulating the dispersion of contaminants around point-source introduction into the environment. The quality of water can, thus, potentially be improved by favoring adsorption reactions as occur, for example, in artificial bank filtration or land disposal of liquid wastes. Important unknowns still exist such as the extent or total capacity and irreversibility or durability of the trapping and, thus, decontaminating power of solids in a given environment.

The rational management of clean water and the safe disposal of wastes necessitates a good understanding of the processes regulating the sorption of contaminants by solids in natural aquatic systems. A short review of the current knowledge of the adsorption of trace organic and inorganic contaminants is presented in this chapter.

II. INVESTIGATING AND MODELING THE ADSORPTION OF MICROPOLLUTANTS

A. Adsorption Isotherms

A large majority of investigations of adsorption on natural solids use the isotherm approach. Known quantities of a solid are equilibrated with solutions of the compound under investigation. In batch experiments, the adsorption of various concentrations of the substance of interest is measured after equilibrium is reached in the presence of constant or varying concentrations of solid. The results are typically reported as the amount of compound sorbed per unit mass of solid vs. the equilibrium dissolved concentration (Figure 1). Adsorption isotherms can usually be classified into three types. A brief description of each is given below. For a more thorough discussion the reader is referred to, for example, the review of sorption models of Voice and Weber[3] or Sposito[4] and to the references cited therein.

1. Langmuir Isotherm

The isotherm presents a hyperbolic shape and, therefore, also an adsorption maximum (Figure 1), even though it might not be within the concentration range investigated. Langmuir[5] developed a model, initially for the adsorption of gasses into solids but later successfully extended to solid-liquid systems. This model makes the following assumptions:

1. The energy of adsorption is constant and independent of surface coverage.
2. Adsorption occurs on localized surface sites.
3. There is no interaction between adsorbed compounds.
4. The maximum adsorption corresponds to only a complete monolayer.

The Langmuir isotherm is usually written as:

$$\bar{c} = \frac{\bar{c}_{max}bc}{1 + bc} \tag{1}$$

where $\bar{c}$, $\bar{c}_{max}$, c, and b stand for the amount of solute adsorbed per unit weight of adsorbent, the maximum adsorption possible, the liquid-phase concentration at equilibrium, and an

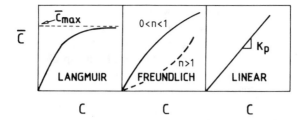

FIGURE 1. The most common types of adsorption isotherms ($\bar{c}$ and c are the equilibrium concentrations of adsorbed compound, on a solid weight basis, and dissolved compound, on a liquid volume basis, respectively).

empirical parameter named the adsorption coefficient (related to the enthalpy of adsorption), respectively.

Identification of the hyperbolic nature of the isotherm is carried out by checking whether the experimental data fit a linear form of Equation 1, such as, for example:

$$\frac{c}{\bar{c}} = \frac{1}{\bar{c}_{max}b} + \frac{c}{\bar{c}_{max}} \qquad (2)$$

The value of $\bar{c}_{max}$ can thus be determined even if the concentration range investigated does not approach the asymptotic x-axis of the hyperbola (c_{max} is the ordinate of this horizontal axis).

The Langmuir model was extended by Brunauer et al.[6] to include the adsorption of multiple molecular layers (BET model).

2. Freundlich Isotherm

Even though the Langmuir and BET models have a sound theoretical basis, they often fail to describe adsorption isotherms. Freundlich[7] proposed that the isotherm has a parabolic shape (Figure 1) and that adsorption can be described by the equation:

$$\bar{c} = Kc^n \qquad (3)$$

where K and n are empirical parameters, representing relative indicators of sorption capacity and of sorption intensity, respectively.[8]

The Freundlich equation is usually regarded as an empirical expression without chemical significance even though it often has a remarkable ability to fit experimental adsorption data. The model is consistent with a heterogeneous surface. If it is assumed that each class of surface sites adsorbs individually according to the Langmuir equation, the Freundlich isotherm can be shown (1) to correspond to a unique distribution of relative adsorption site affinities which is essentially log-normal and (2) that the empirical parameters K and n may be used to mathematically characterize the site distribution, thus providing information about surface heterogeneity of the adsorbent.[9]

Here, also, the identification of the parabolic shape of the experimental data is carried out using a linear form of the isotherm described by Equation 3:

$$\log\bar{c} = \log K + n\log c \qquad (4)$$

3. Linear Isotherm

The simplest isotherm model is a constant partitioning

$$\bar{c} = K_p c \qquad (5)$$

where K_p (sometimes referred to as K_D) is the partition coefficient (Figure 1). The distribution of an ion or a molecule between the surface and the aqueous phase is independent of the total concentration of the adsorbate studied. Because of its mathematical simplicity, this model is often used to describe adsorption equilibria. This is especially true when introducing a solid-solution exchange term in hydrodynamic transport models. However, extreme caution should be observed as the partition coefficient is often not constant, especially over wide ranges of concentration. This model has, nevertheless, been used successfully for hydrophobic organic micropollutants (see Section IV).

4. Drawbacks and Usefulness of the Adsorption Isotherm Approach

The adsorption isotherm methodology has been, and still is, widely used to study the adsorption of contaminants in aquatic systems. However, even in the case of the Langmuir model, it does not give much information about the actual adsorption mechanism. Moreover, it is in some cases a very conditional approach.

For contaminants for which dissolved speciation is strongly dependent on the chemistry of the aqueous phase, adsorption can be significantly affected by environmental parameters. This is especially true for trace elements, whether they are present under cationic or anionic form. It is, in fact, well known that isotherms are pH dependent.[10-13] For example, the parameters of the Langmuir equation corresponding to the adsorption of selenate by amorphous aluminum hydroxide vary by a factor of 3 to 4 between pH 5 and 9 (Figure 2). It is, however, less recognized that other solution parameters can be very significant. The presence of complexing agents or of an adsorption competitor may decrease the amount of pollutant adsorbed.[12,14-18] In some cases, such as for chromium, redox conditions may also be of prime importance. Organic hydrophobic micropollutants, if nonionic, are little affected by the composition of the aqueous phase, and the isotherm approach can therefore be used for modeling their uptake by solids under varying water chemistry. In other specific situations (i.e., when the chemical composition of the aqueous phase is constant and when the concentration of adsorbate is low, thus corresponding to almost a straight line in the adsorption isotherm of Langmuir or Freundlich), it is also possible to use the linear isotherm for trace metals. The effect of the water composition on adsorption will be discussed in more detail in Sections III.B and IV.D.

B. Adsorption Conceptual Models

The mechanisms most often invoked for the sorption of contaminants by model and natural solids are adsorption (physical or chemical) and ion exchange. There does not seem to be any difference between adsorption and coprecipitation.[1,19] For example, the removal of dissolved cadmium by amorphous iron oxyhydroxide is essentially identical whether the iron is precipitated before the trace metal is added (adsorption) or afterward (coprecipitation).[20]

The physical adsorption of a dissolved substance on a solid surface is usually due to either nonspecific forces (van der Waals attraction or hydrogen bonding) or electrostatic interaction. Chemical adsorption involves the formation of chemical associations between dissolved species and specific surface entities.

The ion exchange theory is not really a conceptual model. It only states that the uptake mechanism is characterized by the replacement of one ion, close to or on the surface, by another. This exchange can, however, be due to either physical (electrostatic) or chemical (surface coordination) adsorption.

The various models which have been developed to describe the adsorption of trace elements (both cationic and anionic) on hydrous oxide surfaces can be classified in two categories. In the chemical approach, dissolved substances react chemically with specific surface sites, while physical processes (e.g., electrostatic forces or steric restrictions) are included only as correction factors. In the physical approach, the adsorption is considered in terms of the

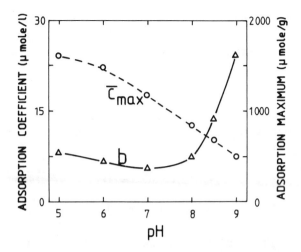

FIGURE 2. Variations of the parameters $\bar{c}_{max}$ (adsorption maximum) and b (adsorption coefficient) of the Langmuir equation corresponding to the adsorption of arsenate on amorphous aluminum hydroxide. (From data of Anderson, M. A., Ferguson, J. F., and Gavis, J., *J. Colloid Interface Sci.*, 54, 391, 1976.)

combination of electrostatic interactions between ions and the surface charge and ion-solvent interactions, while chemical effects are introduced as correction.

Both models can provide a good fit of experimental data.[21,22] The chemical approach is, however, often preferred because cations and anions seem to form inner layer surface complexes.[23,24] The adsorption of dissolved substances is thus simply interpreted by a chemical reaction, which simplifies the inclusion of adsorption phenomena in speciation calculations. This idea of explaining adsorption by regular solution theory is not new. It was proposed by Christ and Truesdell[25] as early as 1963.

Surface chemical reactions are, in fact, slightly different from reactions among simple solutes. They should be considered as reactions with polymers for which the free energy change and the equilibrium constant must be subdivided into an intrinsic adsorption term and an energy of interaction (mostly electrostatic) with neighboring sites:

$$\Delta G_{adsorption} = \Delta G_{intrinsic} + \Delta G_{coulombic} \tag{6}$$

$$\beta_n^{surface} = \beta_{n(intrinsic)}^{surface} \cdot \exp\left(\frac{F\psi z}{RT}\right) \tag{7}$$

where ψ is the electrical potential at the plane of adsorption and Δz is the net change in charge number of the surface species due to the adsorption reaction.

The various chemical models differ mainly by the set of surface species and surface reactions, the mathematical expression of the "mass action law" as a function of surface site concentration, and the formulation of the coulombic term.[26] The choice of surface species and reactions in a model is the result of a compromise between simplicity and the necessity to fit experimental data.

More details about the various chemical models can be found elsewhere.[23,24,26-30]

III. ADSORPTION OF CATIONIC AND ANIONIC MICROPOLLUTANTS: THE SURFACE COMPLEXATION MODEL

A. The Model: From Hydrous Oxides to Sediments and Soils

The surface complexation model is a chemical model which was developed originally through studies of the adsorption of trace-metal cations on solid hydrous oxides.[31-33] It was later extended to anionic adsorbates (AsO_4^{3-}, AsO_3^{3-}, CrO_4^{2-}, PO_4^{3-}, SeO_4^{2-}, SeO_3^{2-}, $H_2SiO_4^{2-}$, F^-, SO_4^{2-}, acetate, and various organic acids) interacting with the same model surfaces.

1. Surface Reactions

The surface of hydrous oxides contains ionizable functional groups. In an aqueous environment these surface groups behave like amphoteres, capable of accepting or giving out protons (depending on the pH):

$$\overline{SOH} \rightleftarrows \overline{SO^-} + H^+ \tag{8}$$

$$\overline{SOH} + H^+ \rightleftarrows \overline{SOH_2^+} \tag{9}$$

where the overbar identifies surface species and S is a hydrolyzed surface atom (such as Si, Al, Fe, Mn).

The surface sites SOH can react with solution ions other than protons, resulting in the formation of surface complexes. Cations are exchanged for surface protons according to the equation:

$$(\overline{SOH})_n + M^{z+} \rightleftarrows (\overline{SO_nM})^{(z-n)+} + nH^+ \tag{10}$$

where for steric reasons n is equal to only 1 or 2. Similarly, the adsorption of inorganic and organic anions can be explained by a competition with OH^- for surface sites:

$$(\overline{SOH})_n + A^{z-} \rightleftarrows (\overline{S)_nA}^{(z-n)-} + nOH^- \tag{11}$$

Here n can have values from 1 to 3. Just as for cation adsorption, the formation of multidentate surface complexes can be prevented for geometric reasons.[24]

The model is summarized in Table 1. Equilibrium constants as described in Equation 7 can be derived from Equations 8 to 11.[31-34] Briefly, acidity and adsorption constants (or surface complex formation constants) can be evaluated from experimental studies of the fraction of element (including H^+ or OH^-) adsorbed as a function of pH.

The value of the electrical potential (ψ) cannot be measured directly. For acid-base surface reactions, the intrinsic constants can be evaluated by extrapolating to conditions of zero surface charge, the values of the microscopic constants measured for different surface charges.[24] For surface-complex formation reactions it is often assumed that the electrostatic term is equal to unity[31,32] or is constant at a given ionic strength (and is included in the value of the intrinsic constant which is therefore an apparent constant). A more rigorous treatment is, however, possible.[33,34]

In practice, the modeling involves choosing one or two reasonable adsorption reactions and adjusting the stability constants for these reactions until the model calculations match the experimental results.

The surface complexation model is easy to manipulate because of the analogy between adsorption and regular solution chemistry (see Section II.B). The concept has, moreover, been very successful in quantitatively predicting the effect of pH on adsorption and of adsorption on the surface charge.[37]

The adsorption characteristics of natural solids have been studied by numerous investigators. Very few of them have provided conceptual interpretations. The sediments, suspended particulate matter, and soils studied at different pH values exhibit the same adsorption patterns as simple hydrous oxides[38-43] (Figure 3). The full applicability of the surface complexation

Table 1
REACTIONS OF SURFACE
COMPLEXATION

SOH	+	M^{2+}	$\rightleftarrows$	SOM^+	+	H^+
$(SOH)_2$	+	M^{2+}	$\rightleftarrows$	$(SO)_2M^\circ$	+	$2H^+$
SOH	+	M^{2+}	$\rightleftarrows$	SOMOH	+	$2H^+$
SOH	+	A^{2-}	$\rightleftarrows$	SA^-	+	OH^-
SOH	+	HA^-	$\rightleftarrows$	SHA	+	OH^-
SOH	+	HB^{2-}	$\rightleftarrows$	SHB^-	+	OH^-
SOH	+	H_2B^-	$\rightleftarrows$	SH_2B	+	OH^-

Note: $M^{2+} = Cd^{2+}, Co^{2+}, Zn^{2+}, Cu^{2+}, Pb^{2+}; A^{2-} = SeO_4^{2-},$
$SeO_3^{2-}, CrO_4^{2-}; B^{3-} = AsO_3^{3-}, AsO_4^{3-}, PO_4^{3-}.$

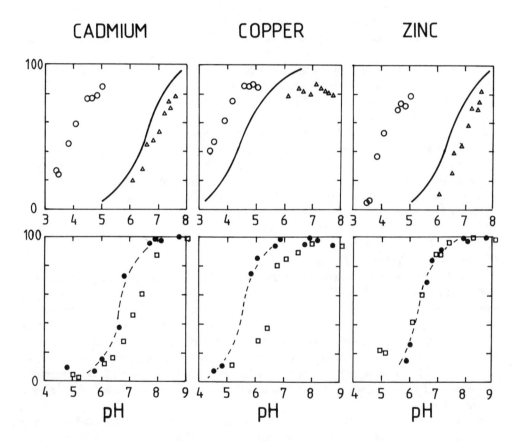

FIGURE 3. Adsorption of trace metals on hydrous oxides and on natural solids. (- : Meuse River bottom sediment; Δ: Rhone River bank sediment; O: Gironde Estuary suspended matter; □: silica; --●--: alumina.) (After Bourg, A C. M., in *Trace Metals in Sea Water,* Wong, C. S., Burton, J. D., Boyle, E., Bruland, W., and Goldberg, E. D., Eds., Plenum Press, New York, 1983, 195; and Mouvet, C. and Bourg. A. C. M., *Water Res.,* 17, 641, 1983.)

model to natural composite solids still needs, however, further investigation, especially for anionic adsorbates.

2. Surface Sites

For solid hydrous oxides[24] and also for some natural composite solids,[38,40,44] the total number of surface sites ($\bar{c}_{max}$) and the pH dependence of the surface charge can be determined by an acid-base titration. However, for many natural solids, dissolution occurs readily when

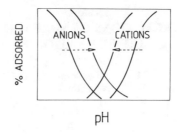

FIGURE 4.   Adsorption edges of anions and cations. (--▷: increasing adsorbent.)

the pH of the titration medium is brought away from the range of natural waters.[41] It can, however, be measured by a tritium exchange method[45] or by a classical cation exchange capacity measurement (displacement of major cations by NH_4^+). These two methods gave values in good agreement for deep ocean interfacial sediments (2.7 ± 0.2 and 3.0 ± 0.2 eq/kg of solids, respectively).[46]

Studies at various adsorption densities and competitive adsorption experiments are consistent with the existence of several distinct types of sites on the surfaces of γ-FEOOH, γ-Al_2O_3, $Fe_2O_3 \cdot H_2O$ (amorphous).[47-49] At very low adsorption densities, all sites are available in excess, and the surface behaves as though it were composed of identical sites. However, when only a few percent of the surface sites are occupied, the average binding constant decreases.[49] This site heterogeneity might not necessarily be a significant problem for the study of natural systems because, as demonstrated by Stumm et al.[24] for iron oxides in lake water, adsorbed trace elements should occupy, under ordinary conditions, only a very small fraction of surface sites, and, if their binding energy is much greater than that of other species, competitive side effects should be negligible.

B. The pH Adsorption Edge

1. Adsorption of Cations and Anions

Typical examples of the adsorption of trace elements in systems with only one strongly binding adsorbate (noncompetitive systems) are shown in Figure 4. The adsorption curves present a sharp increase for a 1- to 1.5-pH range. This part of the curve is called the pH-adsorption edge.[48] Its position is characteristic for each metal, but it depends on adsorbent and also, even if less, on adsorbate concentrations[47] and ionic strength.[50] Typically, a tenfold increase in surface site concentration shifts the adsorption edge by 1 pH unit toward the acidic range. Observations such as "cadmium adsorbs onto iron oxide at pH 7" are, therefore, meaningless if the adsorbate concentration and the chemical characteristics of the system considered are not given.[20]

Inspection of Equations 10 and 11 permits the prediction of the effects on adsorption of changing pH or adsorbent concentration: increasing adsorbent concentration increases the fraction of adsorbed element; increasing pH increases the fraction of adsorbed cation and, conversely, decreases the fraction of adsorbed anion (Figure 4).

For most trace elements, fractional adsorption (amount of element adsorbed/total element in the system) decreases with increasing total trace-element concentration in the system, even when surface sites are available in excess. The trend is more pronounced for cationic than for anionic adsorbates.[35] This phenomenon has been interpreted as indicative of the presence of multiple types of surface sites.[47]

2. Adsorption in Competitive Systems

Complexation of metal cations by inorganic and organic ligands can significantly increase or decrease adsorption compared to a ligand-free system.[36] Complexed species can be either "cation-like" or "anion-like", depending on whether adsorption of the complex increases or decreases with increasing pH (Figure 4). Complexed metal cations behave like cations if they are attached to the surface through the metal. Conversely, they behave like ligands if they are attached to the surface through the ligand. These three-component surface entities are called ternary surface complexes:[51] surface-metal-ligand or surface-ligand-metal.

The adsorption of metal anions can be increased in the presence of trace metal cations. The uptake of chromate and selenate by $Fe_2O_3 \cdot H_2O$(amorphous) was enhanced by Cd, Co, Cu, or Zn, but this was only when the metal formed a surface precipitate, presumably $M(OH)_2$.[52]

C. Reversibility and Kinetics

Whether a reaction is regarded as fast or not actually depends on the time frame of the processes in the context of which the reaction is investigated. For many natural aquatic systems, chemical reactions occurring in a few minutes to a few hours and even, in some cases, up to a few days, can be considered as having reached equilibrium with the surrounding environment. The rate of surface reactions can be limited or controlled by diffusion, especially in low turbidity environments.

Studies of adsorption rates involving agitation and high solution-to-solid ratios usually indicate extremely fast surface interactions with trace elements.[13] Such a result might be extrapolated to natural aquatic environments. There is, however, increasing evidence that sometimes considerably slower rates are involved in coastal and open-ocean systems[49,53,54] and in soils.[55] This time dependence has been attributed to evolution with time of the sorbent or to adsorption-desorption equilibrium followed by a slow diffusion into small pores or into the solid structure.

If uptake reactions are usually reversible, it is often only partially so. Christensen[15] did, however, observe full reversibility for Cd adsorption on soils. Poor desorption can be due to the slower diffusion rate outward than inward in small pores or in the solid lattice structure. In some cases it may well be due to a side reaction. Incomplete desorption of chromate on a subsoil was attributed to reduction of the Cr(VI) to the more adsorbed Cr(III).[39]

The mineralogical form of the bulk sorbent and, even more so, of particle coatings is especially significant in the extent and reversibility of adsorption. Shuman showed that the adsorption capacity and the specific surface of amorphous forms of Fe and Al oxides are larger by one order of magnitude than for the crystalline forms geothite and gibbsite.[56] As particles grow in size, the metals originally adsorbed on the surface can become occluded and are thus no more in equilibrium with the aqueous phase.[57] The formation and growth of particles favors, therefore, the trapping, at least temporarily, of adsorbed metals.

IV. ADSORPTION OF TRACE ORGANIC CONTAMINANTS: THE CONSTANT PARTITION COEFFICIENT MODEL

A. Sorption of Nonpolar Trace Organics

Many organic micropollutants are hydrophobic. This means that even though they are readily soluble in many nonpolar solvents, they tend to repel the water phase (i.e., the solid-solution and solution-air interfaces). For the most hydrophobic organics, the affinity of the solid may even play only a subordinate role in comparison to the repulsion of the water solvent.[2]

Several investigators have observed linear isotherms for the uptake of nonionic hydrophobic organic compounds by aquifer materials, soils, and sediments (Figure 5).[58-65] These isotherms are therefore described by Equation 5 where the distribution coefficient K_p is characteristic of the organic substance and of the environmental solid of interest.

The partition coefficient K_p usually correlates well with the fraction of organic carbon

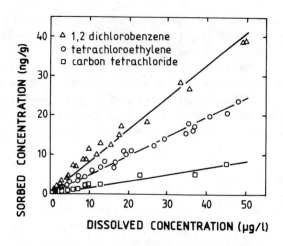

FIGURE 5.   Sorption isotherms for alluvial aquifer material. (From Curtis, G. P., Roberts, P. V., and Reinhard, M., in *A Natural Gradient Experiment on Solute Transport in a Sand Aquifer,* Tech. Rep. No. 22, Roberts, P. V. and Mackay, D. M., Eds., Department of Civil Engineering, Stanford University, Stanford, CA, 1986, 124. With permission).

(f_{oc}) of the solid.[66] The normalization of K_p into K_{oc}, an organic-carbon partition coefficient,

$$K_{oc} = K_p/f_{oc} \tag{12}$$

reduces the variations of K_p from many orders of magnitude to variations of K_{oc} of two- to threefold, suggesting that K_{oc} for a given solute is only weakly related to the nature of the sorbent.[66]

K_{oc} is also well correlated with the aqueous solubility S of the solute:[58]

$$\log K_{oc} = \alpha \log S + \beta \tag{13}$$

In addition, K_{oc} correlates well with the hydrophobicity of the solute.[59] This property can be evaluated by the tendency of the substance to dissolve in a nonpolar solvent. The widely cataloged and easily measured octanol-water distribution coefficient K_{ow} is generally used as the reference of hydrophobicity.

$$\log K_{oc} = \alpha \log K_{ow} + \beta \tag{14}$$

These correlations, in addition to the fact that adsorption isotherms show no indication of curvature even at high relative concentrations, have led to the concept that the uptake of an organic solute by a natural solid proceeds by partitioning into the organic matter fraction of the sorbent rather than by adsorption on the mineral surface.[58,59] The mineral surfaces are likely to be weak and, thus, insignificant adsorbents for nonpolar organics because they are hydrophilic and so should preferentially bind water.[58] The organic matter is expected to be a more powerful sorbent for the nonpolar organics because it has a marked hydrophobic character.[65]

The hypothesis that partition to the soil-organic phase is the primary process for the uptake of nonionic organic compounds from water by soil is further supported by the observation that the simultaneous uptake of such compounds (1,3-dichlorobenzene and 1,2,4-trichlorobenzene) exhibit no competitive effect.[64]

For solids poor in organic matter (f_{oc} <0.001), interactions of the trace-organic solute with the inorganic components of the solid may, however, become significant.[67]

B. Reversibility and Kinetics

Schwarzenbach and Westall[63] conducted laboratory batch and column experiments to elucidate the sorption behavior of nonpolar organic substances (chlorinated and alkyl benzenes and tetrachloroethylene) in a riverwater-groundwater infiltration system. Sorption was found to be reversible. The corresponding material-mass balance confirmed that the organic compounds investigated are removed by sorption and not by biological degradation.

Constant partition coefficients (linear adsorption isotherms) can be estimated from column experiments or field (aquifer) observations of retardation factors according to the equation

$$R_f = 1 + K_p \rho (1 - \epsilon)/\epsilon \qquad (15)$$

where R_f is the retardation factor (equal to the ratio of the time of migration, between an injection well and a sampling well, of the solute to the time of migration of water), ρ is the density of the aquifer material, and ϵ the total porosity.

Partition-coefficient values determined from column experiments run at velocities slower than 10^{-3} cm/s were quite similar to those determined from 18-h equilibrium batch experiments.[63] Column experiments run at a velocity of around 10^{-2} cm/s showed the occurrence of slow sorption kinetics.[63] Thus, if fast equilibrium can be assumed in many natural systems, the sorption kinetics may have a significant effect on organic-micropollutant transport in aquatic systems of fast suspended-matter settling or in fast-flowing aquifers (for velocities faster than 1 to 10 m/d for the solutes and sorbents investigated by Schwarzenbach and Westall.)[63]

Curtiss et al.[65] conducted laboratory rate studies to quantify the approach to equilibrium of five solutes on aquifer solids. An initial rapid sorption step, accounting for about 50% of the total uptake, occurred in less than 2 h. It was followed by a gradually declining rate for 2 to 3 d.

C. Sorption of Ionized Trace Organics

The simple partition model used to describe the uptake of neutral hydrophobic organic substances by natural solids (Equations 12 to 15) is applicable only to a limited extent to chemicals which are fully or partially ionized at environmental pH values. Since the partition of a solute between water and sorbent depends on the relative solubilities of the solute between the two phases, the water pH is expected to have a strong influence on the value of K_p for acids and bases.[58]

This is demonstrated in Figure 6, by the example of 2,3,4,5-tetrachlorophenol.[68] The sorption of chlorinated phenols by sediments and aquifer materials occurs not only for the nondissociated phenols, but also for their conjugate bases (phenolates).[69] The sorption reaction can be described by:

$$\bar{c} = Dc \qquad (16)$$

where D is the overall distribution ratio (Figure 6). In natural waters of low ionic strength (<$10^{-3}M$) and of pH not exceeding the pK_a of the phenol by more than one unit, the sorption of phenolate can be neglected, and D can be expressed by the equation:

$$D = K_p Q \qquad (17)$$

where K_p is the partition coefficient of the nonionized phenol and Q, the degree of protonation.[69]

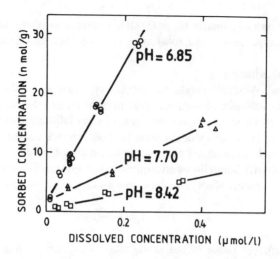

FIGURE 6.   Sorption isotherms of 2,3,4,5-tetrachlorophenol (pK_a = 6.35) on river sediment (f_{oc} = 0.026). (From Schwarzenbach, R. P. and Westall, J., in *Proc. Sem. Degradation, Retention, and Dispersion of Pollutants in Groundwater*, International Association of Water Pollution Research Control, 1984, 39. With permission.)

When the sorption of phenolate has to be considered (e.g., for tetra- and pentachlorophenol since, at ambient pH values, these chemicals are present predominantly in the ionized form), there is strong evidence that this uptake is also predominantly a partitioning process between the water and the organic matter present in a natural sorbent.[69] The extent of phenolate adsorption is dependent on the ionic strength of the aqueous phase.[69]

Partition coefficients corrected for pH can also be used.[70]

$$\log K_{p(a)} = \log K_p - \log(1 + K_a \cdot 10^{pH}) \tag{18}$$

(with K_a: acidity constant of the organic substance).

D. Water-Solubility Enhancement of Trace Organics

The enhancement of water solubility observed for relatively insoluble organic substances in the presence of dissolved high molecular weight organic matter can be accounted for by a partition-like interaction of the solute with dissolved organic matter (DOM).[71] The amplitude of the solubility enhancement is dependent on the type of solute and on the concentration and source (nature) of DOM. A DOM partition coefficient, K_{dom}, can be calculated.[71] It increases with a decrease of the polarity of the DOM. The K_{dom} values for given solutes with soil humic acids are about four times greater than the values with aquatic humic and fulvic acids.[71] For relatively water-soluble compounds such as lindane and 1,2,3-trichlorobenzene, DOM concentrations >100 mg/l would be necessary to observe a solubility enhancement. However, for highly insoluble solutes, solubility can be significantly increased at moderate DOM concentrations.[71]

Valid quantitative data on the interaction of trace-organic pollutants with concentrations of dissolved, or colloidal, organic matter and dissolved/colloidal organic matter are indispensable for assessing the mobility and the transport in aquatic systems of the most hydrophobic organics.

E. Applicability of the Partitioning Model

The direct correlation between K_p and either K_{ow} or S has led to empirical methods of evaluating K_p for a given solute-sorbent system. Equations 13 and 14 are simple, but they may lead to erroneous estimates of sorption, especially since they may not apply when the organic carbon content is lower than 0.1%.[63]

Reported isotherms are frequently linear but not always. Weber et al.[72] observed Freundlich-type isotherms (with n>1) for the sorption of polychlorinated biphenyls (PCBs) on a variety of sediments, soils, suspended matter, and microorganisms. It is, however, very possible that this is an artifact due to the very low concentrations of solid used in the batch experiments. Increased PCB uptake would increase f_{oc} and, thus, would promote further sorption, possibly explaining the isotherm type (Freundlich, with n >1) observed by these authors.

The same artifact could explain the variation of K_p with the solid-to-solution ratio observed by Voice et al.[73]

Other deviations to the simple reversible partition model described here are indicated by the several studies reporting that desorption may be hysteretic. Laboratory experiments indicated that the sediment-adsorbed 2,4,5,2′,4′,5′-hexachlorobiphenyl (HCBP) fractions may be comprised of both reversibly and strongly bound or resistant components.[74] The reversible component exhibited no hysteresis in response to changes in aqueous concentrations, whereas the resistant component reacted only to increases in aqueous concentrations exceeding the initial adsorption concentration.

It was recently shown, however, that these deviations could be experimental artifacts due to changes in the composition of the aqueous phase caused by the presence of suspended organic and inorganic colloidal material that is not removed during centrifugation.[73,75]

V. IMPLICATIONS FOR AQUATIC ECOTOXICOLOGY

A. Surface Interactions with Living Organisms

The surface complexation model of adsorption can be applied to uptake by living organisms and, further, to the understanding of pollutant toxicity. Indeed, the initial step in any biotic process is an interaction with cell walls or with cell membranes.

Grist et al.[76] studied the extent and nature of bonding between metal ions and the cell walls of the alga *Vaucheria* sp. The results were generally very similar for both cell wall fragments and intact cells. A classical acid base titration indicated a total number of surface sites of *circa* 1000 μmol/g. Metal adsorption ranged from 500 μmol/g for Cu to 100 for Na. The strength of surface binding follows the order Cu > Sr > Zn > Mg > Na. Protons displaced by metal adsorption gave ratios for (H displaced)/(M adsorbed) of 1.2 for Cu to 0 for Na. These observations are consistent with a model where metals interact with the protein and polysaccharides of the algal cell wall. The bonding may be of covalent nature (amino and carbonyl groups) and/or of ionic nature (carboxyls and sulfates), Cu and Na representing extremes in these two types of bonding. The surface charge of the cell walls is thought to be generated by increasing pH and covalent bonding from constituent proteins.

The observation that the free copper ion and the triethanolamine-copper complex are about equally adsorbed by the algal wall[76] indicates here also the occurrence of ternary surface complexes (adsorption of trace metal complexes).

Toxicity tests and bioassays commonly give widely scattered results. Moreover, these results are often poorly reproducible. Part of the scatter comes from the large variability in biological response of even a single biological species. However, bioassays would be better understood in quantitative terms if the chemical speciation of the toxic element was taken into account. Indeed, in a Cu bioassay, for example, the various dissolved species of the metal, such as $CuOH^+$, $Cu(OH)_2^{\circ}$, $CuCO_3^{\circ}$, etc., do not necessarily present the same toxicity. It is also difficult to fully control the chemical composition of the aqueous phase during the course of the bioassay. Thus, the dissolved speciation of the toxic element may vary from one batch to the other, leading to an observed, but apparent, poor reproducibility. Pagenkopf[77] developed a chemical model which utilizes competitive equilibria to calculate the chemical activity (effective toxicant concentration) of contaminant species associated with gill surfaces.

The gill membranes consisting of phospholipids can well provide the necessary sites for the formation of Lewis acid-base complexes with metal and hydrogen ions. The interaction may be written similar to Equation 10, as shown for copper, by:

$$Cu^{2+} + \overline{S}^{n-} \underset{}{\overset{K_{Cu}}{\rightleftarrows}} \equiv \overline{SCu}^{2-n} \qquad (19)$$

with the corresponding equilibrium constant, assuming rapid and reversible reaction:

$$K_{Cu} = [\equiv\overline{SCu}^{2-n}]/[Cu^{2+}][\equiv\overline{S}^{n-}] \qquad (20)$$

The surface concentration of Cu on the gill is thus related, not to the total copper, but only to the free species:

$$[\equiv\overline{SCu}^{2-n}] = K_{Cu}[Cu^{2+}][\equiv\overline{S}^{n-}] \qquad (21)$$

Rearrangement of Equation 21, using the conventional definition of α (the ratio of a species concentration to the total element concentration) gives:

$$[\equiv\overline{SCu}^{2-n}] = K_{Cu}\alpha_{Cu}[Cu_T]\alpha_S[\equiv\overline{S}_T] \qquad (22)$$

where the subscript T stands for total concentration.

In a given system (i.e., for constant $[Cu_T]$ and $[\equiv\overline{S}_T]$), varying conditions will result in changes in the metal load associated with the gill surfaces. Increase in dissolved complexing agent or in competitors for surface sites will reduce α_{Cu} or α_S, respectively, provoking a lowering in gill-adsorbed metal, thus resulting in lowering of the copper toxicity. Pagenkopf[77] successfully used his Gill Surface Interaction Model to explain the variability in the toxicity to fish of trace metals at different values of alkalinity, hardness, and pH.

B. Water Chemistry and Adsorption of Polar and Nonpolar Trace Contaminants

Exposure of biological species to trace contaminants is related to the dissolved concentration of these elements and, therefore, to their adsorption characteristics. The uptake of nonpolar hydrophobic organic pollutants should not be dependent on the water constituents since the process is mostly a physical partitioning. The presence of high molecular dissolved (or colloidal) organics such as humic and fulvic acids can, however, increase the solubility of hydrophobic trace organics. The adsorption of ionic trace pollutants is a function of their dissolved speciation, which in turn is very much dependent on some system parameters. Moreover, the extent of adsorption of a given element depends on the presence of other adsorbates. For example, increasing pH or surfaces and decreasing other adsorbates or solution complex formers all tend to increase trace cation adsorption.

As briefly summarized above and as shown in the preceding section, the adsorption and the toxicity to biological species is dependent on the chemical parameters of the aquatic environment. One could readily assume that the conditions favorable to adsorption should reduce the toxicity of trace pollutants. However, these same conditions also increase cell-wall uptake and may, therefore, promote detrimental effects. It would certainly be interesting to make model calculations in systems including biological surfaces to ascertain what conditions, if any, might increase adsorption while not significantly increasing cell-wall uptake.

VI. CONCLUSIONS

The adsorption of contaminants is fairly well understood as far as basic processes under equilibrium conditions are concerned. The models presented here are useful in exploring the problem in order to gain a better comprehension of the adsorption processes. However, we are still quite far from deriving models for full prediction of the effect of the phenomena under a variety of local and environmental chemical conditions.

ACKNOWLEDGMENTS

This paper was written with partial support of the program on "Migration of Micropollutants" of the French Geological Survey. Many thanks to Anna Kay Bourg (AKB) for editing the English and to José Bénichon and AKB for typing the manuscript.

REFERENCES

1. **Jenne, E. A.**, Controls on Mn, Fe, Co, Ni, Cu and Zn concentrations in soils and water: the significant role of hydrous Mn and Fe oxides, *Am. Chem. Soc. Symp. Ser.*, 93, 337, 1968.
2. **Stumm, W.**, Surface chemical theory as an aid to predict the distribution and the fate of trace constituents and pollutants in the aquatic environment, *Water Sci. Technol.*, 14, 481, 1982.
3. **Voice, T. C. and Weber, W. J., Jr.**, Sorption of hydrophobic compounds by sediments, soils and suspended solids. I. Theory and background, *Water Res.*, 17, 1433, 1983.
4. **Sposito, G.**, The chemical forms of trace metals in soils, in *Applied Environmental Geochemistry*, Thornton, I., Ed., Academic Press, London, 1983, chap. 5.
5. **Langmuir, I.**, The adsorption of gases on plane surfaces of glass, mica and platinum, *J. Am. Chem. Soc.*, 40, 1361, 1918.
6. **Brunauer, S., Emmett, P. H., and Teller, E.**, Adsorption of gases in multimolecular layers, *J. Am. Chem. Soc.*, 60, 309, 1938.
7. **Freundlich, H.**, *Colloid and Capillary Chemistry*, Metheun, London, 1926.
8. **Weber, W. J., Jr.**, *Physicochemical Processes for Water Quality Control*, Wiley Interscience, New York, 1972.
9. **Sposito, G.**, Deviation of the Freundlich equation for ion exchange reactions in soils, *Soil Sci. Soc. Am. J.*, 44, 652, 1980.
10. **Anderson, M. A., Ferguson, J. F., and Gavis, J.**, *J. Colloid Interface Sci.*, 54, 391, 1976.
11. **Garcia-Miragaya, J. and Page, A. L.**, Sorption of trace quantities of cadmium by soils with different chemical and mineralogical composition, *Water Air Soil Pollut.*, 9, 289, 1978.
12. **Gerth, J., Schimming, C. G., and Brümmer, G.**, Einfluss der Chloro-Komplexbildung auf Löslichkeit und Adsorption von Nickel, Zink und Cadmium, *Mitt. Dtsch. Bodenkdl. Ges.*, 30, 19, 1981.
13. **Christensen, T. H.**, Cadmium soil sorption at low concentration. I. Effect of time, cadmium load, pH, and calcium, *Water Air Soil Pollut.*, 21, 105, 1984.
14. **Mattigod, S. V., Gibali, A. S., and Page, A. L.**, Effect of ionic strength and ion pair formation on the adsorption of nickel by kaolinite, *Clays Clay Miner.*, 27, 411, 1979.
15. **Christensen, T. H.**, Cadmium soil sorption at low concentration. II. Reversibility, effect of changes in solute composition, and effect of soil aging, *Water Air Soil Pollut.*, 21, 115, 1984.
16. **Christensen, T. H.**, Heavy metal competition for soil sorption sites at low concentrations, *Proc. Int. Conf. Heavy Metals in the Environment*, Athens, Vol. 2, CEP Consultants, Edinburgh, 1985, 394.
17. **Christensen, T. H.**, Cadmium soil sorption at low concentration. IV. Effect of waste leachates on distribution coefficients, *Water Air Soil Pollut.*, 26, 265, 1985.
18. **Schmitt, H. W. and Sticher, H.**, Long term trend analysis of heavy metal content and translocation in soils, *Geoderma*, 38, 195, 1986.
19. **Laxen, D. P. H. and Sholkovitz, E. R.**, Adsorption (co-precipitation) of trace metals at natural concentrations on hydrous ferric oxide in lake water samples, *Environ. Technol. Lett.*, 2, 561, 1981.
20. **Benjamin, M. M., Hayes, M. M., and Leckie, J. O.**, Removal of toxic metals from power-generation waste streams by adsorption and co-precipitation, *J. Water Pollut. Control Fed.*, 54, 1472, 1982.
21. **Vuceta, J.**, Adsorption of Pb(II) and Cu(II) on Quartz from Aqueous Solutions: Influence of pH, Ionic Strength and Complexing Ligands, Ph.D. thesis, California Institute of Technology, Pasadena, CA, 1978.
22. **Bourg, A. C. M.**, Metals in aquatic and terrestrial systems: sorption, speciation and mobilization, in *Chemistry and Biology of Solid Waste: Dredged Material and Mine Tailings*, Salomons, W. and Förstner, U., Eds., Springer-Verlag, Berlin, 1988, 3.

23. **Stumm, W., Hohl, H., and Dalang, F.,** Interaction of metal ions with hydrous oxide surfaces, *Croat. Chem. Acta,* 48, 491, 1976.

24. **Stumm, W., Kummert, R., and Sigg, L.,** A ligand exchange model for the adsorption of inorganic and organic ligands at hydrous oxide interfaces, *Croat. Chem. Acta,* 53, 291, 1980.

25. **Christ, C. L. and Truesdell, A. H.,** Cation exchange in clays interpreted by regular solution theory, in *Proc. 76th Annu. Meet. Geol. Soc. Am.,* Geological Society of America, Boulder, CO, 1963, 32A.

26. **Morel, F. M. M., Westall, J. C., and Yeasted, J. G.,** Adsorption models: a mathematical analysis in the framework of general equilibrium calculations, in *Adsorption of Inorganics at Solid-Liquid Interfaces,* Anderson, M. A. and Rubin, A. J., Eds., Ann Arbor Science Publishers, Ann Arbor, MI, 1981, 263.

27. **James, R. O., Stiglich, P. J., and Healy, T. W.,** Analysis of models of adsorption of metal ions at oxide/water interfaces, *Discuss. Faraday Soc.,* 59, 142, 1975.

28. **Anderson, M. A. and Rubin, A. J., Eds.,** *Adsorption of Inorganics at Solid-Liquid Interfaces,* Ann Arbor Science Publishers, Ann Arbor, MI, 1981.

29. **Stumm, W. and Morgan, J. J.,** *Aquatic Chemistry,* 2nd ed., John Wiley & Sons, New York, 1981.

30. **Schindler, P. W.,** Surface complexation, in *Metal Ions in Biological Systems,* Vol. 18, Sigel, H., Ed., Marcel Dekker, New York, 1984, chap. 4.

31. **Hohl, H. and Stumm, W.,** Interaction of Pb^{2+} with hydrous γAl_2O_3, *J. Colloid Interface Sci.,* 55, 281, 1976.

32. **Schindler, P. W., Fürst, B., Dick, R., and Wolf, P. U.,** Ligand properties of surface silanol groups: surface complex formation with Fe^{3+}, Cu^{2+}, Cd^{2+} and Pb^{2+}, *J. Colloid Interface Sci.,* 55, 469, 1976.

33. **Davis, J. A. and Leckie, J. O.,** Surface ionization and complexation at the oxide-water interface. II. Surface properties of amorphous iron oxyhydroxide and adsorption of metal ions, *J. Colloid Interface Sci.,* 67, 90, 1978.

34. **Davis, J. A. and Leckie, J. O.,** Surface ionization and complexation at the oxide-water interface. III. Adsorption of anions, *J. Colloid Interface Sci.,* 74, 32, 1980.

35. **Benjamin, M. M. and Bloom, N. S.,** Effects of strong binding of anionic adsorbates on adsorption of trace metals on amorphous iron oxyhydroxides, in *Adsorption from Aqueous Solutions,* Tewari, P. H., Ed., Plenum Press, New York, 1981, 41.

36. **Benjamin, M. M. and Leckie, J. O.,** Conceptual model for metal-ligand-surface interactions during adsorption, *Environ. Sci. Technol.,* 15, 1050, 1981.

37. **Hohl, H., Sigg, L., and Stumm, W.,** Characterization of surface chemical properties of oxides in natural waters: the role of specific adsorption in determining the surface charge, in *Particulates in Waters,* Adv. in Chem. Ser. 189, Kavanaugh, M. C. and Leckie, J. O., Eds., American Chemical Society, Washington, D.C., 1980, 1.

38. **Brown, D. W. and Hem, J. D.,** Development of a model to predict the adsorption of lead from solution on a natural streambed sediment, U.S. Geological Survey Water-Supply Paper No. 2187, 1984.

39. **Lion, L. W., Altman, R. S., and Leckie, J. O.,** Trace-metal adsorption characteristics of estuarine particulate matter: evaluation of contributions of Fe-Mn oxide and organic surface coatings, *Environ. Sci. Technol.,* 16, 660, 1982.

40. **Mouvet, C. and Bourg, A. C. M.,** Speciation (including adsorbed species) of copper, lead, nickel and zinc in the Meuse River. Observed results compared to values calculated with a chemical equilibrium computer program, *Water Res.,* 17, 641, 1983.

41. **Bourg, A. C. M. and Mouvet, C.,** Adsorption as a control mechanism of dissolved trace metals in the Meuse River, in preparation.

42. **Elliott, H. A.,** Adsorption behavior of cadmium in response to soil surface charge, *Soil Sci.,* 136, 317, 1983.

43. **Cavallaro, N. and McBride, M. B.,** Zinc and copper sorption and fixation by an acid soil clay: effect of selective dissolutions, *Soil Sci. Soc. Am. J.,* 48, 1050, 1984.

44. **Stollenwerk, K. G. and Grove, D. B.,** Adsorption and desorption of hexavalent chromium in an alluvial aquifer near Telluride, Colorado, *J. Environ. Qual.,* 14, 150, 1985.

45. **Yates, D. E. and Healy, T. W.,** The structure of the silica/electrolyte interface, *J. Colloid. Interface Sci.,* 55, 9, 1976.

46. **Balistrieri, L. S. and Murray, J. W.,** Marine scavenging: trace metal adsorption by interfacial sediment from MANOP site H[1], *Geochim. Cosmochim. Acta,* 48, 921, 1984.

47. **Benjamin, M. M. and Leckie, J. O.,** Multiple-site adsorption of Cd, Cu, Zn and Pb on amorphous iron oxyhydroxide, *J. Colloid. Interface Sci.,* 79, 209, 1981.

48. **Benjamin, M. M. and Leckie, J. O.,** Adsorption of metals at oxide interfaces: effects of the concentrations of the adsorbate and competing metals, in *Contaminants and Sediments,* Baker, R. A., Ed., Ann Arbor Science Publishers, Ann Arbor, MI, 1980, 305.

49. **Balistrieri, L. S. and Murray, J. W.,** Apparent equilibrium constants for metal-solid interactions in the marine environment, *Geochim. Cosmochim. Acta,* 47, 1091, 1983.

50. **Bourg, A. C. M.,** Role of fresh water/sea water mixing on trace metal adsorption phenomena, in *Trace Metals in Sea Water,* Wong, C. S., Burton, J. D., Boyle, E., Bruland, W., and Goldberg, E. D., Eds., Plenum Press, New York, 1983, 195.

51. **Bourg, A. C. M. and Schindler, P. W.,** Ternary surface complexes. I. Complex formation in the system silica-Cu(II)-ethylenediamine, *Chimia,* 32, 166, 1978.

52. **Benjamin, M. M.,** Adsorption and surface precipitation of metals on amorphous iron oxyhydroxide, *Environ. Sci. Technol.,* 17, 686, 1983.

53. **Nyffeler, U. P., Li, Y. H., and Santschi, P. M.,** A kinetic approach to describe trace-element distribution between particles and solution in natural aquatic systems, *Geochim. Cosmochim. Acta,* 48, 1513, 1984.

54. **Santschi, P. H., Nyffeler, U. P., Li, Y. H., and O'Hara, P.,** Radionuclide cycling in natural waters: relevance of sorption kinetics, *Proc. 3rd Int. Symp. Interactions between Sediments and Water, Geneva,* CEP Consultants, Edinburgh, 1984, 18.

55. **Murali, V. and Aylmore, L. A. G.,** No-flow equilibration and adsorption dynamics during ionic transport in soils, *Nature (London),* 283, 467, 1980.

56. **Shuman, L. M.,** Adsorption of Zn by Fe and Al hydrous oxides as influenced by aging and pH, *Soil Sci. Soc. Am. J.,* 41, 703, 1977.

57. **Grimme, H.,** Die Adsorption von Mn, Co, Cu und Zn durch Geothit aus verdünnten Lösungen, *Z. Pflanzenernähr. Bodenkd.,* 121, 58, 1968.

58. **Chiou, C. T., Peters, L. J., and Freed, V. H.,** A physical concept of soil-water equilibria for nonionic organic compounds, *Science,* 206, 831, 1979.

59. **Karickhoff, S. W., Brown, D. S., and Scott, T. A.,** Sorption of hydrophobic pollutants on natural sediments, *Water Res.,* 13, 241, 1979.

60. **Means, J. C., Wood, S. G., Hassett, J. J., and Banwart, W. L.,** Sorption of polynuclear aromatic hydrocarbons by sediments and soils, *Environ. Sci. Technol.,* 14, 1524, 1980.

61. **Brown, D. S. and Flagg, E. W.,** Empirical prediction of organic pollutant sorption in natural sediments, *J. Environ. Qual.,* 10, 382, 1981.

62. **Karickhoff, S. W.,** Semi-empirical estimation of sorption of hydrophobic pollutants on natural sediments and soils, *Chemosphere,* 10, 833, 1981.

63. **Schwarzenbach, R. P. and Westall, J.,** Transport of nonpolar organic compounds from surface water to groundwater. Laboratory sorption studies, *Environ. Sci. Technol.,* 15, 1630, 1981.

64. **Chiou, C. T., Porter, P. E., and Schmedding, D. W.,** Partition equilibria of nonionic organic compounds between soil organic matter and water, *Environ. Sci. Technol.,* 17, 227, 1983.

65. **Curtis, G. P., Roberts, P. V., and Reinhard, M.,** Sorption of organic solutes and its influence on mobility, in *A Natural Gradient Experiment on Solute Transport in a Sand Aquifer,* Roberts, P. V. and Mackay, D. M., Eds., Tech. Rep. No. 22, Department of Civil Engineering, Stanford University, Stanford, CA, 1986, 124.

66. **Karickhoff, S. W.,** Organic pollutant sorption in aquatic systems, *J. Hydraul. Eng.,* 110, 707, 1984.

67. **McCarty, P. L., Reinhard, M., and Rittman, B. E.,** Trace organics in ground water, *Environ. Sci. Technol.,* 15, 40, 1981.

68. **Schwarzenbach, R. P. and Westall, J.,** Sorption of hydrophobic trace organics in groundwater systems, in *Proc. Sem. Degradation, Retention, and Dispersion of Pollutants in Groundwater,* International Association of Water Pollution Research Control, 1984, 39.

69. **Schellenberg, K., Leuenberger, C., and Schwarzenbach, R. P.,** Sorption of chlorinated phenols by natural sediments and aquifer materials, *Environ. Sci. Technol.,* 18, 652, 1984.

70. **Xie, T.-M., Abrahamsson, K., Fogelgvist, E., and Josefsson, B.,** Distribution of chlorophenics in a marine environment, *Environ. Sci. Technol.,* 20, 457, 1986.

71. **Chiou, C. T., Malcolm, R. L., Brinton, T. I., and Kile, D. E.,** Water solubility enhancement of some organic pollutants and pesticides by dissolved humic and fulvic acids, *Environ. Sci. Technol.,* 20, 502, 1986.

72. **Weber, W. J., Jr., Voice, T. C., Pirbazari, M., Hunt, G. E., and Ulanoff, D. M.,** Sorption of hydrophobic compounds by sediments, soils and suspended solids. II. Sorbent evaluation studies, *Water Res.,* 17, 1443, 1983.

73. **Voice, T. C., Rice, C. P., and Weber, W. J., Jr.,** Effects of solids concentration on the sorptive partitioning of hydrophobic pollutants in aquatic system, *Environ. Sci. Technol.,* 17, 513, 1983.

74. **Di Toro, D. M. and Horzempa, L. M.,** Reversible and resistant components of PCB adsorption-desorption isotherms, *Environ. Sci. Technol.,* 16, 594, 1982.

75. **Gschwend, P. M. and Wu, S.,** On the constancy of sediment-water partition coefficients of hydrophobic organic pollutants, *Environ. Sci. Technol.,* 19, 90, 1985.

76. **Grist, R. H., Oberholser, K., Shank, N., and Nguyen, M.,** Nature of bonding between metallic ions and algal cell walls, *Environ. Sci. Technol.,* 15, 1212, 1981.

77. **Pagenkopf, G. K.,** Gill surface interaction model for trace metal toxicity to fishes: role of complexation, pH, and water hardness, *Environ. Sci. Technol.,* 17, 342, 1983.

Chapter 7

GEOCHEMISTRY AND BIOAVAILABILITY OF TRACE METALS IN SEDIMENTS

P. G. C. Campbell and A. Tessier

TABLE OF CONTENTS

I. INTRODUCTION

In recent years the fluxes of many trace metals from terrestrial and atmospheric sources to the aquatic environment have increased.[1] Metals introduced into the aquatic environment partition among various compartments: a portion will be associated with dissolved inorganic and organic ligands in solution whereas another fraction will become associated with the particulate matter following adsorption, precipitation, coprecipitation, or uptake by planktonic living organisms that subsequently settle out of the water column. As a result of complex physical, chemical, and biological processes, a major fraction of the trace metals introduced into the aquatic environment is found associated with the bottom sediments, distributed among a variety of physicochemical forms which exhibit different chemical reactivities.

Predicting the impact of particulate trace metal contamination upon aquatic organisms is difficult. Many such organisms live in contact with both dissolved and particulate trace metals and can, in principle, obtain trace metals either directly from the water or through ingestion of the solid phases. Assessing trace metal availability from the solid phases themselves if often difficult, even in laboratory experiments, due to the tendency of trace metals added in particulate forms to establish solute-solid equilibria. Metal uptake from either source will be influenced by physicochemical factors in the aqueous and particulate phases. In particular, bioaccumulation has been shown to depend upon the particular geochemical phases with which the trace metals are associated.[2] With reference to this latter aspect, there is presently no satisfactory method for determining unambiguously the forms of association of particulate trace metals in natural sediments. The great variety of solid phases that can bind trace metals, the fact that some phases may be present in small amounts and yet exert a scavenging action far out of proportion to their own concentration, the amorphous character of many of these important solid phases, and the fact that trace metals are usually only minor constituents, all conspire to render difficult, if not impossible, the direct measurement of trace metals associated with a particular phase.

The purpose of this chapter is to discuss the merits and limitations of the methods presently available for estimating trace metal partitioning in aquatic sediments and to show the importance of partitioning in assessing trace metal bioavailability. Emphasis will be given to surficial oxic sediments, i.e., those that are most relevant to benthic organisms. Such organisms are generally exposed to particles from an oxic environment either because they live above the anoxic zone or because they have siphons or tubes extending to the oxic zone of the sediments, i.e., they can create their own oxic microenvironments.[3]

II. TRACE METAL PARTITIONING IN SEDIMENTS

In principle, the partitioning of sediment-bound trace metals could be determined both by thermodynamic calculations, provided equilibrium conditions prevail, and by experimental techniques.

A. Thermodynamic Calculation

Thermodynamic calculations on the bulk solution of most fresh waters or seawater show undersaturation with respect to known trace metal solid phases.[4-6] Even for interstitial waters of oxic sediments (in which trace metals have longer residence times than in the bulk solution and are, thus, more apt to approach a state of equilibrium with respect to solid phases in the sediments), trace metals are still undersaturated with respect to their least-soluble solid phases. Hence, since solubility equilibrium with pure solids seems unlikely for trace metals under oxic conditions in natural waters, adsorption by solid surfaces has been suggested as an important mechanism for controlling certain trace metal concentrations.[7,8] Possible substrates for adsorption include clays and hydrous oxides of iron, manganese, aluminum, and silicon. Indeed, empirical studies using sequential extractions of oxidized sediments have consistently shown that nondetrital trace metals bind to more than one substrate. Statistical analysis of the distributions of trace metals in the field, and the potential substrates that can bind them, also support this contention.[9]

Partitioning of a given trace metal, M, between the water and various substrates of a sediment that can compete for it, can be written in a simplified manner:[3,10,11]

$$M^{z+} + \equiv S(1) \underset{\rightleftarrows}{\overset{K_a(1)}{}} \equiv S(1) - M$$

$$M^{z+} + \equiv S(2) \underset{\rightleftarrows}{\overset{K_a(2)}{}} \equiv S(2) - M \tag{1}$$

$$\quad . \qquad . \qquad\qquad .$$
$$\quad . \qquad . \qquad\qquad .$$

$$M^{z+} + \equiv S(n) \underset{\rightleftarrows}{\overset{K_a(n)}{}} \equiv S(n) - M$$

and

$$K_a(n) = \frac{\{\equiv S(n) - M\}}{[M^{z+}]\{\equiv S(n)\}} \tag{2}$$

where $\{\equiv S(1)\}$, $\{\equiv S(2)\}$, . . . $\{\equiv S(n)\}$ are the concentrations of free sites for the various substrates capable of binding M in the sediment (e.g., hydrous oxides of Fe, Mn, Al, Si; clays; humic acids); $\{\equiv S(n) - M\}$ is the concentration of M bound to sites on substrate n; $[M^{z+}]$ is the concentration of the free metal ion present in the ambient water; $K_a(1)$, $K_a(2)$, . . . $K_a(n)$ are apparent overall (or average) equilibrium constants for the respective reactions. Mathematically, this model of competitive adsorption is similar to those used to calculate the distribution of a dissolved trace metal among a mixture of ligands. According to this simple model, the concentration of M associated with a given substrate would depend upon $[M^{z+}]$ and both the intensity $[K_a(n)]$ and capacity ($\{\equiv S(n)\}$) of the substrates for binding M. Estimates of the latter two parameters for the various substrates are thus necessary to calculate the partitioning of M between the ambient water and each of the various substrates in the sediment.

1. Problems Associated with the Choice of Equilibrium Constants

Apparent equilibrium constants for the adsorption of trace metals onto various substrates potentially present in natural sediments have been determined in laboratory experiments using well-defined media.[12-20] Extension of these constants obtained for relatively simple systems to natural sediments is difficult, however, as several key factors can influence the value of K_a:

1. Adsorption can vary with temperature; greater[21] or lower[22] adsorption of trace metals onto substrates such as iron oxyhydroxides has been reported with increasing temperature. However, it is difficult to generalize about the direction and the extent of the changes because the data are scarce and cannot be interpreted unambiguously.[23] The temperature in natural sediments ($\sim$ 4 to 20°C) can differ appreciably from that used for most laboratory experiments ($\sim$20 to 25°C).

2. Typically, monitoring of the amount of trace metal adsorbed as a function of time shows a rapid initial uptake (~minutes) followed by a much slower reaction which may last several days. This behavior has been attributed to surface bonding (fast step) and to solid state diffusion (slow step), respectively.[24] Based on such kinetic information, equilibration times ranging from a few minutes[20,25] to a few hours[17,18,26-28] have been chosen for the equilibrium adsorption experiments in the laboratory. Clearly, these equilibration times are much shorter than those prevailing in natural sediments. Studies on the kinetics of radiotracer adsorption to freshwater and seawater sediments have shown that several days were required for most of the metal radioisotopes to reach an apparent equilibrium distribution between the solid and solution phases.[29-32] Longer reaction times would favor an increase in the K_a value.

3. The extent of adsorption of trace metals is expected to depend strongly upon certain characteristics of the adsorbent surface, such as porosity and specific surface area.[23] For example, Crosby et al.[33] have examined the surface areas and porosities of iron oxyhydroxides prepared under various controlled conditions and have reported significant differences in both surface area and porosity between fresh and aged precipitates and between solids obtained from precipitation of Fe(III) salts or oxidation of Fe(II). A wide range of preparative and pretreatment techniques have been used in laboratory experiments. Such experimental conditions tend to be chosen for convenience and do not necessarily correspond closely to conditions in the natural environment. Most laboratory studies involve the precipitation of oxyhydroxides from concentrated Fe(III) solutions (much more concentrated than in natural waters), while natural iron oxyhydroxides in lake sediments are more commonly derived from the oxidation of Fe(II) originating from the underlying anoxic zone, followed by hydrolysis and precipitation.[33] The precipitate prepared for laboratory experiments is normally aged for minutes,[20,25] hours,[26,27,34] or days[16,17] before initiation of the adsorption experiments. Increasing the aging time generally decreases the extent of subsequent adsorption. Whether short or long aging times are more representative of natural environments remains an open question.

4. Studies over wide ranges of metal concentrations strongly suggest that sediment components such as the oxyhydroxides[26,27,35] and, above all, humic or fulvic material[36-40] comprise various binding sites, the binding energies of which may vary by many orders of magnitude. This can result in an important variation of K_a values with surface coverage. In the laboratory, K_a values are usually obtained in experiments involving high trace metal concentrations, whereas low concentrations are generally involved under natural conditions.

5. Laboratory studies have shown that the concentrations of competing cations influence the adsorption of a given trace metal on clays[41] and humic material.[18] For amorphous iron oxyhydroxides, however, the adsorption of trace metals is not appreciably affected by the presence of other cations, unless these are present at very high concentrations.[27] Most of the adsorption studies in the laboratory have been purposely conducted in the absence of competing cations, whereas their presence in natural waters is unavoidable.

2. Difficulties in Obtaining Representative Site-Density Values

In addition to the equilibrium constants, necessary input data for competitive adsorption models include estimations of the free site concentrations for the various substrates (Equations 1 and 2). At low trace metal coverage, the free site concentration of substrate n can be approximated:

$$\{\equiv S(n)\} \simeq \{\equiv S(n)\}_T \tag{3}$$

where $\{\equiv S(n)\}_T$ is the total concentration of sites for substrate n. The free site concentration can then be written:

$$\{\equiv S(n)\} = N(n) \cdot \{A(n)\} \qquad (4)$$

where $\{A(n)\}$ is the concentration of substrate n in the sediment (e.g., moles per gram) and N(n) is the number of sites per unit quantity of that substrate (site density; e.g., moles of sites per mole of substrate). Concentrations of the various substrates in the sediments can be obtained by chemical extraction techniques, at least for abundant sinks such as metal oxyhydroxides and specific organic substances. The determination of the site density of the individual substrates in a multicomponent system such as a natural sediment is, however, more problematic. Methods to measure the binding capacity of individual substrates in one-component systems have been reviewed by Luoma and Davis.[3] Even for these relatively simple systems, estimates obtained for a given substrate by the various methods are not always consistent.[11] This probably reflects both the complexity of the single substrates (e.g., iron oxyhydroxides exist in a continuum of forms ranging from amorphous to crystalline solids; sedimentary organic matter or humic material is a heterogeneous multicomponent mixture) and the analytical limitations of the methods used. More importantly, the techniques developed for single-component systems cannot be used to determine the site density of a particular substrate within a mixture of substrates such as a sediment. For the moment, any calculations with the competitive adsorption models must rely upon site densities obtained in the laboratory for single-component systems.

3. Other Difficulties

In addition to the problems discussed above, related to the determination of representative binding constants and site densities, competitive adsorption models are subject to other potential problems.[3,11] The main hypothesis underlying this approach is that equilibrium conditions prevail for the partitioning of a trace metal between the water and the various substrates that can bind it; kinetic information is presently lacking to support this assumption. Furthermore, the model assumes that all the substrates behave independently, as a mixture of pure substrates; it is well known, however, that sedimentary particles are not pure substrates, but rather aggregates of many substrates (e.g., clays coated with organic matter and/or iron oxyhydroxides). To what extent aggregation influences the binding of a trace metal on a given substrate is not really known.

4. Comparison of Field and Laboratory Data

Few studies have tried to compare theoretical investigations or adsorption experiments in well-defined media with adsorption in natural systems. Such comparisons would be helpful in providing information about the possible extrapolation of results obtained in simple controlled systems to complex multicomponent systems such as natural sediments. Lion et al.[42] have compared the adsorption of Cd and Pb onto oxic estuarine sediments before and after extracting the sediments with chemical reagents. Changes in Cd and Pb adsorption (vs. pH) after extractions of specific substrates suggested that iron and manganese oxides and organic matter were the important phases in controlling the adsorption of these metals on the studied sediments. These results were consistent with the partitioning obtained for these metals in the same sediments, using a sequential extraction procedure.

Apparent overall equilibrium constants (K_a) for the adsorption of Cd, Cu, Ni, Pb, and Zn onto natural iron oxyhydroxides have been calculated with Equation 2 using field data;[43] $\{\equiv S(n) - M\}$ and $\{\equiv A(n)\}$ were obtained by extracting the oxic sediment samples with a reducing agent and dissolved metal concentrations were measured in the associated pore waters collected *in situ*. These data were compared with equilibrium constants obtained from the literature for the adsorption of the same trace metals onto iron oxyhydroxides in well-defined media. The values obtained for the natural sediments were within the range of those measured in the laboratory which, for a given metal, may vary by several orders of magnitude, depending upon the experimental conditions used in the experiments. The field data were found to be consistent with laboratory experiments and with theory. Both the influence of pH upon adsorption and the binding strength sequence (i.e., Pb > Cu > Zn > Cd > Ni) observed for the field data agreed with theoretical predictions.

For 24 estuarine sediments with different physicochemical characteristics, Luoma[11] determined the partitioning of Cu between organic and inorganic substrates (as obtained with chemical extractions — see below) and compared these experimental results with partitioning calculated with a competitive adsorption model. The two chemical extractants used for obtaining empirical Cu partitioning were ammonium hydroxide (Cu bound with organics) and acetic acid (Cu bound to inorganic substrates). For the model calculations, the equilibrium constants and site density values for iron oxyhydroxides, manganese oxides, and extractable organic material were chosen from adsorption experiments with single substrates reported in the literature.[3,44] The concentrations of the substrates in the sediments were obtained by chemical extractions with ammonium oxalate (iron oxyhydroxides), ammonium hydroxide (extractable organic material), or by complete dissolution (manganese oxides). Both the model calculations and the chemical extractions clearly indicated that the distribution of Cu among the organic and inorganic substrates was influenced by the relative concentrations of the substrates. The agreement was best between the two methods when either organic or inorganic substrates were predominant, and the poorest when both substrate types were present and the model calculations indicated a distribution among various substrates.

B. Experimental Techniques

As an alternative to metal partitioning models, methods have been suggested for fractionating sediments physically and/or chemically. The sediment may be fractionated physically, according to grain size or by density gradient separation, and the individual fractions analyzed separately. Alternatively, sequential extractions with appropriate reagents can be devised to leach successive fractions of the metals selectively from the sediment samples. Some of the extraction techniques and their limitations are discussed below.

1. Chemical Extractants

To extract sediment-bound metals selectively from a particular sedimentary phase or component, one may choose among a wide variety of reagents. These reagents fall naturally into classes of similar chemical behavior; for example, concentrated inert electrolytes, weak acids, reducing agents, complexing agents, oxidizing agents, strong mineral acids.

When the reagents have been selected, there remains a decision as to the order in which they are to be employed. The chemical extractants can be applied sequentially to the sediment sample, with each successive treatment being more drastic in chemical action or of a different nature than the previous one. Many such experimental procedures have been proposed and applied to a wide variety of sediment samples (Table 1).

2. Methodological Considerations

Do metal partitioning patterns obtained with partial extractants or with sequential extraction procedures represent some reality relative to the natural sediments? To respond to this key question, one must first consider a number of methodological points concerning the extractions themselves and the sample pretreatment.

The basic requirement of a selective reagent is to extract metals completely from a given substrate while leaving more or less intact the same metals bound to other substrates.[45] It would be unrealistic to think that one could select a sequence of reagents that would extract

Table 1

EXAMPLES OF SEQUENTIAL EXTRACTION PROCEDURES

Methods	1	2	3	4	5	Ref.
			Step			
1	$NH_2OH \cdot HCl/HNO_3$	$NH_2OH \cdot HCl/HOAc$	$Na_2S_2O_4$	$KClO_4/HCl/HNO_3$	HF/HNO_3	46
2	NH_4OAc	$NH_2OH \cdot HCl$	H_2O_2/HNO_3	$Na_2S_2O_4$	HF/HNO_3	47
3	$MgCl_2$	$NaOAc/HOAc$	$NH_2OH \cdot HCl/HOAc$	H_2O_2/HNO_3	$HF/HClO_4$	48
4	$NaOCl$	$NH_2OH \cdot HCl$	$(NH_4)_2C_2O_4/H_2C_2O_4$	$Na_2S_2O_4$	$HNO_3/HClO_4$	49
5	NH_4OAc	$NH_2OH \cdot HCl/HNO_3$	$(NH_4)_2C_2O_4/H_2C_2O_4$	H_2O_2/HNO_3	HNO_3	50
6	$NaOAc/HOAc$	Na dodecyl-sulfonate/$NaHCO_3$	$NH_2OH \cdot HCl/Na$-citrate	HF/HNO_3		51

the individual fractions without influencing the other sediment constituents to some extent. Any sequential extraction procedure will unavoidably suffer from a certain lack of selectivity, as has been shown theoretically[52] and experimentally. Experimental verification of extraction selectivity has been carried out:

1. By using pure solids of known geochemical composition[53]
2. By determining various parameters (inorganic and organic carbon, sulfur, solid phases examined by X-ray diffraction or by electron microscopy/electron probe analysis) either in the extracts and/or in the residual sediment remaining after the various extractions[48,53,54]
3   By statistical analysis (mostly correlation coefficients) of the trace metals extracted from sediments and the major substrates therein[9,55]
4. By using model sediments built up from various solid phases[56-58]

Although tests to date suggest that selectivity is reasonably satisfactory for oxic sediments, this is not the case for sulfide-rich (anoxic) sediments; in effect, several lines of evidence indicate that the various extraction procedures currently in use are compromised by the tendency of metal-containing sulfide phases to be leached progressively rather than selectively from such sediments.[53,59]

Demonstrating the selectivity (or lack of same) for partial extraction procedures is difficult, however, since none of the approaches described above yield unambiguous results. For example, researchers using model sediments to test extraction procedures usually dope a trace metal onto a given phase and then mix this contaminated phase with other solid phases that do not contain this metal. In doing so, they bias their systems towards metal redistribution, as is predicted by the model described by equations 1 and 2. Applying sequential extraction procedures to the model sediments, they do not recover all the trace metal in the "appropriate" fractions (i.e., in the fractions corresponding to the added phases), and generally conclude that the tested procedure is unsuitable for distinguishing the phase-association of metals in real sediments. It should be considered, however, that the model sediments, as prepared, bear little resemblance to natural sediments, in which the metal would have *already* been distributed among the various sediment components, and where the driving force for redistribution during extraction would have been correspondingly much lower.

Once the reagents have been selected, they can be used in a sequential order. As is evident from Table 1, the precise sequence of extractants may vary considerably. For a given set of reagents, the sequence used will influence the partitioning obtained.[60] During the procedure, there is a potential problem related to readsorption[54,61,62] in that when a metal is liberated with a given reagent it may readsorb onto the remaining solid phases, i.e., the extraction procedure itself may cause a shift of the metal distribution pattern. Even with a given set of reagents and a given extraction sequence, the exact experimental conditions employed to extract the different metal fractions (e.g., ratio of extractant to sediment, contact time) may influence the metal distribution.

It follows from the methodological considerations discussed above that the observed distributions of trace metals among various fractions will not necessarily reflect the relative contributions of discrete sediment phases, but rather should be considered as operationally defined by the methods of extraction.

In addition to these analytical problems, there is also a potential problem in preserving sample integrity between sampling and analysis. The techniques used to preserve the sediments before analysis (e.g., wet storage at 1 to 4°C, freezing, freeze-drying, air-drying) influence the partitioning of trace metals obtained by sequential extraction procedures;[59,63] methods involving the drying of the sediments have especially marked effects. As expected, the effects of pretreatment are more important for anoxic samples than for oxic sediments.

III. BIOAVAILABILITY OF TRACE METALS IN SEDIMENTS

A. General Principles

Methods for assessing the relative bioavailability of sediment-bound metals to benthic species can be grouped into three categories.[64]

1. Laboratory (feeding) experiments, carried out with natural or artificial sediments
2. Field experiments involving, for example, the transfer of benthic organisms between contaminated and uncontaminated environments (with or without their original sediments), or the use of field "plots" subject to different experimental manipulations
3. Field surveys of sediment characteristics (e.g., $[M]_{sed}$) and metal burdens in indigenous benthic species

In this final section we shall consider results derived largely from this latter approach. The purpose of this section is to examine relationships between the partitioning of trace metals in sediments, as discussed above, and their bioavailability to benthic organisms. The emphasis will be on *in situ* undisturbed surficial sediments, rather than on dredged or resuspended material.

The least ambiguous criterion for evaluating the "bioavailability" of a particular metal will be its accumulation within one or more target organisms. Benthic organisms can take up metals directly from the sediment ambient water* and/or from the particulate material. For equal metal concentrations in the external medium, uptake from solution is faster than from sediments (i.e., dissolved metals are inherently more "available");[65] opposing this tendency will be the higher metal concentrations ($\rightarrow 10^6 \times$) in the particulate phase. The uptake pathway(s) used by a particular organism can obviously affect the relationships between metal partitioning and metal bioavailability. For type A organisms that live in intimate contact with the sediments but are incapable of ingesting particulate material (e.g., sediment bacteria, epipelic algae, rooted aquatic plants), the chemical composition of the ambient water assumes primary importance.

For most metals a key factor influencing metal bioavailability in the ambient water will be the free aquo-ion concentration,** $[M^{z+}]$ (cf. the "free-ion" equilibrium model of metal toxicity).[66] Factors controlling α_M, the proportion of total metal $[M]_T$ that is present as the aquo ion, include:[66,67] nature and concentrations of the ligands present in the water (L_1, L_2, . . . L_x); concentrations of competing cations, including Ca^{2+} and Mg^{2+}; redox potential; temperature; pH; ionic strength; degree to which chemical equilibrium is attained (reaction kinetics). Clearly these factors will vary from one environment to another, and, thus, the *total* metal concentration in the sediment pore water will be an imperfect indicator of metal bioavailability. Ideally, one would determine $[M^{z+}]$ directly in the ambient water, but this has only very rarely been attempted;[68,69] the inherent sampling and analytical problems (small sample volumes, lack of method selectivity, inadequate detection limits) impose severe limits on this approach. A possible compromise would involve measurement of $[M]_T$ together with

---

*   The nature of the "ambient water" encountered by different benthic species will depend on their burrowing behavior and feeding strategies. It may include, in different cases: anoxic interstitial waters for organisms in contact with deeper sediments; oxic interstitial waters near the surface for organisms living within the sediment, but whose primary environmental contact is above the redox interface; or waters at the sediment-water interface for suspension-feeders.

**   Metals existing in alkylated forms (e.g., Hg, Pb, As, Se, Sn) constitute an exception to this general rule; for such metals the relative contribution of the organometallic species must be considered.

such complementary parameters as pH and the dissolved organic carbon concentration [DOC].[70] Since metal-metal interactions at the level of biological membranes may also be of importance, measurement of the concentrations of possible competing metals (e.g., Ca^{2+}, Mg^{2+}, Fe^{2+}) could also be used to refine estimates of the bioavailability of metal M.

An alternative to the direct measurement of ambient water characteristics would be to determine metal concentrations in those sediment forms in equilibrium with the ambient water (see Section II.A, Equations 1 and 2). In such an approach the relevant parameter will not be the *total* metal concentration in the particulate material, but rather the concentration of the metal bound to the solid surfaces ($\{\equiv S(1) - M\}$, $\{\equiv S(2) - M\}$, . . . , $\{\equiv S(n) - M\}$), together with the concentrations of sites, $\{\equiv S(n)\}$, responsible for binding the metal in these forms. In this approach the ratio $\{\equiv S(n) - M\}/\{\equiv S(n)\}$ represents a surrogate measure of $K_a (n) \cdot [M^{z+}]$ (see Equation 2) and thus should be a useful predictor of metal bioaccumulation. Note that the concentration of adsorbed or "exchangeable" metal is not in itself sufficient to estimate $[M^{z+}]$ or to estimate metal bioavailability. In effect, high levels of adsorbed metal in a given sediment may simply reflect an abundance of adsorbing surfaces, rather than a high value of $[M^{z+}]$ in the sediment pore water.

The reasoning described to this point for type A organisms also applies to type B organisms, i.e., those capable of ingesting sediments and taking up metals from particulate phases. To these considerations, however, must be added an evaluation of the possible uptake of metals during digestion of the sediments in the gut.

Controlled laboratory feeding experiments, often involving the use of synthetic sediments, have shown that accumulation of a given metal by type B organisms may differ by as much as 1000-fold among different sediment types (e.g., iron oxyhydroxides; manganese oxyhydroxides; organic debris; suspended bacteria; carbonates[2,65,71,72]). The ranking of these different solid phases, as a function of the relative "bioavailability" of their associated metals, may, however, vary from one metal to another. Differences in metal availability seem to be related to the strength of metal binding to the particulate phase; more weakly bound metals are generally more available, whereas more strongly bound metals tend to be less available.[71] Changes in overall metal-binding intensity are, however, not the only factor involved in alterations of metal availability from ingested particles. For example, Harvey and Luoma[65] examined the effects of adherent bacteria and bacterial extracellular polymer upon the uptake of particle-bound Cd by deposit-feeding clams. Adsorption of exopolymer on synthetic iron oxide particles had little or no effect on the extent of Cd binding, but did lead to an increase in the availability of particle-bound Cd.

The bioavailability of particulate metals will also vary among different type B species, as a function of their respective digestive chemistries. Of obvious importance will be the pH and redox conditions prevailing in the digestive tract of the target organism. Luoma[71] suggests that the digestive pH of deposit-feeding organisms is generally in the range 6 to 7, but supporting data are scarce. Even less is known about conditions inside digestive vesicles or amoebocytes in the many invertebrate species that employ intracellular digestion. The residence time of the sediments within the organism will also be important, in that longer residence times will favor more complete extraction of the particle-bound metal (since the desorption and/or dissolution reactions involved in the digestive process may well be slow). In addition particle size must be considered, since type B organisms may be selective in their feeding/ingestion habits, and digestive extraction will likely be more efficient as particle size decreases.

With both type A and B organisms, metal uptake from solution may be influenced by metal-metal interactions, for example, competition among dissolved metals for a given uptake site on a cell membrane:

$$
\begin{array}{c}
\text{M} \\
\\
\quad + \text{L-membrane} \\
\\
\text{M}'
\end{array}
\begin{array}{c}
\nearrow \text{M — L-membrane} \\
\\
\searrow \text{M}' \text{— L-membrane}
\end{array}
\quad
\begin{array}{c}
\text{transport of M} \\
\longrightarrow \\
\end{array}
\cdots
$$

where M' = competing metal; L-membrane = metal transport site on the cell membrane. For type B organisms, knowledge of the concentration of other metals solubilized during the digestive process might, thus, be used to improve the prediction of the bioavailability of M, the metal of concern. Alternatively, competition may occur *for* solubilized metal, between the uptake sites, on the one hand, and the undigested sediment, on the other (i.e., those solid phases least affected by the digestive process, including clays; Fe-oxyhydroxides; Mn-oxyhydroxides; humic substances):

$$
M \quad
\begin{array}{l}
+ \text{ L-membrane} \rightleftarrows \text{M} - \text{L-membrane} \quad \xrightarrow{\text{transport of M}} \quad \dots \\[8pt]
+ \equiv S(n) \qquad \rightleftarrows \equiv S(n) - M
\end{array}
$$

where $\equiv S(n)$ = a potential adsorption site on the undigested sediment. If such competition does occur, the bioavailability of metal M will be lower in sediments rich in one or more of these "ameliorative" phases.

B. Relationships between the Partitioning of Trace Metals in Sediments and their Bioavailability

Despite the inherent limitations of sediment extraction procedures, as discussed in Section II.B, one might intuitively expect metal bioavailability to be higher for those fractions more readily removed from the sediment (with "milder" reagents). Indeed, evidence from laboratory microcosms does suggest that metal bioavailability is affected by metal partitioning among sedimentary compartments.[71] With this in mind, we shall consider what empirical (field) evidence exists to indicate that certain metal fractions may indeed be more "available" than others, or that metal concentrations in certain sediment fractions may be better predictors of metal bioavailability than others.

In the following discussion, type A and type B organisms are considered separately, as are freshwater and estuarine/marine habitats. Rather than attempt an exhaustive review for the different types of organisms, we have chosen to present a number of representative examples for macrophytes (type A) and for benthic filter- or deposit-feeders (type B). Results are presented from field studies in which sediment-dwelling organisms have been collected along a spatial gradient in metal concentrations in the sediments, $[M]_{sed}$.

1. Aquatic Macrophytes

Two pathways exist for the accumulation of trace metals in the stems and leaves of rooted aquatic plants: (1) direct uptake from the water column or (2) uptake from the interstitial water in the sediments, followed by acropetal translocation from the roots to the stems and leaves. This same duality of pathways exists for other solutes (e.g., phosphorus), and both accumulation routes may be operative in the same plant.[73] For *submerged* rooted species in a field situation, it clearly is difficult to distinguish between the two possible uptake vectors; in such cases the least ambiguous "integrator" of metal bioavailability in the sediments will be the metal levels in the underground parts of the plant, including the roots and rhizome.[74-76] For *emergent* species, the metal content of the aboveground parts of the plant constitutes an indicator of the bioavailability of the metal of concern in the host sediment, provided that any contribution of airborne particulates to plant leaf surfaces can be discounted[77] and that uptake via the submerged portion of the stem in negligible.

Literature concerning the prediction of metal accumulation by aquatic macrophytes is scarce. If uptake from sediment interstitial water were widespread, one might expect the concentration of "available" metal in the sediment environment to be related to the metal concentration in the plant. In particular, correlations would be anticipated between $[M^{z+}]$ in the interstitial water and metal levels in plants. Somewhat surprisingly, however, no one seems to have tested this prediction in a field setting. A recent literature review did turn up some 105 cases where (total) metal concentrations had been determined both in rooted freshwater plants and in the underlying sediments.[78,105] However, simple relationships between metal concentrations in the plant, $[M]_p$, and the total metal concentrations in the sediments, $[M]_{sed}$, were rarely observed. As discussed in Section III.A, one might anticipate an improved prediction of metal accumulation by aquatic plants if metal partitioning in the host sediment were considered. This approach is discussed below.

Freshwater plants — Two field studies were found where attempts had been made to predict metal concentrations in the underground portion of submerged freshwater plants from metal partitioning data in the adjacent sediments; for emergent species a single such study was identified. This latter example was not a true field study; highly contaminated sediments were collected in the field, returned to the laboratory and used as a substrate for growing *Cyperus esculentus*, a freshwater marsh plant, under controlled conditions simulating either reduced ("flooded") or oxidized ("upland") situations.[79,80]

Extractants used to evaluate bioavailable metal for rooted aquatic plants have included a concentrated electrolyte ($MgCl_2$), dilute acid (HC1), and DTPA (diethylenetriaminepentaacetate), an organic chelator commonly used to estimate the concentration of bioavailable nutrients and metals in soils.[81] In three cases the "extractable" metal proved a better predictor of plant metal content than did "total" metal (Table 2: cases 2 and 3a and b), in one case there was no significant improvement (Table 2: case 1), and in nine cases no positive relationship could be found between $[M]_p$ and $[M]_{sed}$ even for "extractable" metal (see NRCC[70] for a more detailed summary of these results).

Estuarine/marine plants — For rooted marine plants, a single field study was identified where attempts had been made to refine estimates of sediment metal bioavailability by extraction techniques.[77] Four extractants (distilled water; NH_4OAc; DTPA; HCl/H_2SO_4) were tested and the concentrations of extracted metal were used to predict metal uptake by salt marsh plants at existing dredged material disposal sites. Of these, DTPA showed good potential for predicting accumulation of Cd, Cu, and Zn (plus, to a lesser extent, Cr and Pb) in two of the three species studies, *Spartina alterniflora* and *Distichlis spicata* (Table 2: cases 4 and 5). Prediction of plant uptake of Hg and Ni was, however, consistently poor.

The following points should be emphasized with respect to the bioavailability of sediment-bound metals to type A organisms, as represented by rooted aquatic plants, and to the use of sediment extraction techniques to estimate this bioavailability:[70]

- Information concerning the prediction of metal accumulation by aquatic macrophytes from sediments is scarce. Published studies are often restricted to a limited number of sites in a single geographical (geological) area; the transposability of such results to other regions is unknown.

- The extraction procedures have often been performed on dried sediments. The drying step may profoundly influence metal partitioning,[59] and the results of a given extraction procedure performed after drying generally differ markedly from those obtained with the original fresh sediment. This is of particular importance for those metals originally present in the interstitial water or loosely held at adsorption sites.

- Results published to date suggest that, for a given metal, marked species-to-species differences may exist (cf. Table 2: DTPA-extractable metal was a good predictor of metal accumulation by *S. alterniflora* and *D. spicata*, but not by *S. cynosuroides*). Similarly, for a particular plant species different metals may yield contrasting results. These limited observations would dictate caution in the search for a single extraction procedure that will have broad applicability to several metals and all rooted aquatic plants.

Table 2
EXAMPLES OF USE OF SEDIMENT EXTRACTION TECHNIQUES TO PREDICT THE BIOACCUMULATION OF SEDIMENT-BOUND METALS BY ROOTED AQUATIC PLANTS

Environment	Species	Plant part	Metal	N[a]	Predictor of plant metal accumulation[b]		Ref.
					Fraction	r^2	
Fresh water	1. *Nuphar lutea*	Rhizome	Cu	7	[Cu] ext. with HCl (0.5 *M*) ([Cu]$_T$)	0.99 (1.00)	82
	2. *N. variegatum*	Rhizome	Cu	13	[Cu] ext. with MgCl$_2$ ([Cu]$_T$)	0.30 (0.20)	83
					[Cu]/[Fe] ext. with MgCl$_2$	0.88	
	3. *Cyperus esculentus*	Above-ground	a. Cd	15	[Cd] ext. with DTPA ([Cd]$_T$)	0.27 (0.07)	80,84
			b. Pb	14	[Pb] ext. with DTPA ([Pb]$_T$)	0.26 (0.01)	
Estuary	4. *Spartina alterni-flora*	Leaf	a. Cd	79	[Cd] ext. with DTPA	0.81	77
			b. Cu	79	[Cu] ext. with DTPA	0.30	
					[Cu] ext. with HCl/H$_2$SO$_4$ (dilute)	0.32	
			c. Zn	79	[Zn] ext. with DTPA	0.67	
	5. *Distichlis spicata*	Leaf	a. Cd	25	[Cd] ext. with DTPA	0.83	77
			b. Cu	25	[Cu] ext. with DTPA	0.32	
			c. Zn	25	[Zn] ext. with DTPA	0.69	

Note: Table has been adapted from Reference 70.

[a] N = number of sites.
[b] r^2 = coefficient of determination.

- Published studies to this point (Table 2) have all attempted to relate metal concentrations in plants, $[M]_p$, to one or more extractable metal fractions in the host sediments. As discussed in Section III.A, the concentration of extractable metal is not in itself a reliable estimate of $[M^{z+}]$; high concentrations of extractable metal in a particular sediment may simply reflect an abundance of adsorbing surfaces rather than a high value of $[M^{z+}]$ in the sediment pore water. One might anticipate better predictions of metal bioaccumulation from the ratio $\{\equiv S(n) - M\}/\{S(n)\}$, which should vary as a function of K_a (n) $\cdot [M^{z+}]$. The usefulness of this approach remains to be tested.
- Although such sediment characteristics as pH, redox potential, organic matter content, and amorphous iron concentration have been shown qualitatively to influence metal accumulation by aquatic macrophytes, few attempts have been made to incorporate these variables into predictive equations applicable in a field setting.

2. Benthic Invertebrates

Deposit-feeding invertebrates are directly exposed to sediment-bound metals and are thus capable of accumulating metals from the ambient water and/or from ingested sediment.[64,65,71,72] Concentration factors with respect to sediments are generally higher for sessile macroinvertebrates than for other aquatic organisms.[85] For this reason, and because of their limited mobility, macroinvertebrates have often been proposed as biological monitors of pollution by metals. Metal concentrations found in benthic invertebrates are affected by both geochemical and biological factors (for a review of these latter variables, e.g., size, age, sex, molting cycles, reproductive cycles, detoxification mechanisms, see References[64,71,86]). In the following discussion we shall emphasize the geochemical factors, but the influence of "biological variability" should not be neglected.

On the basis of the results of controlled laboratory feeding experiments and in accord with the earlier discussion in this chapter, one might anticipate the existence of consistent relationships between metal accumulation in benthic invertebrates and metal partitioning in the adjacent or host sediment. Indeed, attempts to relate metal concentrations in various benthic species to those in the adjacent sediment are rather more numerous (and more successful) than for the aquatic macrophytes discussed earlier.

In this section we consider results from field surveys in which (1) surficial sediment samples and benthic invertebrates were collected along gradients of sediment metal concentrations and (2) where attempts were made to predict metal accumulation in the organisms from the metal content of the host sediment. Since simple relationships between $[M]_{organism}$ and the total metal concentration in the sediment are seldom found in such studies,[64,87,88] the present compilation has been limited to those investigations for which metal partitioning data are reported for the sediments.

a. Freshwater Invertebrates

For freshwater systems three field surveys were identified which satisfied the criteria outlined above (Table 3). Extractants used to evaluate metal partitioning included concentrated electrolytes ($MgCl_2$; NH_4OAc), weak and strong acids (acetic acid, HOAc; HCl), a reducing agent designed to solubilize amorphous Fe and Mn oxyhydroxides ($NH_2OH \cdot HCl$), and an acidic oxidizing solution (H_2O_2, HNO_3).

For two filter-feeding mollusks, *Elliptio complanata* and *Anodonta grandis*, collected from a restricted geographical area downstream from a mining/smelting complex, Cu, Pb, and Zn levels in various tissues (or in the whole organism, without the shell) were best related not to the total metal concentrations in the adjacent sediment, but rather to one or more of the relatively easily extracted fractions present in the fine sediments (particle diameter <70 μm, i.e., the ingestible size fraction). Accumulation of Cu, Pb, and Zn was also influenced by the protective or competitive effect of certain other sediment constituents,

Table 3
EXAMPLES OF USE OF SEDIMENT EXTRACTION TECHNIQUES TO PREDICT THE BIOACCUMULATION OF SEDIMENT-BOUND METALS TO FRESHWATER BENTHIC INVERTEBRATES

Species	Organism part	Metal	N^a	Fraction[c]	r^2	Ref.
				Predictor of plant metal accumulation[b]		
				Filter-Feeders		
1. *Elliptio complanata*	Whole	Cu	8	[Cu]/[Fe] ext. with $NH_2OH \cdot HCl$	0.95	89
	Whole	Zn	8	[Zn]/[Fe] ext. with $NH_2OH \cdot HCl$	0.66	
				[Zn] ext. with $H_2O_2 \div$ [Fe] ext. with $NH_2OH \cdot HCl$	0.79	
	Gills	Pb	8	[Pb]/[Fe] ext. with $NH_2OH \cdot HCl$	0.83	
2. *Anodonta grandis*[d]	Whole[e]	Cu	9	[Cu]/[Fe] ext. with $NH_2OH \cdot HCl$	0.87	78
	Whole[e]	Zn	9	[Zn] ext. with $H_2O_2 \div$ [Fe] ext. with $NH_2OH \cdot HCl$	0.90	
	Gills	Pb	10	[Pb]/[Fe] ext. with $NH_2OH \cdot HCl$	0.82	
				Deposit-Feeders		
3. Oligochaetes (unidentified; mainly *Limnodrilus hoffmeisten* and *Tubifex tubifex*)	Whole	Cu	12	[Cu] ext. with $NH_2OH \cdot HCl$	0.85	90,91
	Whole	Pb	12	[Pb] ext. with $NH_2OH \cdot HCl$	0.35	
				$([Pb]_{IT})$	(0.46)	

Note: Table has been adapted from Reference 70.

[a] N = number of sites.
[b] r^2 = coefficient of determination.
[c] Fractions extracted from the sediment in a sequential extraction scheme.
[d] Age class 5 to 6 years.
[e] For gravid females, eggs were removed.

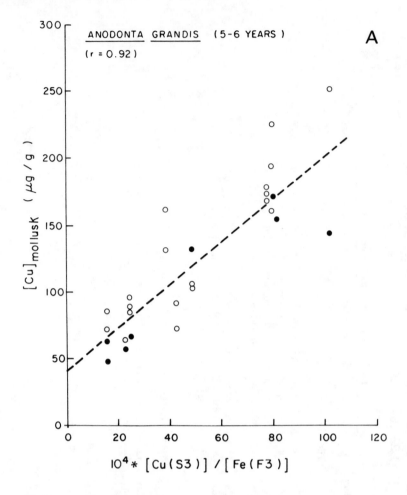

FIGURE 1.   Relationship between the concentration of metal M in the freshwater mollusk *Anodonta grandis* and the ratio [M(S3)]/[Fe(F3)] in the sediments. (A) M = Cu ([Cu]$_{mollusk}$ = (1.6×10^4) [Cu(S3)]/[Fe(F3)] + 41); (B) M = Zn ([Zn]$_{mollusk}$ = (2.1×10^4) [Zn(S3)]/[Fe(F3)] − 140); (C) M = Pb ([Pb]$_{gill}$ = (3.8×10^3) [Pb(S3)]/[Fe(F3)] − 4.3). Note that [M(S3)] corresponds to the metal concentrations (Cu, Pb, Zn) removed from the sediment by sequential extraction with $MgCl_2$, NaOAc/HOAc (pH 5), and $NH_2OH \cdot HCl$, where [Fe(F3)] corresponds to the amorphous iron oxide fraction extracted by $NH_2OH \cdot HCl$. The filled circles correspond to gravid specimens (eggs removed). (See References 78 and 89 for details.)

notably the amorphous iron oxyhydroxides. The best predictions of metal levels in the mollusks were obtained when metal concentrations extracted from the sediments were normalized with respect to the amorphous iron concentration[78,89] (see Figure 1).

Results for deposit-feeding oligochaetes were less consistent than those for the mollusks (Table 3). Copper concentrations in the easily reducible phases ($NH_2OH \cdot HCl$) were the best predictor of Cu bioaccumulation, whereas total lead in the sediments was the best predictor for lead levels in the organisms. For zinc, no significant relationships were observed between metal accumulation in the organism and zinc concentrations in any of the sediment extracts, suggesting a certain homeostatic regulation of internal zinc concentrations.[64] No attempts were made[90,91] to improve the predictions by normalizing Cu, Pb, or Zn concentrations in the sediment with respect to other sediment constituents.

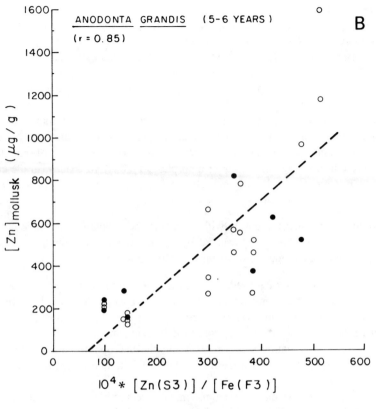

FIGURE 1B.

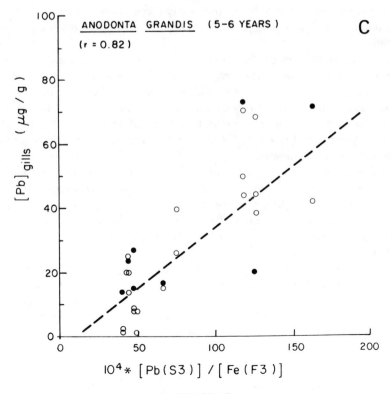

FIGURE 1C.

b. *Estuarine/Coastal Invertebrates*

Studies relating metal bioaccumulation by macroinvertebrates to metal levels in sediments are more numerous for the estuarine environment than for freshwater systems. Many extractants have been explored (Table 4); the individual reagents have usually been applied in nonsequential fashion to the fine sediment fraction (<100 or <250 μm). Three invertebrate species have been studied extensively: two deposit-feeding clams occupying similar ecological niches (*Scrobicularia plana*, *Macoma balthica*) and a polychaete (*Nereis diversicolor*). Metal accumulation in these organisms was best predicted from metal concentrations in one or more of the more readily extracted fractions present in the sediments. Normalization of these sediment metal concentrations with respect to the concentrations of one of the major metal sinks present in the sediment, either amorphous Fe oxyhydroxides (in the case of As, Pb, and Zn) or organic matter (Hg), frequently led to marked improvements in the predictions of metal accumulation (Table 4).

The following points merit attention with respect to the bioavailability of sediment-bound metals to type B organisms, as represented by benthic macroinvertebrates, and to the use of sediment extraction schemes to estimate this bioavailability:

* Information concerning the prediction of metal accumulation is far more abundant for estuarine benthic invertebrates than for freshwater species. The available field data for estuaries are derived from a variety of geological systems, both in Britain and in North America, whereas the freshwater data base is much smaller.

* In contrast to the situation prevailing for aquatic macrophytes, most sediment extractions have been performed on fresh samples; artifacts due to sediment drying should thus be minimal.

* In both freshwater and marine systems, metal levels in benthic invertebrates are better related to metal concentrations in one or more relatively easily extracted fractions, present in the fine sediments, than to total metal concentrations. Although a single extraction procedure cannot be unequivocally recommended, treatment of sediments with dilute hydrochloric acid (HC1, 3 *M*) has proven consistently useful in estuarine systems.

* Metal concentrations in benthic organisms decrease in the presence of appreciable concentrations of various metal sinks, notably amorphous Fe oxyhydroxides and organic matter. Similar conclusions regarding the protective or competitive role of these phases have been derived from experimental manipulations[65,72,97,98] and from multiple regression analyses of metal levels in benthic invertebrates as a function of various sediment geochemical factors.[91,96] However, the specific factors affecting metal bioavailability differ among metals.

* A number of laboratory investigations suggest that metal uptake from solution may be important for some benthic invertebrates, notably the annelids.[64,99,100] In field surveys, however, very little attention has been paid to metal concentrations and metal speciation in the sediment interstitial water. Future studies of this type could be markedly improved if such concentration values were obtained (e.g., by *in situ* dialysis with pore-water "peepers".[100-104]

* In this connection, it should be noted that a strong relationship between trace metal levels in a given benthic species and certain sediment characteristics does not require that the main route of entry of the trace metal be via ingestion of the sediments themselves. Such relationships can also be explained by adsorptive control of the dissolved trace metal concentrations in the ambient water to which the organisms are exposed, i.e.,

$$[M(\text{organisms})] \propto [M^{z+}] \left(= \frac{\{\equiv S(n) - M\}}{\{\equiv S(n)\}} \cdot \frac{1}{K_a(n)} \right) \qquad (5)$$

Table 4
EXAMPLES OF USE OF SEDIMENT EXTRACTION TECHNIQUES TO PREDICT THE BIOACCUMULATION OF SEDIMENT-BOUND METALS TO MARINE BENTHIC INVERTEBRATES

Species	Organism part	Metal	N[a]	Predictor of plant metal accumulation[b]		Ref.
				Fraction[c]	r²	
Filter-Feeders						
1. *Scrobicularia plana*	Whole	As	75	[As]/[Fe] ext. with HCl	0.93 (0.82)*	92
	Whole	Hg	78	[Hg] ext. with HNO₃ ÷ % organic matter	0.63	93
	Whole	Pb	37	[Pb]/[Fe] ext. with HCl	0.98 (0.88)*	94
	Whole	Zn	40	[Zn] ext. with NH₄OAc	0.38	95
2. *Macoma balthica*	Whole	Hg	29	[Hg] ext. with HNO₃ ÷ % organic matter	—	93
	Whole	Ag	15	[Ag] ext. with HCl	0.68*	64
		Cd	14	[Cd] ext. with HCl	0.21*	
		Pb	15	[Pb]/[Fe] ext. with HCl	0.58*	
		Zn	16	[Zn] ext. with HCl	0.31*	
Deposit-Feeder						
3. *Nereis diversicolor*	Whole	As	82	[As]/[Fe] ext. with HCl	0.55*	92
	Whole	Ag	37	[Ag] ext. with HCl	0.41*	96
		Co	37	[Co] ext. with HCl	0.66*	
		Cu	37	[Cu] ext. with HNO₃	0.72*	
		Pb	37	[Pb] ext. with HNO₃	0.54*	
	Whole	Ag	38	[Ag] ext. with HCl	0.62*	64

Note: Table has been adapted from Reference 70.

[a] N = number of sites.
[b] r² = coefficient of determination; an asterisk indicates that the statistical analysis was performed on log-transformed data.
[c] Fractions extracted from the sediment (nonsequential extraction).

Equation 5 clearly shows that an apparent relationship between $[M]_{organism}$ and $\{\equiv S(n) - M\}/\{\equiv S(n)\}$ may occur even though the main route of metal uptake by the organisms is via the dissolved trace metal, $[M^{z+}]$.

- The chemical conditions prevailing in the digestive tract of many benthic invertebrates, the kinetics of the digestive process, and the residence time of sediments in the digestive system are matters of considerable speculation. Similarly, the feeding habits of these organisms (particle size ingested; depth of sediment encountered) are often poorly understood. These would be fruitful areas for future research.

- Despite the complexity of the environmental factors involved, results obtained to date "offer some optimism that metal concentrations in (estuarine) organisms generally can be predicted and explained using relatively simple chemical measurements."[96]

ACKNOWLEDGMENTS

This review has benefited from the constructive criticism of many of our colleagues. In particular, many of the ideas presented in Section III were developed and refined while one of us (P.G.C.C.) worked on a review document for the National Research Council of Canada[70] in collaboration with P. M. Chapman, A. Crowder, W. K. Fletcher, B. Imber, A. G. Lewis, S. N. Luoma, P. M. Stokes, and M. Winfrey. Financial support from the Quebec Ministry of Education (Fonds FCAR), the Natural Sciences and Engineering Research Council of Canada (strategic grant), and the World Wildlife Fund Canada is gratefully acknowledged.

REFERENCES

1. **Förstner, U. and Wittmann, G. T. W.,** *Metal Pollution in the Aquatic Environment,* 2nd ed., Springer-Verlag, Berlin, 1981.
2. **Luoma, S. N. and Jenne, E. A.,** The availability of sediment-bound cobalt, silver, and zinc to a deposit-feeding clam, in *Biological Implications of Metals in the Environment,* Wildung, R. E. and Drucker, H., Eds., U.S. Energy Research and Development Administration, Washington, D.C., 1977, 213.
3. **Luoma, S. N. and Davis, J. A.,** Requirements for modeling trace metal partitioning in oxidized estuarine sediments, *Mar. Chem.,* 12, 159, 1983.
4. **Schindler, P. W.,** Heterogeneous equilibria involving oxides, hydroxides, carbonates, and hydroxide carbonates, in *Equilibrium Concepts in Natural Water Systems,* Adv. Chem. Ser. 67, Stumm, W., Ed., American Chemical Society, Washington, D.C., 1967, 196.
5. **Hem, J. D.,** Chemistry and occurrence of cadmium and zinc in surface water and groundwater, *Water Resour. Res.,* 8, 661, 1972.
6. **Hem, J. D.,** Geochemical controls on lead concentrations in stream water and sediments, *Geochim. Cosmochim. Acta,* 40, 599, 1976.
7. **Krauskopf, K. B.,** Factors controlling the concentrations of thirteen rare metals in sea-water, *Geochim. Cosmochim. Acta,* 9, 1B, 1956.
8. **Schindler, P. W.,** Removal of trace metals from the oceans: a zero order model, *Thalassia Jugosl.,* 11, 101, 1975.
9. **Luoma, S. N. and Bryan, G. W.,** A statistical assessment of the form of trace metals in oxidized estuarine sediments employing chemical extractants, *Sci. Total Environ.,* 17, 165, 1981.
10. **Oakley, S. M., Nelson, P. O., and Williamson, K. J.,** Model of trace-metal partitioning in marine sediments, *Environ. Sci. Technol.,* 15, 474, 1981.
11. **Luoma, S. N.,** A comparison of two methods for determining copper partitioning in oxidized sediments, *Mar. Chem.,* 20, 45, 1986.
12. **Loganathan, P. and Burau, R. G.,** Sorption of heavy metal ions by a hydrous manganese oxide, *Geochim. Cosmochim. Acta,* 37, 1277, 1973.
13. **Murray, J. W.,** The interaction of metal ions at the manganese dioxide-solution interface, *Geochim. Cosmochim. Acta,* 39, 505, 1975.

14. **Davis, J. A. and Leckie, J. O.,** Surface ionization and complexation at the oxide/water interface. II. Surface properties of amorphous iron oxyhydroxide and adsorption of metal ions, *J. Colloid Interface Sci.,* 67, 90, 1978.

15. **Farrah, H. and Pickering, W. F.,** pH effects in the adsorption of heavy metal ions by clays, *Chem. Geol.,* 2, 317, 1979.

16. **Swallow, K. C., Hume, D. N., and Morel, F. M. M.,** Sorption of copper and lead by hydrous ferric oxide, *Environ. Sci. Technol.,* 14, 1326, 1980.

17. **Balistrieri, L. S. and Murray, J. W.,** The adsorption of Cu, Pb, Zn, and Cd on goethite from major ion seawater, *Geochim. Cosmochim. Acta,* 46, 1253, 1982.

18. **Beveridge, A. and Pickering, W. F.,** Influence of humate-solute interactions on aqueous heavy metal ion levels, *Water Air Soil Pollut.,* 14, 171, 1980.

19. **Bourg, A. C. M.,** Role of fresh water/sea water mixing on trace metal adsorption, in *Trace Metals in Sea Water,* Wong, C. S., Boyle, E., Bruland, K. W., Burton, J. D., and Goldberg, E. D., Eds., Plenum Press, New York, 1983, 195.

20. **Millward, G. E. and Moore, R. M.,** The adsorption of Cu, Mn, and Zn by iron oxyhydroxide in model estuarine solutions, *Water Res.,* 16, 981, 1982.

21. **Tewari, P. H., Campbell, A. B., and Woon, L.,** Adsorption of Co^{+2} by oxides from aqueous solution, *Can. J. Chem.,* 50, 1642, 1972.

22. **Dyck, W.,** Adsorption and coprecipitation of silver on hydrous ferric oxide, *Can. J. Chem.,* 46, 1441, 1968.

23. **Kinniburgh, D. G. and Jackson, M. L.,** Cation adsorption by hydrous metal oxides and clay, in *Adsorption of Inorganics at Solid-Liquid Interfaces,* Anderson, M. A. and Rubin, A. J., Eds., Ann Arbor Science Publishers, Ann Arbor, MI, 1981, 91.

24. **Leckie, J. O., Benjamim, M. M., Hayes, K., Kaufman, G., and Altman, R. S.,** Adsorption/coprecipitation of trace elements from water with iron oxyhydroxide, Electric Power Research Institute Rep. CS-1513, Palo Alto, CA, 1980.

25. **Millward, G. E.,** The adsorption of cadmium by iron(III) precipitates in model estuarine solutions, *Environ. Technol. Lett.,* 1, 394, 1980.

26. **Benjamin, M. M. and Leckie, J. O.,** Multiple-site adsorption of Cd, Cu, Zn, and Pb on amorphous iron oxyhydroxide, *J. Colloid Interface Sci.,* 79, 209, 1981.

27. **Benjamin, M. M. and Leckie, J. O.,** Competitive adsorption of Cd, Cu, Zn, and Pb on amorphous iron oxyhydroxide, *J. Colloid Interface Sci.,* 83, 410, 1981.

28. **Kinniburgh, D. G. and Jackson, M. L.,** Concentration and pH dependence of calcium and zinc adsorption by iron hydrous oxide gel, *Soil Sci. Soc. Am. J.,* 46, 56, 1982.

29. **Nyffeler, U. P., Li, Y.-Y., and Santschi, P. H.,** A kinetic approach to describe trace-element distribution between particles and solution in natural aquatic systems, *Geochim. Cosmochim. Acta,* 48, 1513, 1984.

30. **Santschi, P. H., Nyffeler, U. P., O'Hara, P., Buchholtz, M., and Broecker, W. S.,** Radiotracer uptake on the sea floor: results from the MANOP chamber deployments in the eastern Pacific, *Deep Sea Res.,* 31, 451, 1984.

31. **Santschi, P. H., Nyffeler, U. P., Anderson, R. D., Schiff, S. L., and O'Hara, P.,** Response of radioactive trace metals to acid-base titrations in controlled experimental ecosystems: evaluation of transport parameters for application to whole-lake radiotracer experiments, *Can. J. Fish. Aquat. Sci.,* 43, 60, 1986.

32. **Balistrieri, L. S. and Murray, J. W.,** Marine scavenging: trace metal adsorption by interfacial sediment from MANOP site H, *Geochim. Cosmochim. Acta,* 48, 921, 1984.

33. **Crosby, S. A., Glasson, D. R., Cuttler, A. H., Butler, I., Turner, D. R., Whitfield, M., and Millward, G. E.,** Surface area and porosities of Fe(III)- and Fe(II)-derived oxyhydroxides, *Environ. Sci. Technol.,* 17, 709, 1983.

34. **Davis, J. A.,** Adsorption of Trace Metals and Complexing Ligands at the Oxide/Water Interface, Ph.D. thesis, Stanford University, Stanford, CA, 1977.

35. **Balistrieri, L. S. and Murray, J. W.,** Metal-solid interactions in the marine environment: estimating apparent equilibrium binding constants, *Geochim. Cosmochim. Acta,* 47, 1091, 1983.

36. **Buffle, J., Greter, F.-L., and Haerdi, W.,** Measurement of complexation properties of humic and fulvic acids in natural waters with lead and copper ion-selective electrodes, *Anal. Chem.,* 49, 216, 1977.

37. **Shuman, M. S., Bradley, J. C., Fitzgerald, P. J., and Olson, D. L.,** Distribution of stability constants and dissociation rate constants among binding sites on estuarine copper-organic complexes: rotated disk electrode studies and an affinity spectrum analysis of ion-selective electrode and photometric data, in *Aquatic and Terrestrial Humic Materials,* Christman, R. F. and Gjessing, E. T., Eds., Ann Arbor Science Publishers, Ann Arbor, MI, 1983, 349.

38. **Gamble, D. S., Underdown, A. W., and Langford, C. H.,** Copper(II) titration of fulvic acid ligand sites with theoretical, potentiometric, and spectrophotometric analysis, *Anal. Chem.,* 52, 1901, 1980.

39. **Fish, W., Dzombak, D. A., and Morel, F. M. M.,** Metal-humate interactions. II. Application and comparison of models, *Environ. Sci. Technol.,* 20, 676, 1986.

40. **Turner, D. R., Varney, M. S., Whitfield, M., Mantoura, R. F. C., and Riley, J. P.,** Electrochemical studies of copper and lead complexation by fulvic acid. I. Potentiometric measurements and a critical comparison of metal binding models, *Geochim. Cosmochim. Acta,* 50, 289, 1986.

41. **Farrah, H. and Pickering, W. F,** Influence of clay-solute interactions on aqueous heavy metal ion levels, *Water Air Soil Pollut.,* 8, 189, 1977.

42. **Lion, L. W., Altman, R. S., and Leckie, J. O.,** Trace-metal adsorption characteristics of estuarine particulate matter: evaluation of contribution of Fe/Mn oxide and organic surface coatings, *Environ. Sci. Technol.,* 16, 660, 1982.

43. **Tessier, A., Rapin, F., and Carignan, R.,** Trace metals in oxic lake sediments: possible adsorption onto iron oxyhydroxides, *Geochim. Cosmochim. Acta,* 49, 183, 1985.

44. **Davies-Colley, R. J., Nelson, P.O., and Williamson, K. J.,** Copper and cadmium uptake by estuarine sedimentary phases, *Environ. Sci. Technol.,* 18, 491, 1984.

45. **Chao, T. T.,** Use of partial dissolution techniques in geochemical exploration, *J. Geochem. Explor.,* 20, 101, 1984.

46. **Chao, T. T. and Theobald, P. K.,** The significance of secondary iron and manganese oxides in geochemical exploration, *Econ. Geol.,* 71, 1560, 1976.

47. **Engler, R. M., Brannon, J. M., Rose, J., and Bigham, G.,** A practical selective extraction procedure for sediment characterization, in *Chemistry of Marine Sediments,* Yen, T. F., Ed., Ann Arbor Science Publishers, Ann Arbor, MI, 1977, 163.

48. **Tessier, A., Campbell, P. G. C., and Bisson, M.,** Sequential extraction procedure for the speciation of particulate trace metals, *Anal. Chem.,* 51, 844, 1979.

49. **Hoffman, J. S. and Fletcher, W. K.,** Selective sequential extraction of Cu, Zn, Fe, Mn and Mo from soils and sediments, in *Geochemical Exploration 1978,* Watterson, J. R. and Theobald, P. K., Eds., Association of Exploration Geochemists, Rexdale, Ontario, 1979, 289.

50. **Förstner, U.,** Accumulative phases for heavy metals in limnic sediments, *Hydrobiologia,* 91, 269, 1982.

51. **Robbins, J. M., Lyle, M., and Heath, G. R.,** A Sequential Extraction Procedure for Partitioning Elements Among Co-existing Phases in Marine Sediments, Rep. 84-3, College of Oceanography, Oregon State University, Corvallis, 1984.

52. **Sigg, L., Stumm, W., and Zinder, G.,** Chemical processes at the particulate-water interface: implications concerning the form of occurrence of solute and adsorbed species, in *Complexation of Trace Metals in Natural Waters,* Kramer, C. J. M. and Duinker, J. C., Eds., Martinus Nijhoff and Junk, The Hague, 1984, 251.

53. **Rapin, F. and Förstner, U.,** Sequential leaching techniques for particulate metal speciation: the selectivity of various extractants, in *Proc. Int. Conf. Heavy Metals Environ.,* CEP Consultants, Edinburgh, 1983, 1074.

54. **Tipping, E., Hetherington, N. B., and Hilton, J.,** Artifacts in use of selective chemical extraction to determine distributions of metals between oxides of manganese and iron, *Anal. Chem.,* 57, 1944, 1985.

55. **Tessier, A., Campbell, P. G. C., and Bisson, M.,** Particulate trace metal speciation in stream sediments and relationships with grain size: implications for geochemical exploration, *J. Geochem. Explor.,* 16, 77, 1982.

56. **Guy, R. D., Chakrabarti, C. L., and McBain, D. C.,** An evaluation of extraction techniques for the fractionation of copper and lead in model sediment systems, *Water Res.,* 12, 21, 1978.

57. **Miguellati, N., Robbe, D., Marchandise, P., and Astruc, M.,** A new chemical extraction procedure in the fractionation of heavy metals in sediments. Interpretation, in *Proc. Int. Conf. Heavy Metals Environ.,* CEP Consultants, Edinburgh, 1983, 1090.

58. **Nirel, P., Thomas, A. J., and Martin, J. M.,** A critical evaluation of sequential extraction techniques, in *Proc. Semin. Speciation Fission Activation Products in the Environment,* Oxford, U.K., 1985.

59. **Rapin, F., Tessier, A., Campbell, P. G. C., and Carignan, R.,** Potential artifacts in the determination of metal partitioning in sediments by a sequential extraction procedure, *Environ. Sci. Technol.,* 20, 836, 1986.

60. **Miller, W. P., Martens, D. C., and Zelazny, L. W.,** Effect of sequence in extraction of trace metals from soils, *Soil Sci. Soc. Am. J.,* 50, 598, 1986.

61. **Rendell, P. S., Batley, G. E., and Cameron, A. J.,** Adsorption as a control of metal concentrations in sediment extracts, *Environ. Sci. Technol.,* 14, 314, 1980.

62. **Roger, B.,** Comparative study of two selective extraction procedures. Readsorption phenomenon during mineralisation, *Environ. Technol. Lett.,* 7, 539, 1986.

63. **Thompson, E. A., Luoma, S. N., Cain, D. J., and Johansson, C.,** *Water Air Soil Pollut.,* 14, 215, 1980.

64. **Bryan, G. W.,** Bioavailability and effects of heavy metals in marine deposits, in *Wastes in the Sea,* Vol. 6, Ketchum, B., Capuzzo, J., Burt, W., Duedall, I., Park, P., and Kester, D., Eds., John Wiley & Sons, New York, 1985, 41.

65. **Harvey, R. W. and Luoma, S. N.,** Effect of adherent bacteria and bacterial extracellular polymers upon assimilation by *Macoma balthica* of sediment-bound Cd, Zn and Ag, *Mar. Ecol. Prog. Ser.,* 22, 281, 1985.

66. **Morel, F. M. M.,** *Principles of Aquatic Chemistry,* John Wiley & Sons, New York, 1984.

67. **Stumm, W. and Morgan, J. J.,** *Aquatic Chemistry. An Introduction Emphasizing Chemical Equilibria in Natural Waters,* 2nd ed., John Wiley & Sons, New York, 1981.

68. **Batley, G. E. and Giles, M. S.,** A solvent displacement technique for the separation of sediment interstitial waters, in *Contaminants and Sediments,* Vol. 2, Baker, R. A., Ed., Ann Arbor Science Publishers, Ann Arbor, MI, 1980, 101.

69. **Abdullah, M. I. and Reusch-Berg, B.,** Metal species in sediments and interstitial water, in *Proc. Int. Conf. Heavy Metals Environ.,* CEP Consultants, Edinburgh, 1981, 669.

70. **NRCC,** Biologically Available Metals in Sediments, Campbell, P. G. C. and Lewis, A. G., Eds., National Research Council of Canada, Associate Committees on Marine Analytical Chemistry/Scientific Criteria for Environmental Quality, NRCC Rep. No. 27694, Ottawa, Ontario, 1988.

71. **Luoma, S. N.,** Bioavailability of trace metals to aquatic organisms — a review, *Sci. Total Environ.,* 28, 1, 1983.

72. **Harvey, R. W. and Luoma, S. N.,** Separation of solute and particulate vectors of heavy metals uptake in controlled suspension feeding experiments with *Macoma balthica, Hydrobiologia,* 121, 97, 1985.

73. **Denny, P.,** Solute movement in submerged angiosperms, *Biol. Rev.,* 55, 65, 1980.

74. **Chiaudani, G.,** Contenuti normali ed accumuli di rane in *Phragmites communis* L come risposta a guelli nei sedimenti de sei laghi italiani, *Mem. Ist. Ital. Idrobiol.,* 25, 81, 1969.

75. **Welsh, R. P. H. and Denny, P.,** The uptake of lead and copper by submerged aquatic macrophytes in two English lakes, *J. Ecol.,* 68, 443, 1980.

76. **Schierup, H. H and Larsen, V. J.,** Macrophyte cycling of zinc, copper, lead and cadmium in the littoral zone of a polluted and non-polluted lake. I. Availability, uptake and translocation of heavy metals in *Phragmites australis* (Cav.) Trin., *Aquat. Bot.,* 11, 197, 1981.

77. **Lee, C. R., Smart, S. M., Sturgis, T. C., Gordon, R. N., and Landin, M. C.,** Prediction of heavy metal uptake by marsh plants based on chemical extraction of heavy metals from dredged material, Dredged Material Research Program, Tech. Rep. D-78-6, Environmental Effects Laboratory, U.S. Army Engineer Waterways Experimental Station, Vicksburg, MS, 1978.

78. **Tessier, A., Campbell, P. G. C., Auclair, J. C., Bisson, M., and Boucher, H.,** Évaluation de l'impact de rejets miniers sur des organismes biologiques, INRS-Eau, Rapport scientifique No. 146, Université du Québec, Ste.-Foy, Québec, 1982.

79. **Folsom, B. L. and Lee, C. R.,** Zinc and cadmium uptake by the freshwater marsh plant *Cyperus esculentus* grown in contaminated sediments under reduced (flooded) and oxidized (upland) disposal conditions, *J. Plant Nutr.,* 3, 233, 1981.

80. **Lee, C. R., Folsom, B. L., and Bates, D. J.,** Prediction of plant uptake of toxic metals using a modified DTPA soil extraction, *Sci. Total Environ.,* 28, 191, 1983.

81. **Lindsay, W. and Norvell, W.,** Development of the DTPA soil test for zinc, iron, manganese and copper, *Soil Sci. Soc. Am. J.,* 42, 421, 1978.

82. **Aulio, K.,** Accumulation of copper in fluvial sediments and yellow water lilies *(Nuphar lutea)* at varying distances from a metal processing plant, *Bull. Environ. Contam. Toxicol.,* 25, 713, 1980.

83. **Campbell, P. G. C., Tessier, A., and Bisson, M.,** Accumulation of copper and zinc in the yellow water lily, *Nuphar variegatum:* relationships to metal partitioning in the adjacent lake sediments, *Can. J. Fish. Aquat. Sci.,* 42, 23, 1985.

84. **Lee, C. R., Folsom, B. L., and Engler, R. M.,** Availability and plant uptake of heavy metals from contaminated dredged material placed in flooded and upland disposal environments, *Environ. Int.,* 7, 65, 1982.

85. **Waldichuk, M.,** Some biological concerns in heavy metals pollution, in *Pollution and Physiology of Marine Organisms,* Vernberg, F. J. and Vernberg, W. B., Eds., Academic Press, New York, 1974, 1.

86. **Phillips, D. J. H.,** The use of biological indicator organisms to monitor trace metal pollution in marine and estuarine environments. A review, *Environ. Pollut.,* 13, 281, 1977.

87. **Neff, J. W., Foster, R. S., and Slowey, J. F.,** Availability of sediment-adsorbed heavy metals to benthos with particular emphasis on deposit-feeding infauna, Dredged Material Research Program, Tech. Rep. D-78-42, Environmental Effects Laboratory, U.S. Army Engineer Waterways Experimental Station, Vicksburg, MS, 1978.

88. **Marquenie, J. M., de Kock, W. Chr., and Dinnden, P. M.,** Bioavailability of heavy metals in sediments, in *Proc. Int. Conf. Heavy Metals Environ.,* CEP Consultants, Edinburgh, U.K., 1983, 944.

89. **Tessier, A., Campbell, P. G. C., Auclair, J. C., and Bisson, M.,** Relationships between the partitioning of trace metals in sediments and their accumulation in the tissues of the freshwater mollusc *Elliptio complanata* in a mining area, *Can. J. Fish. Aquat. Sci.,* 41, 1463, 1984.

88. **Marquenie, J. M., de Kock, W. Chr., and Dinnden, P. M.,** Bioavailability of heavy metals in sediments, in *Proc. Int. Conf. Heavy Metals Environ.,* CEP Consultants, Edinburg, U.K., 1983, 944.

89. **Tessier, A., Campbell, P. G. C., Auclair, J. C., and Bisson, M.,** Relationships between the partitioning of trace metals in sediments and their accumulation in the tissues of the freshwater mollusc *Elliptio complanta* in a mining area, *Can. J. Fish. Aquat. Sci.,* 41, 1463, 1984.

90. **Bindra, K. S. and Hall, K. J.,** Geochemical partitioning of trace metals in sediments and factors affecting bioaccumulation in benthic organisms, Internal Report, Westwater Research Centre, University of British Columbia, Vancouver, 1977.

91. **Hall, K. J. and Bindra, K. S.,** Geochemistry of selected metals in sediments and factors affecting organism concentration, in *Proc. Int. Conf. Management Control Heavy Metals Environ.,* CEP Consultants, Edinburgh, 1979, 337.

92. **Langston, W. J.,** Arsenic in U.K. estuarine sediments and its availability to benthic organisms, *J. Mar. Biol. Assoc. U.K.,* 60, 869, 1980.

93. **Langston, W. J.,** The distribution of mercury in British estuarine sediments and its availability to deposit-feeding bivalves, *J. Mar. Biol. Assoc. U.K.,* 62, 667, 1982.

94. **Luoma, S. N. and Bryan, G. W.,** Factors controlling the availability of sediment-bound lead to the estuarine bivalve *Scrobicularia plana, J. Mar. Biol. Assoc. U.K.,* 58, 793, 1978.

95. **Luoma, S. N. and Bryan, G. W.,** Trace metal bioavailability: modeling chemical and biological interactions of sediment-bound zinc, in *Chemical Modelling in Aqueous Systems* , ACS Symp. Ser. 93, Jenne, E. A., Ed., American Chemical Society, Washington, D.C., 1979, 577.

96. **Luoma, S. N. and Bryan, G. W.,** A statistical study of environmental factors controlling concentrations of heavy metals in the burrowing bivalve *Scrobicularia plana* and the polychaete *Nereis diversicolor, Estuarine Coastal Shelf Sci.,* 15, 95, 1982.

97. **Breteler, R. J., Valiela, I., and Teal, T. M.,** Bioavailability of mercury in several northeastern U.S. *Spartina* ecosystems, *Estuarine Coastal Shelf Sci.,* 12, 155, 1981.

98. **Breteler, R. J. and Saksa, F. I.,** The role of sediment organic matter on sorption-desorption reactions and bioavailability of mercury and cadmium in an intertidal ecosystem, in *Aquatic Toxicology and Hazard Assessment,* 7th Symp., ASTM STP 854, Cardwell, R. D., Purdy, R., and Bahner, R. C., Eds., American Society for Testing and Materials, Philadelphia, 1985, 454.

99. **Bryan, G. W. and Hummerstone, L. G.,** Adaptation of the polychaete *Nereis diversicolor* to estuarine sediments containing high concentrations of heavy metals. I. General observations and adaptation to copper, *J. Mar. Biol. Assoc. U.K.,* 51, 845, 1971.

100. **Prosi, F.,** Bioavailability of heavy metals in different freshwater sediments: uptake in macrobenthos and biomobilization, in *Proc. Int. Conf. Management Control Heavy Metals Environ.,* CEP Consultants, Edinburgh, 1979, 288.

101. **Hesslein, R. H.,** An *in situ* sampler for close interval pore water studies, *Limmnol. Oceanogr.,* 21, 912, 1976.

102. **Mayer, L. M.,** Chemical water sampling in lakes and sediments with dialysis bags, *Limnol. Oceanogr.,* 21, 909,1976.

103. **Carignan, R.,** Interstitial water sampling by dialysis: methodological notes, *Limnol. Oceanogr.,* 29, 667, 1984.

104. **Carignan, R., Rapin, F., and Tessier, A.,** Sediment porewater sampling for metal analysis: a comparison of techniques, *Geochim. Cosmochim. Acta,* 49, 2493, 1985.

105. **Campbell, P. G. C.,** unpublished compilation, INRS-Eau, Université du Québec, 1986.

In Situ Ecotoxicological Studies: Bases and Methodologies

Chapter 8

THE POLLUTION OF THE HYDROSPHERE BY GLOBAL CONTAMINANTS AND ITS EFFECTS ON AQUATIC ECOSYSTEMS

François Ramade

TABLE OF CONTENTS

I. INTRODUCTION

Over 75,000 chemicals are in common use today, and several thousands of new compounds are added to this figure each year.[1,2] Fortunately, few categories among them are used in large amounts or are released unintentionally into the environment to such an extent that they really threaten the aquatic ecosystems on a wide scale or, worse, on a global one.

Presently, freshwater and brackish habitats as well as marine ecosystems are currently polluted by persistent organic compounds (organochlorine pesticides and polychlorinated biphenyls (PCBs) for example), oil slicks or chronic hydrocarbon release, and, to a smaller extent, by heavy metals and radioactive wastes. Furthermore, some atmospheric contaminants through the so-called acid rain phenomenon are impinging more and more severely on freshwater ecosystems of the whole Northern Hemisphere and even on some industrialized areas south of the equator.[3] This phenomenon of acid precipitation has aroused deep concerns during the last decade for it has been spreading without disruption since the end of the 1950s.

As a consequence, not only the direct impact of this chemical pollution on the hydrosphere, but also its effect on the physical nature of the aquatic environment has to be assessed. Ultimately, even nontoxic contaminants can indirectly exert their ecotoxicological effects on aquatic ecosystems on a rather wide scale. The most documented example is the one of continental and coastal water dystrophication by organic matter from domestic and industrial origin discharged by sewage pipes.

The present extent of contamination by these various pollutants is a matter of serious concern. For example, over 3 million tons of dichlorodiphenyltrichloroethane (DDT) have been manufactured since its discovery in 1939, of which over 1 million tons had already been transferred to the oceanic ecosystem at the beginning of the of the last decade.[4]

Though figures for PCBs are known with less accuracy, it may be estimated that more than 1 million tons of these compounds are still in use all around the world. As a consequence of their widespread use in electrical devices such as plasticizers and additives of various chemical products, PCBs can presently be detected in the most remote area of the biosphere,[5] i.e., in the benthic fishes Trematomidae living in the deep waters of the antarctic continental shelf.[6]

Among the other major aquatic contaminants must be listed the oil spills and other causes of hydrocarbon release, of which 4.5 million tons are expected to be discharged annually into continental waters and ultimately into the oceans of the world.

If we turn now to atmospheric pollution, it has been estimated that 200 million tons/year of sulfur dioxide was released into the atmosphere at the beginning of the present decade,[8] a significant part of it appearing as the SO_4 anion in continental water bodies through the water cycle, causing their acidification.

II. DISPERSION AND CIRCULATION MECHANISMS OF CONTAMINANTS IN THE BIOSPHERE

The pollutants mentioned above and some others may be listed among the most serious types of contaminants that threaten aquatic ecosystems, for they are widespread, released in sufficient quantities and over a wide-enough area that significant pollution of large ecosystems and even of the whole biosphere, could result in the long run. Moreover, even point spills of sufficient duration and amount can be the origin of widespread environmental contamination.

The emission of pollutants into the environment is a complex phenomenon and cannot be limited to the deceptively fixed image of a waste pipe spilling out its effluents into a lake. In almost all cases, substances discharged into the environment are going to be carried a very long way from their source. Atmospheric and hydrological circulation systems will then disperse them progressively throughout the biosphere.

A. Atmospheric Transport of Pollutants

Many pollutants of aquatic ecosystems are not directly discharged into them, but are primarily from terrestrial origin or behave as secondary contaminants. As a consequence, atmospheric transport plays a fundamental role in the dispersion of pollutants and their distribution into different aquatic biotopes. Any organic or mineral compound, even if it is solid, can theoretically be carried by the air. For gases, entry into the atmosphere is direct; for liquids with a weak vapor pressure, it is in the form of aerosols; and for nonsublimable solids, it takes the form of fine particles. Even low-vapor-pressure compounds, such as various organic pesticides, can dissipate into the atmosphere through codistillation with water vapor.[9]

Some of the major contaminants released by man into the atmosphere are natural constituents of it. Sulfur dioxide, nitrogen dioxide, carbon dioxide, and even mercury add to the normal quantities present in the atmosphere which come from various biogeochemical processes and natural phenomena such as volcanism.

It has been possible to calculate the average retention time for microscopic nonsedimentable particles as being 1 week at 3000 m, 2 months at the tropopause level, 1 year in the lower stratosphere, and 2.5 years at an altitude of 340 km.[10]

B. Transfer of Pollutants from the Atmosphere into Water

Apart from a few rare exceptions, atmospheric pollutants do not remain in the air ad infinitum, but precipitation introduces them into the hydrosphere or indirectly via the runoff and the leaching of soils. Solid particles are carried mechanically or by dissolving; gaseous substances are dissolved in rainwater. Snowfalls which filter through the atmosphere are also a very efficient process for extracting contaminants from air and bringing them into continental water bodies and the sea.[11] The pollutants then circulate on the continental surface, trickling into the soils and contaminating groundwater. Processes of leaching and water erosion play an essential part in the transfer of pollutants to the hydrosphere. Finally, geochemical phenomena will eventually introduce the majority of man-made pollutants into the oceans of the world, which are the ultimate receptacles for toxic agents and other pollutants produced by technological civilization.

The fundamental role of the water cycle in the transfer of pollutants was proved by studies made of radioactive fallout following the A- and H-bomb tests in the 1950s.[12]

Various analytical studies have confirmed that the combined processes of atmospheric circulation and precipitation play a major role in the contamination of aquatic habitats and could transport pollutants a very long way from their point of emission. For example, in the most remote rural areas of Great Britain, rainfall averages 14 ppt of PCBs, which is a rather high concentration.[13]

Far worse, at the end of the 1960s it was even shown that the snow falling in the central region of the Antarctic continent is polluted by DDT and other organochlorine compounds![14] In areas of the world where the ocean is very far from industrialized countries, marine mammals are significantly contaminated by PCBs, DDT, and hexachlorocyclohexanes (HCHs).[15] Analysis of the pH of rainwater indicates that the pH level has dropped seriously over Europe (Figure 1) as a result of the use of fossil fuels rich in sulfur.[16-18] At the end of the 1970s, it was shown that the entire land mass of the Northern Hemisphere, as far north as 10° latitude was exposed to acid precipitation.[19]

Similarly, rainfall can play a major role in the contamination of oceanic waters by pollutants. Evidence of the role of atmospheric transport in the pollution of the continents and oceans by halogenated hydrocarbons was provided by Ware and Addison,[20] These scientists demonstrated a positive correlation between the amount of PCBs contained in marine phytoplankton from the Gulf of St. Lawrence and the average amount of precipitation in the same zone during the day preceding the sampling.

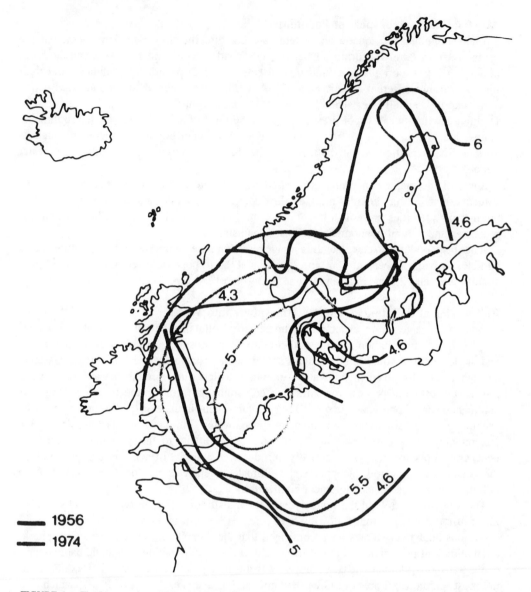

FIGURE 1.   The increase in acidity of precipitation over Western Europe from 1956 to 1974. (Modified from Reference 17.)

More recently, after the Chernobyl accident, a clear correlation was found in Western Europe, at the local level, between the amount of rainfall during the days following the failure of this nuclear reactor and the degree of contamination by radioactive fallout.[21]

The runoff and leaching of the soil, the circulation of surface waters, and, subsequently, the discharge of rivers into the sea play a major role in the contamination of the hydrosphere.

Impressive evidence of this role is given by comparison of the average levels of concentration of various organohalogen compounds (Figure 2) in estuarine and coastal waters, on one hand, and of seawater, on the other hand.[22]

C. Uptake by Biota

The contamination of continental and oceanic waters by polluting agents will sooner or later be transferred to living beings. Chemical contaminants, either mineral or organic, will

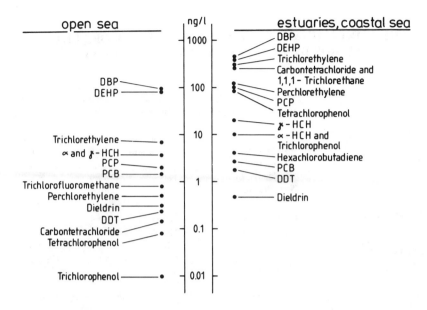

FIGURE 2.   Concentration of various organochlorine pollutants in water of the open sea and of estuaries and coastal areas. (From Ernst, W., *Helgol. Wiss. Meeresunters,* 33, 302, 1980. With permission.)

be absorbed or ingested by aquatic organisms following two major pathways: a direct one, through the cells walls (in the case of phytoplankton), the integument, and the branchial circulation; and an indirect one, through the alimentation and trophic chains.

The persistence of nonbiodegradable polluting agents in the ecosystem is bound to help their passage into aquatic plants and then into animal communities; that is, into all the trophic systems of each biocenosis. The systematic study of contaminated freshwater and marine animals with either carnivorous or ichthyophagous diets has revealed for a score of years the extent of the pollution of the hydrosphere by nonbiodegradable substances, especially organohalogen compounds such as DDT, PCBs, dioxins, and heavy metals.

An analysis of various pelagic birds of the order Procellariiforms (petrels and shearwaters) which live in the most remote areas of the oceans of the world[23] showed the extent of the contamination of the biosphere by these substances (Table 1). Many species of birds belonging to this group are threatened with extinction as their exposure to such strong doses of organohalogen compounds have partially or totally sterilized them.[24]

The extent of the contamination of aquatic ecosystems by various pollutants lies in their great stability and consequently in their weak biodegradability. For example, the theoretical half-life of DDT in pure water inferred from its rate constant is 81 years;[25] the actual one of dieldrin in natural waters is more than 20 years.

1. Bioconcentration by Living Organisms

Most living things can absorb the pollutants contained in the environment and concentrate them in their organism. This phenomenon has been known for a long time due to the existence of species which are capable of accumulating natural substances in concentrations many tens of thousands times more than those usually found in water. For instance, the ability of algae of the genus *Fucus* or *Laminaria* to concentrate bromine or iodine from seawater is used in the industrial extraction of these elements.

The same phenomenon of concentration is seen with various pollutants released into the aquatic environment by man. For example, the plutonium dumped into the ocean in diluted waste products from nuclear reprocessing plants can be concentrated by phytoplankton at

Table 1

THE CONTAMINATION OF VARIOUS SPECIES OF PELAGIC SEABIRDS BY DDT, ITS METABOLITES, AND PCBs

Species	Place of capture (place of reproduction)	Tissues analyzed	DDT and metabolites (ppm)	PCBs (ppm)
Fulmarus glacialis	California (Alaska)	Whole bird	7.1	2.3
Puffinus creatopus	Mexico (Chile)	Whole bird	3.0	0.4
P. griseus	California (New Zealand)	Fats	11.3	1.1
		Fats	40.9	52.6
P. gravis	New Brunswick (Southern Atlantic)	Fats	70.9	104.3
Pterodroma cahow	Bermudas (same)	Whole bird	6.4	—
Oceanodroma leuchor-hoea (Leach's Petrel)	California (same)	Fats ex ovo	953!	351!
Oceanites oceanicus (Wilson petrel)	New Brunswick (Antarctica)	Fats	199!	697!

Modified from Rizebrough, R. W., in *Impingement of Man on the Ocean,* Hood, D. W., Ed., John Wiley & Sons, New York, 1971, 262.

levels 3000 times greater than that of plutonium in the seawater, and in concentrations of up to 1200 times greater by benthic algae.[26]

Ecotoxicological studies have also shown that in each aquatic community there exist true bioconcentrators capable of literally sucking up minute traces of contaminants present in water and accumulating them in their organism. Phytoplankton has an astonishing capacity for accumulating various mineral and organic pollutants. Although PCBs rarely exceed a concentration of 0.1 ppb in the surface waters of the North Atlantic, they occur at a concentration of 200 ppb in the phytoplankton collected from those waters; up to 3050 ppb have been found in phytoplankton samples off the Gulf of St. Lawrence corresponding to a concentration factor (CF) exceeding 30,000.

At the higher levels of aquatic food chains, certain aquatic animals are also capable of a surprising accumulation of pollutants. Invertebrates with a microphagous diet, especially bivalve mollusks, can achieve very high CFs, reaching 690,000 for both DDT and PCBs.[27] With radioisotopes such as ^{54}Mn, ^{65}Zn, ^{55}Fe, concentration factors over 10,000 are currently observed in these species.[28] Fish can also directly accumulate pollutants present in water. The uptake is paradoxically not so much by ingestion (as long as concentration in the food remains low), but mainly through the skin and also through the gills as a consequence of the rapid circulation of water at the level of these organs in order to ensure sufficient oxygenation of the blood. In this way, an American minnow (*Pimephales promelas*) after a few months in water contaminated with weak traces of endrin can concentrate the insecticide at levels 100,000 times stronger than that in water.

It may be concluded from the extensive body of literature in the field of ecotoxicology that the most important factors in predicting the dynamic of retention of chemicals by living organisms in natural ecosystems are the water insolubility (Figure 3) and its reverse picture — lipid solubility — both of them being reliably predicted by simple measurements such as water-*n*-octanol partition coefficient,[29] molecular size, and degree of ionization (Figure 4).

There is, for example, a good correlation between the CF of organochlorine compounds in marine organisms and the degree of insolubility of these compounds in water.[30]

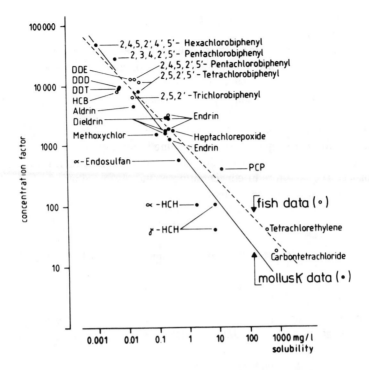

FIGURE 3. Concentration factor, referring to wet weight, of various chlorinated hydrocarbons in mollusks and fish as a function of water solubility. Data from various studies are compared with the water solubility of the applied compound. A clear negative correlation exists between accumulation and water solubility. (From Ernst, W., *Helgol. Wiss. Meeresunters,* 33, 304, 1980. With permission.)

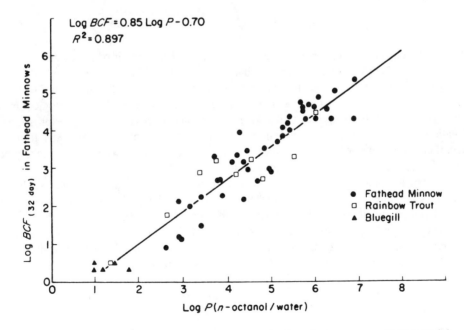

FIGURE 4. Correlation between the biological concentration factor in three species of freshwater fish and the *n*-octanol/water partition coefficient corresponding to various organic chemicals. (From Veith, G. D., Defoe, D. L., and Bergstedt, B. V., *J. Fish. Res. Board Can.,* 36, 1040, 1979. With permission.)

2. The Transfer and Concentration of Pollutants in Aquatic Trophic Chains and Food Webs in Natural Ecosystems

As a consequence of the process of bioaccumulation, every aquatic food chain can be the site of transfer and biological magnification of the pollution within the contaminated biocenose as far as any nonbiodegradable substance is concerned.

Miller[31] proposed a mathematical expression making it possible to quantify the transfer process of pollutants with bioaccumulation in the trophic systems.

Assume that a predator situated at the trophic level n + 1 with a body weight b1 eats a1 grams of the prey from the trophic level n containing a concentration x0 of pollutant. Assume further that the predator absorbs a fraction f1 of the pollutant which it then excretes at a daily rate of k1. The equilibrium concentration of the given pollutant, x1, in the body of the predator will be given by the relation:

$$x1 = Ft(0,1)xo \tag{1}$$

in which Ft(0,1), the transfer factor from the trophic level 0 to 1 has the value:

$$Ft(0,1) = a1 * f1/b1 * k1 \tag{2}$$

If the predator is then himself prey to a higher carnivore from the trophic level n + 2, the equilibrium concentration of the pollutant in its organism will be given by the relation:

$$x2 = Ft(1,2) * x1 \tag{3}$$

where:

$$Ft(1,2) = a2 * f2/b2 * k2 \tag{3A}$$

Finally, we will have the relation:

$$x2 = a2 * f2/b2 * k2 * a1 * f1/b1 * k1 * xo \tag{4}$$

This can be repeated several more times as a function of the length of the chain.

As the coefficient k is always much less than 1 for substances that are slightly or completely biodegradable (k = 0.01 for substances for which the biological half-life is 70 d, for example), then the transfer factor will clearly be higher than 1, and so a biological magnification that can be manyfold will take place at each trophic level. The expression in Equation 4 makes it easy to verify that with an average transfer factor equal to 10, a magnification on the order of 100 times for the concentration of pollutant will take place between an herbivore and a secondary predator.

An extensive body of literature is available in the field of biomagnification studies in natural ecosystems. An historical example is given by the pollution of Clear Lake by tetrachlorodiphenylethane (TDE).[32,33] This California lake was treated repeatedly between 1949 and 1957 with TDE, an insecticide related to DDT, in order to destroy a little midge (*Chaoborus astictopus*), frequent outbreaks of which were annoying bathers in this resort area. Consequently, the DDT accumulated in the limnic food chain:

Water → phytoplankton → zooplankton → microphagous fish →

TL   I   II   III

→ macrophagous fish → fish-eating birds

TL   IV   V

Table 2
THE CONCENTRATION OF TDE IN
THE TROPHIC CHAINS OF CLEAR
LAKE

Element or species	Trophic level	Concentration (ppm)
Water	0	0.014
Phytoplankton	I	59
Plankton-eating fish	II & III	7—9
Predatory fish	III & IV	
Micropterus salmoides		22—25
Ameirus catus		22—221
Fish-eating birds (grebes)	V	2500 (in fats)

Table data taken from various authors in *Ecotoxicology*, Ramade, F., Ed., John Wiley & Sons, Chichester, 1987.

Table 3
CONCENTRATION FACTORS REPORTED FOR
CHLORINATED PESTICIDES IN WATER AND FISH

DDT	Dieldrin	Endrin	Chlordane	Lindane
61,600[a]	5,800[a]	4,100[a]	11,400[a]	550[a]
260,000—4,400,000[b]	63,800[c]	6,200[d]	37,800[e]	180[e]
	12,000[f]		16,900[g]	
1,100,000[h]				

p,p′-DDE	Endosulfan		Heptachlor	Toxaphene
181,000[i]	330[j]		17,400[a]	10,000[k]
			9,500[e]	52,000[l]

[a] Fathead minnow, 28-d exposure.
[b] Lake whitefish, natural population (Great Lakes).
[c] Lake trout, 152-d exposure.
[d] Black bullhead, 32-d exposure.
[e] Fathead minnow, 32-d exposure.
[f] Guppy, 32-d exposure.
[g] Sheepshead minnow, 189-d exposure.
[h] Yellow perch, natural population (Great Lakes).
[i] Rainbow trout, 108-d exposure.
[j] Sheepshead minnow, 28-d exposure.
[k] Brook trout, 140-d exposure.
[l] Fathead minnow, 98-d exposure.

Table data after Reference 34.

Despite the rather low concentration of this insecticide in lake waters (0.014 ppm), the biomagnification reached such an extent that the amount of TDE sampled from the fats of grebes reached 2500 ppm with a CF of 178,500 compared to the waters of the lake (Table 2).

Some organochlorine insecticides, especially DDT, have the greatest CF ever observed in natural food webs (Table 3). A CF of 4,400,000 has been measured in natural populations of lake whitefish (*Coregonus clupeaformis*) from the Great Lakes.[34]

Table 4
CONCENTRATIONS OF ORGANOCHLORINE COMPOUNDS IN PELAGIC ORGANISMS

Organisms[a]	Trophic level	Concentrations (μg/kg dry weight) p,p′DDE	PCBs	Concentration factor for PCBs
Microplankton[a]	1	<0.5	4,500	170,000
Meganyctiphanes norvegica[b]	2	26	620	50,000
Sergestes arcticus[c]	3	15	470	47,000
Pasiphaea sivado[c]	3	5	210	20,000
Myctophus glaciale[d]	3—4	1	50	6,000
Surface water		N.D.	0.0025	

Note: Data were collected within 2 h in November 1974, at depths of 0—100 m in the Mediterranean, 5 km off the coast at Villefranche-sur-Mer, France.

[a] Principally copepods, chaetognaths, phytoplankton, and seston.
[b] Carnivorous euphausiid.
[c] Carnivorous shrimps.
[d] Fish feeding principally on pelagic crustaceans.

From Fowler, S. W. and Elder, D. L., *Bull. Environ Contam. Toxicol.*, 19, 244, 1978. With permission.

In spite of a vast array of experimental evidence suggesting that persistent organic pollutants and heavy metals usually undergo biomagnification processes through the trophic chains, this concept is perhaps too simple an idea for aquatic ecosystems and cannot be considered as an absolute law.

For aquatic habitats, the available experimental evidence suggests that intake from food is unlikely to be the major source of residues for persistent contaminants. As long as concentrations of these pollutants in food remain low, the major source of contamination in aquatic species is the direct intake from water. As a consequence, various examples do exist[35] which show a significant decrease in the concentration of contaminants occurring at the upper levels of trophic webs (see, for example, Table 4).

At the opposite, CFs observed in experimental food chains either for organic compounds or for heavy metals are usually far lower than the highest values of CFs monitored in natural aquatic trophic webs (see, for example, Table 4).

The various examples cited there emphasize the relevance of *in situ* studies on the contamination of aquatic biota. Though experimental models and laboratory microcosms are not only useful, but even invaluable in the field of aquatic ecotoxicology, their use can be misleading as they cannot really reconstruct the complexity of natural ecosystems.

A good demonstration of the necessity of field studies and of their comparison with ecotoxicological data obtained from experimental microcosms has been provided by Cronan and Schofield[36] and Johnson et al.[37] Early studies on the effects of water acidification on freshwater organisms, utilizing simple aquatic bioassays on macroinvertebrates or fishes, showed detectable effects at sublethal levels, and mortality occurred at pH level below 4.5. In the natural freshwater ecosystems, however, the resultant acidification of aquatic habitats by acid deposition produced mortality effects on similar biota at significantly higher pH values in the range 4.5 to 5.5 (Figure 5). Additional research on natural ecosystems revealed the cause. As a consequence of water acidification, aluminum was mobilized from soils in the aquatic environment and adsorbed onto fish or invertebrate gills, leading to mortality at

pH

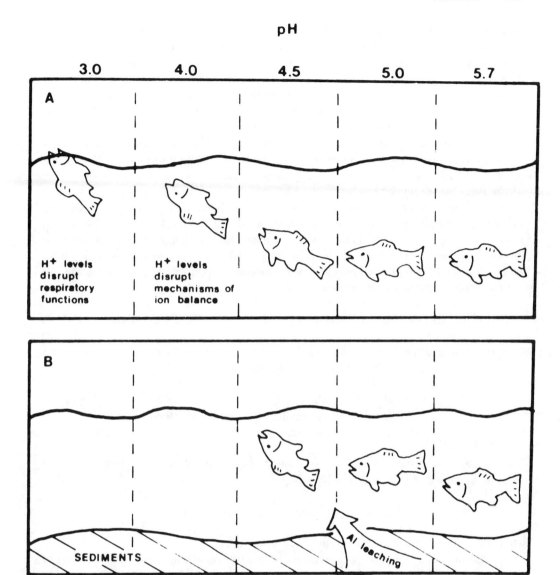

FIGURE 5. Difference in mortality as determined by pH and the availability of aluminum in the environment. Trout succumb at a lower pH under laboratory conditions without sediments. (A) Compared with field conditions; (B) where toxic levels of aluminum can be leached from the watershed soils and lake sediments. (From Levin, S. A. and Kimball, K. D., *Environ. Pollut. Manage.*, 8, 385, 1984. With permission.)

pH levels that were not by themselves directly toxic to the fish. As Levin and Kimball[38] point out, simplified laboratory bioassays failed to predict the indirect terrestrial effects of aluminum toxicity on aquatic biota because they ignored the biogeochemical linkage between the terrestrial and aquatic ecosystems and the resultant complexity of the natural environment.

Similarly, comparison of the effects of methyl parathion applied in outdoor ponds and in laboratory tests showed that various secondary effects occurred that could not be predicted from the latter ones. An increase in population of *Diaptomus* in treated ponds was possibly caused by mortality of predators and competitors. A bloom of filamentous algae which then collapsed, leading to severe depletion of dissolved oxygen and fish deaths, may have been triggered by mortality of herbivorous Ephemeroptera and daphnids. The growth of juvenile rainbow trout in treated ponds was significantly less than in untreated ponds. Moreover, their growth in laboratory aquaria was not affected when rainbow trout were exposed to higher concentrations of methyl parathion that occurred in the outdoor ponds.[39]

In another instance, even more complex laboratory microcosms containing natural sediments failed to be adequate for the prediction of CFs in aquatic food chains leading to CF estimation well below the ones actually occurring in natural aquatic biota. For example, the radioconcentration of ^{104}Ru predicted for the affluents discharged into the sea by the French nuclear reprocessing plant of La Hague proved to be about 100-fold lower than the CF effectively observed in lobster caught in the British Channel.[112] More generally, discrepancies in values of CF for radionuclides reaching up to 1000-fold between real aquatic ecosystems and artificial laboratory food chains have been observed.[40]

The preceding examples emphasize the relevance of *in situ* studies and of environmental monitoring in aquatic ecotoxicology. The reductionist laboratory approach of problems is, of course, useful, but it must be constantly confronted with the results from field studies in order to improve the value of its models. These examples also demonstrate the necessity of an ecosystem perspective in aquatic ecotoxicology as laboratory measurements of effects up on individual do not translate easily into potential effects upon natural populations and, all the more, upon complex communities.

III. THE ECOSYSTEM PERSPECTIVE IN AQUATIC ECOTOXICOLOGY

A misleading concept, still widespread among policymakers and even environmental toxicologists, can be summarized by the relationship:

$$\text{Aquatic contaminant} \rightarrow \text{Individuals of susceptible species} + \text{some ``side effects''}$$

As a matter of fact, such an approach of ecotoxicological problems is very unrealistic and must be discarded, the real relationship being

$$\text{Aquatic contaminant} \rightarrow \text{the whole ecosystem}$$

(In both relationships, the arrow means ''acts upon''.)

As Pimentel et al.[41,42] point out, relatively little information exists about most aspects of the effects of pesticides on whole ecosystems including species diversity, ecosystem stability, nutrient cycling, energy flow, genetics of organisms, and physical resources. Such a conclusion can be easily extended to the deleterious effects of any pollutant occurring in aquatic ecosystems. Studies on the ecotoxicological impact of xenobiotics on aquatic communities have to be initiated at several organizational levels.

First of all, ecotoxicological effects may be divided into two major categories: demoecological and biocenotical.[43]

A. Demoecological Effects

Demoecological effects result from the acute or long-term action of pollutants at population level. These effects will be displayed through immediate or premature death, reduced reproductive success and recruitment, reduced growth, and (or) increased losses at juvenile stages. Ultimately, they are reflected in the lower abundance and perturbated distribution of the exposed populations of sensitive species.

Although immediate and short-term mortality is the most obvious ecotoxicological impact of a given contaminant on populations, other manifestations of toxicity occurring at sublethal concentration can be far more noxious, in the long run, for the exposed species. Moreover,

Table 5
SOME EFFECTS OF POLLUTANTS IMPINGING ON POPULATIONS AT SUBLETHAL DOSAGES

Vital process	Critical effects of pollutants
Development of gametes	Gene damage, abnormal development of gametes, reduced fertility in both male and female
Fertilization	Interference with homing of spermatozoa to ova, impairment hatchability of spermatozoa to fertilize the ova
Embryo development	Cellular abnormalities linked to chromosome alterations, interference with the egg metabolism; induction of failures in organogenesis
Hatching	Failure to hatch, high mortalities of newly hatched larvae or young, teratogenic abnormalities
Growth	Physiological damages inducing growth retardation and deficiencies, behavioral alterations especially related to avoidance and feeding; increased sensitivity to parasite and diseases; impaired metamorphosis
Courting and mating	Destruction of spawning and mating grounds: inappropriate courting and mating behavior leading to reduced mating success

From Sheehan, P. J., in *Effects of Pollutants at the Ecosystem Level,* Sheehan, P. J., Miller, O. R., Butler, G. C., and Bourdeau, P., Eds., John Wiley & Sons, New York, 1984, 40. With permission.

exposure to sublethal concentrations is by far the most frequent ecotoxicological problem in the study of freshwater pollution as a consequence of the general and chronic deterioration of aquatic habitats, especially in industrialized countries.

Table 5 presents a summary of critical sublethal effects leading to population declines in contaminated aquatic ecosystems.[44]

Among the most detrimental altered performance may be growth retardation, impairment of invertebrates' metamorphosis, and reproductive failure.

The importance of reproductive damages occurring even at very low environmental concentrations of pollutants has stimulated a vast array of research. Reproductive failure can be observed during a number of steps: development of gametes, courtship mating, fecundation, embryo development, hatching.

The inability of a population to successfully complete any one step would result in a reduced reproductive fitness of the population. Reproductive performance of plant organisms may also be affected by sublethal concentrations of pollutants. It was shown, e.g., that various pesticides inhibited zygospore germination of microphytic algae.[45,46] Similarly, acid rain has noxious effects on pollen germination of higher plants.[47]

Behavioral effects resulting from exposure to contaminants, even in exceedingly low concentrations, can elicit behavioral reponses for species living in a polluted ecosystem. For example, laboratory and *in situ* studies with sublethal concentrations of an organophosphorus insecticide, Temephos, showed that this insecticide used to control mosquito larvae in brackish waters and estuarine areas reduced significantly the density of fiddler crabs (*Uca pugnax*). *In situ* experiments comparing the survival of crabs in caged plots and in open plots demonstrated that in test plots, caged in order to prevent predation by marsh birds, the Temephos remained nontoxic to the fiddler crabs. It was concluded from laboratory experiments (Figure 6) that these sublethal levels of Temephos impaired the space reactions of fiddler crabs, the population of which undergoes an increase of predation pressure resulting from the exposure to this mosquito larvicide.[48]

Other investigations carried out on vertebrates, especially waterfowl, have provided a vast body of evidence on the occurrence of disturbed behavioral responses resulting from exposure to various environmental pollutants. For example, the avoidance behavior of mallard ducklings was strongly affected by methylmercury exposure,[49] and the same effect was observed

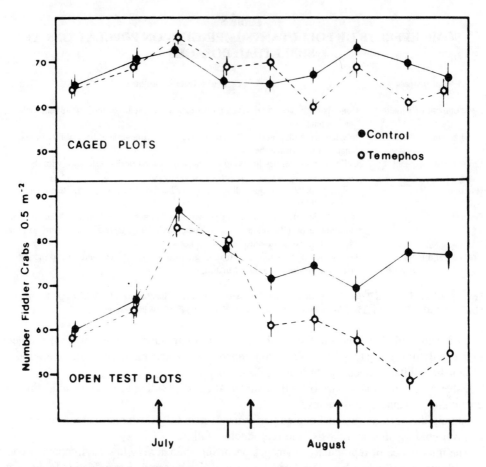

FIGURE 6. Effects of the insecticide Temephos on fiddler crab densities in caged plots to reduce natural predation and open-field plots. Arrows represent the application of Temephos. Symbols are the mean values, and bars are the 95% least significant intervals. (From Ward, D. V., Howes, B. L., and Ludwig, D. F., *Mar. Biol.,* 35, 119, 1976. With permission.)

on young black ducks treated with cadmium[50] or chromium.[51] Although fish safely avoid and escape in the aquatic environment, even at sublethal concentrations of toxicants,[52] this behavior often changes the distribution of fish and, subsequently, affects the aquatic ecosystem and fisheries.[53] Moreover, sublethal dosages of pollutants frequently elicit, in one way or another, changes in the reproductive behavior. The avoidance of contaminated areas during reproduction, although it would selectively protect a given species, is also generally detrimental to the population involved, particularly when it is a migratory species prevented from reaching its spawning or breeding grounds (or its nurseries). It was demonstrated in the early 1960s that water pollution by copper of the Northwest Miramichi River (New Brunswick, Canada) prevented the spawning of Atlantic salmon in the upper part of its watercourse.[54] Identically, it has been observed in the early 1970s that thermal pollution of rivers by the cooling systems of nuclear power plants could prevent migrating fish species from reaching their spawning areas or their nurseries.[55]

B. Effects on Community Structure and Dynamic

Natural communities are dynamic assemblages from hundreds to hundreds of thousands of species which interact with the complex physicochemical components of their ecosystem. The biota specific to a given ecosystem play a major role in fundamental ecological processes and modify their physical and chemical environment. conversely, the latter influences the composition and diversity of community. Therefore, it is of the utmost importance in eco-toxicology when interest is focused at ecosystem level to assess the effect of stresses on community structure and dynamic. Such studies will afford us specific methodologies enabling us to appraise the impact of water pollution on aquatic living resources.[56]

1. Reduction in Population Size and Loss of Species

The most impressive effects of pollutants, obviously, are the total elimination of all species from the impacted area. Several examples of such dramatic pollution have occurred in the past. In 1969, for example, a tank of endosulfan accidentally fell into the Rhine River downstream from Bingen, West Germany. It caused the biggest fish kill ever observed in Western Europe: all fish in the contaminated section of this river — 400 km long — were poisoned.[57]

Even in a case of less dramatic contamination, ample evidence indicates that the introduction of foreign, toxic chemicals into aquatic habitats results in reduction of density and, sometimes, in the complete extinction of populations of sensitive species.

Large mortalities in aquatic biota are frequently observed after pesticide spraying, particularly with broad-spectrum insecticides. Though recent advances in pesticide research have allowed the manufacturing of compounds more selective and less detrimental to non-target organisms, lakes, or running water, pollution through spray drift or runoff still occurs often, everywhere in the world. For example, the widespread use of pyrethroids to eradicate various vectors of diseases, like the tsetse fly in Africa, displayed a drastic impact on aquatic organisms. In Nigeria, aerial applications of Cypermethrin and Decamethrin contaminated the waters of numerous tributaries of the Karami River.[58,59] Acute mortality occurred in many species of aquatic arthropods, notably water beetles, Corixidae water bugs, and crustaceans. Freshwater prawns (*Macrobrachium* sp.) and mayfly larvae disappeared from river benthos samples for up to 1 year, the severe reduction in aquatic arthropods representing a hazard to fish populations through reduced food supply. The decrease in the density of the most susceptible species may be the result of direct mortality when concentrations reach the acute toxic level, as shown in the previous example, but more currently the consequence of long-term exposure that impairs postembryonic growth, reproduction, predator avoidance behavior, and various other demoecological parameters of the contaminated population.[60]

Apart from direct lethal effects, strong mortalities have been observed in aquatic animals following water pollution by xenobiotics. A vast body of literature describes indirect poisoning throughout bioamplification in the food chain. Though less documented, other examples have shown mortalities occurring as a consequence of the disappearance of food species. During the 1960s, extensive studies were conducted in New Brunswick (Canada) watersheds of the Sevogle and Miramichi rivers following aerial forest spraying of DDT to combat the spruce budworm (*Choristoneura fumiferana*). These investigations showed that the majority of aquatic insects were killed (Figure 7), especially larvae of Ephemeroptera, Plecoptera, and Trichoptera.[61] As these species are the prevailing diet of young salmon, the growth of parrs and smolts was impaired, and their density was lowered in polluted creeks and streams,[62] apart from areas where higher DDT concentrations in water exerted direct toxic effects on fish.[63,64]

2. Reduction in Density and Species Richness, the Indicator Species

The reduction in density and species richness of chronically polluted aquatic habitats acts as the primary factor altering community structure. Our own investigations demonstrated that permanent exposure of macroinvertebrate communities to low pesticide concentrations occurring in ponds located among field areas intensively sprayed resulted in a severe decrease

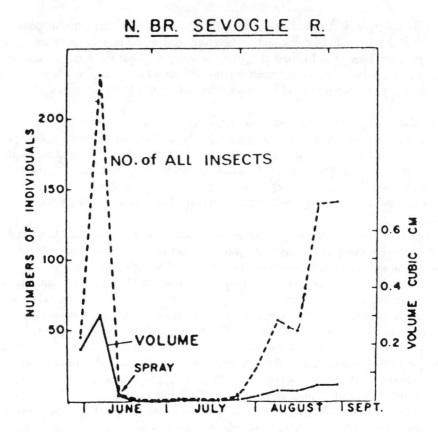

FIGURE 7.   Effects of forest spraying with DDT on densities and biomass (assessed by overall volume) of stream insects from the Sevogle watershed, New Brunswick. (From Ide, F. P., *J. Fish. Res. Board Can.*, 24, 769, 1967. With permission.)

in density, and even in the disappearance, of several families of benthic invertebrates, especially pollutant-sensitive ones.[65,66]

Among fish, a wide body of investigations have shown that pollutant sensitivity increases from cyprinids to percids to salmonids, the last being the first affected when any chronic pollution occurs in a water body.

Studies assessing the impact of acid rain on planktonic communities of oligotrophic lakes demonstrated a sharp decrease of species richness associated with increasing water acidity. A significant drop in the number of species of phytoplankters and zooplankters occurred below pH 6 as a consequence of the acidification of Swedish lakes by acid precipitation.[67] The lakes studied averaged over 40 species of phytoplanktonic algae and 15 species of zooplankton when their pH was above 7, but these numbers dropped to 10 and 5 species, respectively, below pH 4.8 (Figure 8).

As a general rule, stenecious species are significantly more susceptible to any contaminant than euryecious ones.

Initiated with the study of freshwater ecosystems receiving organic sewage, the completion of species list/relative abundance and the use of species or taxonomic groups as ecological indicators of water pollution provide a very useful and invaluable methodology in aquatic ecotoxicology.[56]

The relative susceptibility — or tolerance — of individual species to toxic pollutants forms the fundamental basis for the indicator species concept. Positive bioindicators are the tolerant species, populations of which are found in the contaminated ecosystem. Conversely,

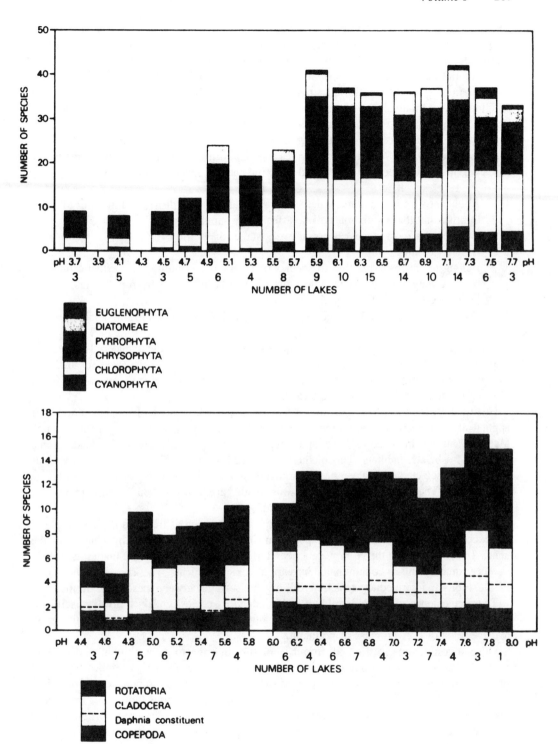

FIGURE 8. Influence of water acidification on the taxonomic richness of phytoplankton and zooplankton from Swedish lakes. (From Almer, B., Dickson, W., Ekstrom, C., et al., *Ambio,* 3, 33, 1974. With permission.)

negative bioindicators are represented by sensitive species found to occur in low abundance, or be absent from polluted areas, or the relative abundance of which in the communities sharply decreases in a gradient of increasing pollution. Several points need to be emphasized regarding the use of indicator species for the assessment of pollution.[68] Detection of changes in populations and communities does not necessarily tell one anything about the effects that pollutants may actually be exerting. These changes may be correlated with natural phenomena linked, for example, to successional processes. It is because of these difficulties that the need for base-line data is frequently urged. Second, many (sometimes conflicting) criteria can affect the choice of bioindicator species for monitoring the level of water contaminants and assessing their effects. As a general rule, indicator organisms should be

1. Abundant, if not dominant, where they occur and widely distributed in order to avoid the risk that the population will be affected by sampling
2. Sedentary at most stages, if not at every stage, of its life cycle
3. Easy to identify
4. Belong as much as feasible to species of economic, scientific, or aesthetic interest, because they are the species that most need to be protected from pollutants and for which we have most ecological information
5. Of a convenient size depending on the projected use of the indicator species

Individuals of small species can be grouped together to give adequate material for residue analysis. Conversely, large animals permit an easy removal of specific tissues for analysis. Moreover, they tend to have stable populations which make it easier to detect significant changes in their population size and structure.

Historically, the use of bioindicator species was the first ecotoxicological criterion ever applied to the detection of any kind of water pollution and was devised by Kolkwitz and Marsson.[69] It was intended to detect and assess the gross level of pollution by dead organic matter in freshwater habitats. This method, so-called saprobies, relied on the use of various bioindicators selected among benthic macroinvertebrates and other components of the aquatic communities. The ranking of these species was decided according to their increasing tolerance to higher degrees in biological oxygen demand.

Among the various invertebrate species, some of them such as mayflies (Ephemeroptera) are exceedingly susceptible to water pollution; others such as aquatic oligochaetes of the genus *Tubifex* can even thrive in waters heavily loaded with dead organic matter as well as in other aquatic habitats contaminated by various toxic chemicals. The isopod *Asellus aquaticus* exhibits an intermediary sensitivity for organic sewage between these two extremes.[70,71]

The convenience of methods relying of bioindicator species renders it always useful for qualitatively detecting a decrease in water quality and even, in some instances, the gross average level of a given class of pollutant.

Other investigations, carried out in our laboratory in order to devise bioindicators useful for the detection of water pollution by pesticides, stem from the comparison of the frequencies of occurrence of several families that proved to be especially good bioindicators of this class of micropollutants.[72] This purpose was achieved by sampling various families of freshwater insects occurring in ponds located among fields heavily sprayed by pesticides where residue analyses had shown the occurrence of a variety of organochlorine compounds in the sediment and an episodic water pollution by some herbicides. The frequency of these families in these ponds was compared with the frequency of their occurrence in samples from control ponds located in natural meadows of the same area (south of Paris) where pesticides had never been used. Several families occurred at a significantly lower frequency in field ponds than in control ponds or were even absent. We studied pollutant-sensitive families occurring in both ponds. As an example, Figure 9 shows the results obtained for two water bug families:

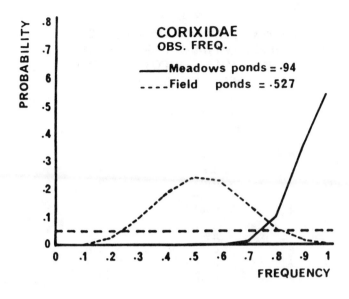

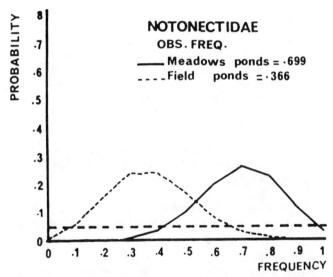

FIGURE 9. Probability of occurrence for the family of Corixidae and Notonectidae in a given number of samples from ponds either polluted with
pesticides (field ponds) or used as control (meadow ponds). (From Ramade,
F., Cosson, R., Echaubard, M., Le Bras, S., Moreteau, J. C., and Thybaud,
E., *Bull. Ecol.*, 15(1), 29, 1984. With permission.)

Corixidae and Notonectidae. For each family the average frequency of occurrence inside
the sample F_i (ratio of the number of samples where the species was present to the total
number of samples taken) was calculated. If we compute the binomial law for ten samples
and for an average frequency corresponding to this frequency F_i, we are able to determine
the probability corresponding to the occurrence of the family inside 1, 2, 3, . . . 10 samples
on 10. Table 6 and Figure 9 illustrate the application of this kind of ecotoxicological criteria
on the Corixidae family, for which we found an average frequency F_i = 0.527 in field
pond samples and F_i = 0.94 in meadow ponds. If F_i is equal or lower than 7, the probability
that waters of this pond are not polluted by pesticides is lower than the 0.05 threshold.
Conversely, if F_i is higher than 8, the probability that waters are not polluted is higher than
the 0.95 threshold.

Table 6
PROBABILITY FOR FINDING 1, 2, ...,
10 TIMES THE OCCURRENCE OF
THE FAMILY CORIXIDAE IN THE
WHOLE SAMPLING

	P_x = probability of F_i = 0.1 X	
X	Polluted ponds	Control ponds
0	0.001	0
1	8.006	0
	0.031	0
3	0.093	0
4	0.182	0
5	0.243	0.001
6	0.225	0.002
7	0.143	0.017
8	0.060	0.099
9	0.015	0.344
10	0.002	0.538

Note: F_i = observed frequency for the family i; F_i = number of samples where i occurs/total number of samples in the sampling; Fi Corixidae from field ponds = 0.0527; from meadow ponds F_i = 0.940; x = number of samples where the family occurs; P_x = probability of occurrence of the family in x samples of 10

From Ramade, F., Cosson, R., Echaubard, M., Le Bras, S., Moreteau, J. C., and Thybaud, E., *Bull. Ecol.,* 15(1), 27, 1984. With permission.

Another use of indicator species is bound to the assessment of pollutant distribution in the environment and of the overall level of contamination of a given area. Such use relies on accumulator species, those species which may be considered positive bioindicators of pollution as they acquire high levels of pollutants.

Among the major advantages for analysis of biotic vs. abiotic samples, the following must be cited:

1. The occurrence of some pollutants, notably heavy metals and organochlorine compounds, in much higher concentrations within aquatic organisms, sometimes by as much as a factor of 10^6, can make chemical analysis much easier.
2. Analysis of organisms measures the pollutant availability, which is more important for biological effects than the measure of the total amount of pollutant in the environment.
3. Organisms integrate the amount of pollutant present during a period of time. Each combination of species and pollutant is unique with its own degree of integration over time.

Whatever the use of bioindicator species in the field area of ecotoxicology, some points need to be stressed. As even species with very similar exposures differ in one way or another in some relevant feature of their ecological characteristics, it may be concluded that there is no universal indicator of environmental contamination. Accordingly, Bryan et al.[73] suggest that one should analyze several species with different types of exposure. Moreover, a variety of species have to be screened in order to check their suitability for the detection of a given class of pollutants. Ultimately, it is not a single species but an assemblage which is more likely to give a significant assessment of the actual state of pollution when effects are monitored.[74]

3. Species Diversity and the Use of Bioecological Indices

The concept of ecological diversity integrates both the number of species (richness) and their relative frequency in a given biota.

Various bioecological indices have been devised which constitute a first attempt at application — at the qualitative level — of the diversity concept in aquatic ecotoxicology.

The biotic index method has been devised by the Trent Water Authority in the U.K. and by Verneaux and Tuffery[75,76] in France during the 1960s in order to provide a convenient method for people committed to monitoring water quality of freshwater bodies and river watersheds.

The basic principle of the method lies in the use of empirical indices describing the quality and the degree of diversity of benthic communities of macroinvertebrates. Because of the difficulty of determination of the various taxonomic classes, the level of taxonomic accuracy required for the index calculation has been arbitrarily set for the species, genus, or family in order to increase the flexibility of the method. Table 7 shows the standard table for calculation of the biotic indices. It can easily be observed that biotic index decreases from top to bottom as the proportion of families of bioindicators of water pollution by organic matter increases.

The use of diversity indices as indicators of water pollution has been proposed by many ecotoxicologists since the turn of the 1950s.[77] Decreased diversity has been used to assess gross environmental degradation in aquatic ecosystems.

Among the most frequently applied to the detection of water pollution, we shall cite the following diversity index: first of all, in the rather uncommon instance where all the individuals of an aquatic community could be numbered, Pielou has proposed the use of the Brillouin's index of which the variant of Margaleff is more known:

$$H = (1/N)Log_2N! - \sum_{i=1}^{S} Log_2ni! \tag{5}$$

(N = total number of individual; S = total number of species; ni = number of individuals of the ith species.)

For practical applications, ecotoxicologists refer to other indices that may be applied to samples, because communities cannot usually be numbered entirely.

Coste[78] has applied the index of Lyod, Zad, and Karr to the study of the pollution of the Seine River in its Parisian area section (Figure 10):

$$H' = C/N\left(NLog_{10}N - \sum_{i=1}^{S} niLog_{10}ni\right) \tag{6}$$

where C is the number of class of frequency expressed in bits (for 10 class, $C = 3.321928$). However, Shannon's index has been by far, up to now, the most widely used in aquatic ecotoxicology:

$$H' = -\sum_{i=1}^{S} ni/NLog_2ni/N \tag{7}$$

Table 7
TABLE STANDARD OF DETERMINATION OF THE BIOTIC INDEX

Taxonomic class	Taxonomic units found in the sample	Subclass according to the number of taxonomic units (T.U.)	Biotic Index (Overall number of taxonomic units found in the samples)				
			(0—1)	(2—5)	(6—10)	(11—15)	(16+)
Plecoptera of Ecdyonuridae	1	More than 1 T.U.	—	7	8	9	10
	2	Only 1 T.U.	5	6	7	8	9
Trichoptera or (caddis-flies)	1	More than 1 T.U.	—	6	7	8	9
	2	Only 1 T.U.	5	5	6	7	9
Ancylidae or Ephemeroptera (except Ecdyonuridae)	1	More than 2 T.U.	—	5	6	7	8
	2	2 or less T.U.	3	4	5	6	7
Aphelocheirus or Odonata or Gammaridae or Mollusca (except Sphaeridae)	0	All the T.U. thereupon absent	3	4	5	6	7
Asellus or Hirudinae or sphaeridae or Hemiptera (except Aphelocheirus)	0	All the T.U. absent	2	1	4	5	/
Tubificidae or Chironomidae of the group thummi and plumosus	0		1	2	3	/	/
Eristalinae	0		1	2	3	/	/

From Verneaux, J. and Tuffery, G., in *La Pollution des Eaux Continentales*, Pesson, P., Ed., Gauthier-Villars, Paris, 1976, 210. With permission.

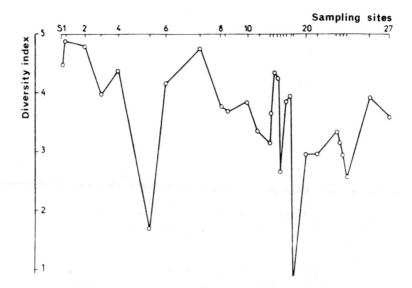

FIGURE 10.   Influence of sewage discharge on the diversity index computed for diatom community from the Seine River, sampled in different locations of its Paris section. (Modified from Reference 78.)

It must, nevertheless, be pointed out that there are limitations hampering the effectiveness of diversity indices in distinguishing stressed and unstressed communities. Shannon's index, for example, gives the same weight to species of the same abundance, whatever their taxonomic affinities. In order to avoid such criticism, Osborne et al.[79] have proposed the use of a hierarchical diversity index devised by a formula expanded from Pielou[80] to include three taxonomic levels (familial, generic, and specific):

$$HDI = H'(F) + H'(G) + H'(S) \tag{8}$$

where $H'(F)$ is the familial component of the total diversity, $H'(G)$ is the generic component of the total diversity, and $H'(S)$ the specific component of the total diversity. Data from five invertebrate communities, known to be affected by limestone strip-mining, were analyzed with this taxonomic-based hierarchical diversity index computed from individual numbers or trophic bases. Both results showed that more information can be obtained using this index.[79]

The most universal criticism of the application of the Shannon diversity measure of aquatic ecotoxicology is the misleading interpretation of data from depauperate communities resulting from the large influence of the evenness component.[81] Another problem in the use of the Shannon index stems from the fact that the community response to a given gradient of increasing pollution is never univocal. During an initial stage, at sublethal exposure, some dominant species among the most pollutant sensitive will decrease their abundance, increasing as a consequence the evenness and, accordingly, the diversity index value quite paradoxically! It is only after the onset of lethal exposures triggering the disappearane of species less and less pollutant sensitive that the index will drop down (Figure 11).[56]

In conclusion, the diversity index reflects changes in water quality only during a period of severe stress. In moderately polluted aquatic ecosystems, changes in dominance strongly affect equitability, thereby hampering the effectiveness of diversity indices in distinguishing degraded communities from unstressed ones.

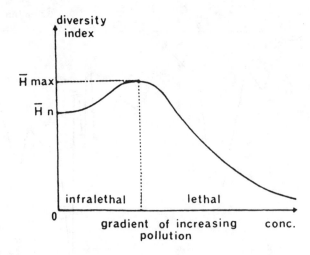

FIGURE 11.   Theoretical variation of the diversity index inside a
gradient of increasing concentration of a given pollutant. (From
Ramade, F., Cosson, R., Echaubard, M., Le Bras, S., Moreteau,
J. C., and Thybaud, E., *Bull. Ecol.*, 15(1), 36, 1984. With permission.)

4. Effects of Pollutants on Frequency Distribution of Species within the Affected Communities
    The means by which species populations are controlled is one of the most important
aspects of its biological niche. In polluted environments, the occurrence of a given contaminant will necessarily impact on the extent of the resource space occupied by each species
in one way or another, according to the level of tolerance or sensitivity of the species. As
a consequence, the balance existing between the various components of the community will
be disturbed as the occurrence of the pollutant will force modification of the interspecific
competition, leading to the vanishing of the most sensitive populations. The frequency
distribution of species in a given community will be more or less distorted by a chronic
pollution of aquatic ecosystems.
    The use of importance value distribution curves afford us, accordingly, an interesting
opportunity from a methodological standpoint to assess the response of a given community.
    Historically, the first attempt at application of the importance value distribution curve was
designed to determine whether the thermal pollution of the Savannah River by an Atomic
Engergy Commission plant had any effect upon the aquatic organism. The survey relied on
the structure and composition of the diatom community between a disturbed and a control
area.[82,83] Since then, the study of frequency distribution of species from this taxonomic
group has proven very useful in assessing the impact of pollution on water quality. For
example, Coste[78] described a strong disturbance in the structure of the diatom community
of the Seine River downstream from Paris resulting from the discharge of sewage and other
polluted effluents into its waters at the level of its Parisian section (Figure 12). This work
has shown that both communities adjust by the Preston frequency distribution curve, but
that it occurs with a decrease in the number of species and an increase in the number of
individuals of the dominant ones in samples from the most polluted section of this river.
    In our own research,[56,66] we have compared the importance value curves computed from
experimental data collected for two lentic communities of benthic macroinvertebrates: one
from ponds contaminated by pesticides and the other from meadow ponds used as control
(Figure 13). The latter fits rather well on the long-normal distribution of Preston whereas
the community from the field ponds fits on an intermediary position between Preston's
distribution and the log-linear one (Motomura). Generally speaking, it may be assumed that

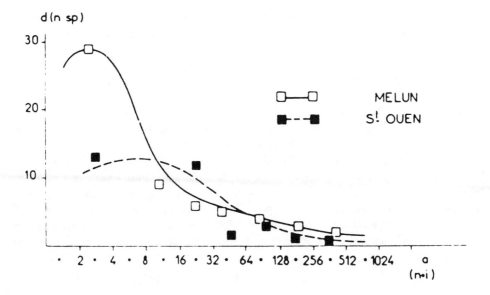

FIGURE 12. Influence of the level of water pollution (according to the sampling site) on the frequency distribution of diatom species in the Seine River. (Modified from Reference 78.)

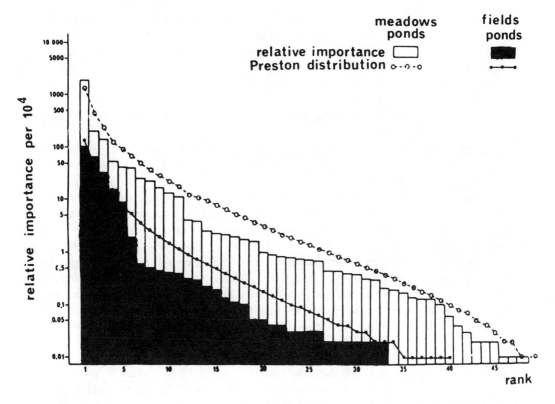

FIGURE 13. Comparison of the importance value curve computed for a community of benthic macroinvertebrates, from ponds chronically polluted by pesticides and from control ponds. (From Ramade, F., Moreteau, J. C., Le Bras, S., and Echaubard, M., *Acta Oecol., Oecol. Appl.*, 6(3), 232, 1985. With permission.)

communities strongly disturbed by any water pollutant exhibit importance value curves which would fit on the log-linear model which is proper for communities living in constraining environments.

5. Biocenotic Methods for the Study of the Pollution Impact on Aquatic Ecosystems

These methods represent another ecotoxicological criterion that is applied more and more to the assessment of the impact of pollution on the quality of continental waters.

The response of a community taken as a whole to a given kind of contaminant affords us an interesting opportunity for such impact assessment.

The most elaborate method providing quantitative ecotoxicological criteria of water pollution at the ecosystem level relies on the use of factorial index. For example, Elouard and Jestin[84] have devised such an index for the assessment of the impact of pesticides on aquatic communities in West African rivers exposed to an organophosphorus insecticide — Temephos — of widespread use for the control of blackflies transmitting ocular onchocerciasis. The researchers' factorial index is computed from the data, collected from samples taken in control rivers or sprayed ones, by the use of the factorial correspondency analysis:

$$INO(j) = \left(\sum_{i=1}^{n} Xi * ai / \sum_{i=1}^{n} Xi \right) c + 5 \qquad (9)$$

where $c = 10/d$, d being the distance from the pollution factorial axis for each species Xi to the location on the same axis of the most pollutant-sensitive one. A weight is given to every taxon according to its position on this axis, expressed between 1 and 10 as a function of its degree of pollutant sensitivity.

6. Succession and Recovery

Few ecotoxicological data exist regarding successional patterns in freshwater ecosystems exposed to a given kind of pollutant.

Historically, the most documented examples are related to the impact of sewage discharge into running water,[71] which describe with much detail the recovery of an aquatic community downstream from the pollution release.

Broadly speaking, the permanent exposure to a toxic pollutant maintains the biota in an early successional stage where only some opportunistic, pollutant-tolerant, and short-lived species of high reproductive potential (r strategist) are able to thrive in this artificially constraining environment.

The oligochaete *Tubifex* exemplifies such a species in temperate freshwater ecosystems. After severe water pollution renders an aquatic habitat depauperate in animal life, this "opportunist" species can colonize an empty space rapidly under conditions of reduced competition. At the opposite, under the condition of chronic pollution, it figures among the few surviving species from the benthic communities.

C. Effects on Ecosystem Functioning

Extreme water pollution exerts a drastic impact on energy flow and element cycles in aquatic ecosystems. At lower levels of pollution complex internal feedback and controls can minimize overall disturbances as various forms of redundancy exist inside the community. For example, pollutant-sensitive species may be replaced by a pollutant-tolerant ecological equivalent without affecting the overall functioning of the ecosystem.

1. Effects on Biomass and Productivity

Water contaminants can act directly on primary production by inhibiting photosynthesis or by hampering the growth of autotrophic organisms (for example, through the reduction of the phytoplankton mitotic rate).

Among the major pollutants impinging on primary producers must be named herbicides and various organochlorine compounds. Investigations carried out during the 1960s had already demonstrated that very low concentrations of pesticides, especially herbicides from the substituted ureas and carbamate groups, strongly inhibited phytoplankton.[85,86]

Organochlorine compounds like DDT have also proven to be detrimental to the growth of freshwater algae.[87] As a general rule, concentrations of PCB between 1 and 10 ppb cause a noticeable decrease in the total biomass as well as cellular size when they are added to cultures of phytoplanktonic populations. The same PCB concentrations also inhibit chlorophyll synthesis from 3 to 4 d without, however, perceptibly changing photosynthetic activity.[88] These effects on biomass have been shown to be a result of inhibiting effects of PCBs on the frequency of cell division.[89]

More recently, other herbicides (for example, atrazine) exhibited a powerful inhibitory action on phytoplankton and other freshwater algae.[90,91] In other experiments, the inhibiting action of the herbicides diuron, Propanil, and Chlorpropham on photosynthesis was observed by Maule and Wright[92,93] in some cyanobacteria (*Anabaena* sp.) and green algae (*Chlamydomonas* sp. and *Anacystis nidulans*).

Studies carried out in our own laboratory have shown that the herbicides used in intensive agriculture, especially Chlortoluron and simazine, impacted heavily on phytoplankton and filamentous algae in adjacent ponds. The concentration in chlorophyll-a in water in these contaminated ponds proved to be far lower than those from control ponds (Figure 14) as a consequence of the strong decrease in algal biomass.[94]

Similar conclusions can be drawn from the study of lakes exposed to acid precipitation where the sharp decrease in phytoplankton diversity was associated with a lowered primary productivity: research carried out in Lake Gardsjon in southern Sweden demonstrated that an extensive impoverishment of the plankton diatom community has taken place concomitant with the acidification during recent decades. The annual accumulation rate of valves of euplanktonic species, mainly *Cyclotella* sp. has decreased to 10 to 40% of the values recorded from the preacidification period, these results indicating a decrease in productivity from the pelagic zone.[95]

Heavy metals can also impinge on primary productivity of phytoplankton and other algae, especially copper which is very toxic to these organisms. For example, the inhibiting effects on the carbon fixation rate of phytoplankter *Tellina tenuis* is already drastic at a copper concentration about 4 ppb higher than controls.[96]

However, in some acidified lakes, the growth of filamentous algae, mosses, and periphyton has been greatly increased so that the gross photosynthetic biomass has not changed very much in spite of the decrease of euplanktonic species and macrophytes.[95,97] Some authors correlate the decrease in planktonic productivity to reduced nutrient supply, whereas the increase of photosynthetic biomass could be the result of the decrease of grazer populations in acidified water bodies. As long ago as the early 1960s, Kiser et al.[98] demonstrated that treatment of fresh water with sodium arsenate, rotenone, and other assorted chemicals not only influenced the target organisms, but also stimulated noxious algae bloom by killing keystone species feeding on phytoplankters in the zooplankton assemblage.

2. Effects on Secondary Productivity

The impact of water pollution on secondary productivity of aquatic ecosystems is a consequence of two distinct processes: the direct long-term effect on invertebrate and vertebrate consumers and the indirect effect which can result from a decrease in primary productivity.

The adverse effects of water acidification, metals and organic contaminants on aquatic

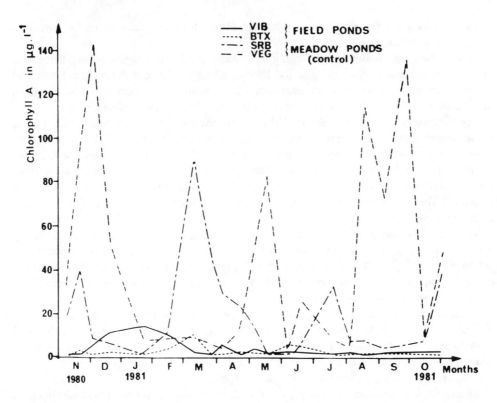

FIGURE 14.   Comparison of chlorophyll content taken as a measure of phytoplanktonic and filamentous algae biomass between meadow ponds (control) and field ponds exposed to an episodic pollution with herbicide sprayed in adjacent fields. (From Goacoulou, J., Influence des Traitements Phytosanitaires sur les Biocenoses Limniques, Ph. D. thesis, Université de Paris-Sud, Orsay, France, 1987 [Also published modified in *Hydrobiologia*, 148, 269, 1987].)

macroinvertebrates encompass all taxonomic groups, though some tolerant species always occur in polluted environment, the trend being always toward a reduced number of species, individuals, and, accordingly, biomass.

In acidified lakes and streams, a sharp decrease in abundance and biomass of zooplankton has been observed.[67,99] There are few species and low abundance of cyclopoid copepods. Few cladocerans are found, but some species may become abundant (e.g., *Bosmina aragoni, Diaphanosoma brachyurum*). Daphnids may be totally missing. One calanoid species usually dominates the grazer community. Thus *Eudiaptomus gracilis* is replaced in Scandinavian lakes by *Diaptomus laciniatus*, whereas in acidified lakes in Northern America it is replaced by *Diaptomus miniatus*.[100,101]

Moreover, as a consequence of fish disappearance because of acidification, some invertebrate predators like *Chaoborus obscuripes* or *Glaenocorixa propinqua* become abundant in lakes. They both consume zooplankton, predominantly cladocerans, which may lead to an increased abundance of copepods. This has actually been observed[102] in acidified water bodies.

In our own investigations of the effects of pesticide use on the functioning of freshwater ecosystems located in field areas intensively sprayed, we have shown a decrease in the secondary productivity of the benthic macroinvertebrate community. This decrease is correlated with the occurrence of insecticide and herbicide residues inside field ponds. Figure 15 represents the comparison of standing biomass and secondary productivity for ponds located in polluted areas and for ponds located in natural meadows, (where no pesticide use ever occurred) used as control.[103]

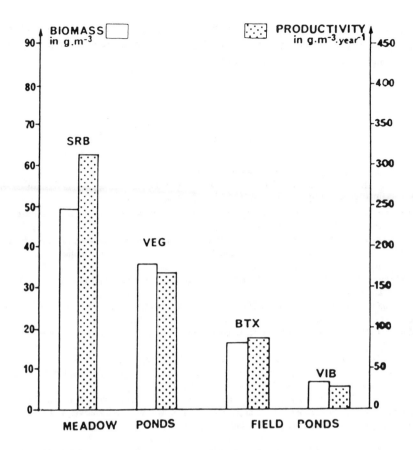

FIGURE 15. Comparison of secondary productivity of the invertebrate community, from ponds exposed to chronic water pollution with pesticides and from control ponds. (After Reference 103.)

3. Decomposition and Element Cycling

Water pollution interferes to a more or less wide extent with the decomposition process and nutrient cycling, both at the producer-consumer level and at the decomposer one.

In the case of acid rain, there is a direct effect on nutrient availability, a reduced phosphorus level, and a decreased nitrification rate in acidified lakes.[104,105] Aquatic ecosystem acidification deeply depresses litter decomposition and nutrient cycling processes as a direct consequence of lowered pH and of the solubilization of various toxic metals (aluminum, mercury, cadmium, and so on) which inhibit microbial decomposition of organic matter.[106]

Heavy metals can obviously interfere drastically with decomposer activity in polluted aquatic ecosystems. As demonstrated for acid rain, excessive metal concentrations disrupt the delicate nitrogen cycle. Low levels of copper, for example, suppress the nitrogen fixation by cyanobacteria *Anabaena* and *Nostoc*.[107]

As previously discussed, various herbicides proved to be highly toxic to the blue green algae *Anabaena* and *Nostoc* and subsequently disrupt the nitrogen cycle.[92,93,108]

Another cause of disturbance in element circulation and cycling throughout an ecosystem is due to the impact of a given pollutant on the trophic web structure.

The role of consumers in regulating ecosystems, especially energy and nutrients, is of the utmost importance. Accordingly, changed predator-prey relations in polluted freshwater ecosystems may be responsible for many of the ecological changes observed and which, up to now, have not been satisfactorily investigated.[41,42,109,110]

Such a topic is of leading importance as the impingement of water contamination on the productivity of freshwater ecosystems, and, therefore, on conservation of living resources, becomes a major issue in aquatic ecotoxicology.

REFERENCES

1. **Cairns, J.**, Estimating hazards, *BioScience,* 30, 101, 1980.
2. **Miller, D. R.**, Chemicals in the environment, in *Effects of Pollutants at Ecosystem Level,* Vol. 22, Sheehan, P. J., Miller, D. R., Butler, G. C., and Bourdeau, P., Eds., John Wiley & Sons, New York, 1984, 15.
3. **McCormick, J.**, *Acid Earth, the Global Threat of Acid Pollution,* International Institute for Environment and Development, London, 1985, 190.
4. **Goldberg, E. D. et al.**, Chlorinated hydrocarbons in the marine environment, in *Man's Impact on the Terrestrial and Oceanic Ecosystem,* Matthews, W. H., Smith, F. E., and Goldberg, E. D., Eds., MIT Press, Cambridge, 1971, 275,
5. **Risebrough, R. W., Reiche, P., Peakall, D. B., et al.**, Polychlorinated biphenyls in the global ecosystem, *Nature (London),* 220, 1098, 1968.
6. **Subramanian, B. R., Tanabe, S., Hidaka, M., and Tatsukawa, R.**, DDT's and PCB isomers and congeners in fish, *Arch. Environ. Contam. Toxicol.,* 12, 621, 1983.
7. **Moore, J. W. and Ramamoorthy, S.**, *Organic Chemicals in Natural Waters,* Springer-Verlag, Berlin, 1984, 289.
8. **Bosang, B.**, Le soufre dans l'atmosphére, *Recherche,* 13, 1132, 1982.
9. **Bowman, M. C., Schechter, M. S., and Carter, R. L.**, Behaviour of chlorinated insecticides in a broad spectrum of soil types, *J. Agric. Food Chem.,* 13, 360, 1965.
10. **Bowen, H. F. M.**, Natural cycles of the elements and their perturbation by man, in *Environment and Man and The Chemical Environment,* Lenihan, J. and Fletcher, W. W., Eds., Blackie, Glasgow, 1977, 1.
11. **Leivestad, H. and Muniz, I. P.**, Fish kill at low pH in Norwegian river, *Nature (London),* 269, 391, 1976.
12. **Fowler, E. B., et al.**, *Radioactive Fallout, soils, plant, food, man,* Elsevier, Amsterdam, 1965, 317.
13. **Wells, D. E. and Johnstone, S. J.**, The occurrence of organochlorine residues in rain water, *Water Air Soil Pollut.,* 9, 271, 1978.
14. **Peterle, T. J.**, DDT in antarctic snow, *Nature (London).,* 224, 620, 1969.
15. **Tanabe, S., Mori, T., Tatsukawa, R., and Miyazaki, N.**, Global pollution of marine mammals by PCBs, DDTs and HCHs, *Chemosphere,* 12, 1269, 1983.
16. **Gorham, E.**, Factors influencing supply of major ions to inland water with special reference to the atmosphere, *Ecol. Soc. Am. Bull.,* 72, 795, 1961.
17. **Oden, S.**, The acidification of air precipitation and its consequences in the natural environment, Ecology Committee Bull. No. 1, Swedish National Science Research Council, Stockholm, 1968.
18. **Oden, S.**, The acidity problem — an outline of concepts, *Water Air Soil Pollut.,* 6, 137, 1976.
19. **Hutchinson, T. C. and Havas, M.**, Effects of acid precipitation on terrestrial ecosystems, in *NATO Conf. Ser., Ecology,* PLenum Press, New York, 1980, 654.
20. **Ware, D. M. and Addison, R. F.**, PCB residues in plankton from the Gulf of St. Lawrence, *Nature (London),* 246, 519, 1973.
21. **Beninson, D. and Lindell, B.**, Chernobyl reactor accident, ICP/CEH 129, Doc. 7246 E, OMS, Regional Office for Europe, Copenhagen, 1986, 36.
22. **Ernst, W.**, Effects of pesticides and related organic compounds in the sea, *Helgol. Wiss. Meeresunters,* 33, 301, 1980.
23. **Rizebrough, R. W.**, Chlorinated hydrocarbons, in *Impingement of Man on the Ocean,* Hood, D. W., Ed., John Wiley & Sons, New York, 1971, 259.
24. **Wurster, C. F. and Wingate, D. B.**, DDT residues and declining reproduction in the Bermuda petrel, in *Man's Impact on Environment,* Dettwyler, T. R., Ed., McGraw-Hill, New York, 1971, 572.
25. **Wolfe, . L., Zepp, D. F., Pavis, G. L., et al.**, Methoxychlor and DDT degradation in water: rates and products, *Environ. Sci. Technol.,* 11, 1977, 1077.
26. **Fraizier, A. and Guary, J. C.**, Recherche d'indicateurs biologiques appropriés au controle de la contamination du littoral par le Plutonium, *Abstr. Int. Symp. Transuranium in the Environment,* San Francisco, 1975, International Atomic Energy Association, Vienna, 1976, 68.
27. **Rizebrough, R. W., De Lappe, B. W., and Schmidt, T. T.**, Bioaccumulation factors of chlorinated hydrocarbons between mussels and sea water, *Mar. Pollut. Bull.,* 7, 225, 1976.

28. **Amiard-Triquet, C. and Amiard, J. C., Eds.,** *Radioécologie des Milieux Aquatiques,* Masson, Paris, 1980, 200.
29. **Veith, G. D., Defoe, D. L., and Bergstedt, B. V.,** Measuring and estimating the bioconcentration factors of chemical in fish, *J. Fish. Res. Board Can.,* 36, 1040, 1979.
30. **Ernst, W.,** Determination of the bioconcentration potential of marine organisms: a steady state approach, *Chemosphere,* 11, 731, 1977.
31. **Miller, D. R.,** Models for total transport, in *Principles of Ecotoxicology,* Butler, G. C., Ed., John Wiley & Sons, New York, 1978, 71.
32. **Hunt, E. G. and Bischoff, A. I.,** Inimical effects on wildlife of periodic DDD application to Clear Lake, *Calif. Fish Game,* 46, 91, 1960.
33. **Craig, R. B. and Rudd, R. L.,** The ecosystem approach to toxic chemicals in the biosphere, in *Survival in Toxic Environments,* Khan, M. and Bederka, J. P., Eds., Academic Press, New York, 1974, 1.
34. **Moore, J. W. and Ramamoorthy, S.,** Chlorinated pesticides, in *Organic Chemicals in Natural Waters,* Moore, J. W. and Ramamoorthy, S., Eds., Springer-Verlag, New York, 1984, 103.
35. **Fowler, S. W. and Elder, D. L.,** PCB and DDT residues in a Mediterranean pelagic food chain, *Bull. Environ. Contam. Toxicol.,* 19, 244, 1978.
36. **Cronan, C. S. and Schofield, C. L.,** Aluminum leaching response to acid precipitations: effects on high level watersheds in the Northeast, *Science,* 204, 304, 1979.
37. **Johnson, N. C., Driscoll, C., Eaton, J., Likens, G., and McDowell, W.,** "Acid rain", dissolved aluminium and chemical weathering at the Hubbard Brook Experimental Forest, New Hampshire, *Geochim. Cosmochim. Acta,* 45, 1421, 1981.
38. **Levin, S. A. and Kimball, K. D.,** New perspectives in ecotoxicology, *Environ. Pollut. Manage.,* 8, 375, 1984.
39. **Crossland, N. O.,** Fate and biological effects of methyl parathion in outdoor ponds and laboratory aquaria. I. Fate. II. Effects, *Ecotoxicol. Environ. Saf.,* 8, 471, 1984.
40. **Patel, B.,** Field and laboratory comparability of radioecological studies, in *Design of Radiotracer Experiments in Marine Biological Systems,* Tech. Rep. Ser. International Atomic Energy Association, Vienna, 1975, 211.
41. **Pimentel, D. and Edwards, C. A.,** Pesticides and ecosystems, *BioScience,* 32, 595, 1982.
42. **Pimentel, D. and Andow, D. A.,** Pest management and pesticide impacts, *Insect Sci. Appl.,* 5, 141, 1983.
43. **Moore, N. W.,** A synopsis of the pesticide problem, in *Advances in Ecological Research,* Vol. 4, Academic Press, New York, 1967, 75.
44. **Sheehan, P. J.,** Effects on individuals and populations, in *Effects of Pollutants at the Ecosystem Level,* Sheehan, P. J., Miller, O. R., Butler, G. C., and Bourdeau, P., Eds., John Wiley & Sons, New York, 1984, 23.
45. **Cain, J. R. and Cain, R. K.,** The effects of selected herbicides on zygospore germination and growth of Chlamydomonas moewusii *(Chlorophyceae, Volvocales), J. Physiol.,* 19, 301, 1983.
46. **Cain, J. R. and Cain, R. K.,** Effects of five insecticides on zygospore germination and growth of the green alga Chlamydomonas moewusii, *Bull. Environ. Contam. Toxicol.,* 33, 571, 1984.
47. **Cox, R. M.,** The response of plant reproductive processes to acidic rain and other pollutants, in *Effects of Acidic Deposition and Air Pollutants on Forest, Wetlands and Agricultural Ecosystems,* Under Press, Toronto, 1985, 24.
48. **Ward, D. V., Howes, B. L., and Ludwig, D. F.,** Interactive effects of predation pressure and insecticide (Temefos) toxicity on populations of the marsh fiddler crab *(Uca pugnax), Mar. Biol.,* 35, 119, 1976.
49. **Heinz, G.,** Effects of low dietary levels of methyl mercury on mallard reproduction, *Bull. Environ. Contam. Toxicol.,* 11(4), 386, 1974.
50. **Heinz, G. H. and Haseltine, S. D.,** Avoidance behaviour of young black ducks treated with chromium, *Toxicol. Lett.,* 8, 307, 1981.
51. **Heinz, G. H., Haseltine, S. D., and Sileo, L.,** Altered avoidance behavior of young black ducks fed cadmium, *Environ. Toxicol. Pharmacol.,* 2, 419, 1983.
52. **Henry, M. G. and Atchison, G. J.,** Behavioral effects of methyl parathion on social groups of bluegill (Lepomis macrochirus), *Environ. Toxicol. Chem.,* 3, 399, 1984.
53. **Hidaka, M. and Tatsukawa, R.,** Avoidance test of a fish, medaka *(Oryzias latipes),* to aquatic contaminants with special reference to chloramine, *Arch. Environ. Contam. Toxicol.,* 14, 565, 1985.
54. **Sprague, J. B., Elson, P. F., and Saunders, R. L.,** Sublethal copper-zinc pollution in a salmon river a field and laboratory study, *Advances in Water Pollution Research,* Pergamon Press, Elmsford, N.Y., 1965, 65.
55. **Merriman, D.,** Does industrial calefaction jeopardize the ecosystem of a long tidal river?, in *Proc. Symp. IAEA, Environmental Aspects of Nuclear Power Stations,* International Atomic Energy Association, Vienna, 261, 507, 1971.
56. **Ramade, F.,** Proposal ecotoxicological criteria for the assessment of the impact of pollution on environmental quality, *Toxicol. Environ. Chem.,* 13, 189, 1987.

57. **Greve, P. A., and Wit, S. L.,** Endosulfan in the Rhine River, *J. Water Pollut. Control. Fed.,* 43, 238, 1971.

58. **Koeman, J. H. et al.,** Three years' observation on side effects of helicopter applications of insecticides used to exterminate Glossina species in Nigeria, *Environ. Pollut.,* 15, 31, 1978.

59. **Smies, M., Evers, R. H. J., Peijnenburg, F. H., and Koeman, J. H.,** Environmental aspects of field trials with pyrethroids to eradicate tse tse fly in Nigeria, *Ecotoxicol. Environ. Saf.,* 4, 114, 1980.

60. **Everts, J. W., Frankenhuyzen, K., Roman, B., and Koeman, J. K.,** Side effects of experimental pyrethroid applications for the control of tse tse files in a riverine forest habitat (Africa), *Arch. Environ. Contam. Toxicol.,* 12, 91, 9183.

61. **Ide, F. P.,** Effects of forest spraying with DDT on aquatic insects of salmon streams in New Brunswick, *J. Fish. Res. Board Can.,* 24, 769, 1967.

62. **Kennleyside, M. H. A.,** Effects of forest spraying with DDT in New Brunswick on food of young Atlantic salmon, *J. Fish. Res. Board Can.,* 24, 807, 1967.

63. **Elson, P. F. and Kerswill, C. J.,** Forest spraying and salmon angling, *Atl. Salmon J.,* 3, 20, 1964.

64. **Elson, P. F.,** Effects on wild young salmon of spraying DDT over New Brunswick forests, *J. Fish. Res. Board Can.,* 24(4), 731, 1967.

65. **Ramade, F., Echaubard, M., Le Bras, S., and Moreteau, J. C.,** Influence des traitements phytosanitaries sur les biocoenoses limniques. I. Comparaison faunistique de mares situées dans des zones de grande culture de la région parisienne, *Acta Oecol., Oecol. Appl.,* 4(1), 3, 1983.

66. **Ramade, F., Moreteau, J. C., Le Bras, S., and Echaubard, M.,** Influence des traitements phytosanitaires sur les biocoenoses limniques. II. Comparaison de la structure des peuplements propres aux biotopes étudiés, *Acta Oecol., Oecol. Appl.* 6(3), 227, 1985.

67. **Almer, B., Dickson, W., Ekstrom, C., et al.,** Effects of acidification on Swedish lakes, *Ambio,* 3, 30, 1974.

68. **Moriarty, F.,** *Ecotoxicology, the Study of Pollutants in Ecosystems,* Academic Press, London, 1983, 233.

69. **Kolkwitz, R. and Marsson, W. A,** Okologie der pflanzlichen saprobien, *Ber. Dtsch. Bot. Ges.,* 26A, 505, 1908.

70. **Gaufin, A. R. and Tarzwell, C. M.,** Aquatic invertebrates as indicators of stream pollution, *Public Health Rep.,* 67(1), 57, 1952.

71. **Hynes, H. B. N.,** *The Biology of Polluted Waters,* Liverpool University Press, Liverpool, 1960, 202.

72. **Ramade, F., Cosson, R., Echaubard, M., Le Bras, S., Moreteau, J. C., and Thybaud, E.,** Détection de la pollution des eaux en milieu agricole, *Bull. Ecol.,* 15(1), 21, 1984.

73. **Bryan, G. W., Langston, W. J., and Hummerstone, L. G.,** The use of biological indicators of heavy metal contamination in estuaries with special reference to an assessment of the biological availability of metals in estuarine sediments from south-west Britain occasion, Publ. No. 1, Marine Biological Association of the U.K., Plymouth, 1980.

74. **Hellawell, J. M.,** *Biological Indicators of Freshwater Pollution and Environmental Management,* Elsevier, London, 1986, 546.

75. **Verneaux, J. and Tuffery, G.,** Une méthode zoologique pratique de détermination de la qualité biologique des eaux courantes. Indices biotiques, *Ann. Sci. Univ. Besancon. Zool.,* 3, 79, 1967.

76. **Verneaux, J.,** Fondements biologiques et écologiques de l'étude de la qualité des eaux continentales, in *La Pollution des Eaux Continentales,* Pesson, P., Ed., Gauthier-Villars, Paris, 229, 1976.

77. **Washington, M. G.,** Diversity, biotic and similarity indices — a review with special reference to aquatic ecosystems, *Water Res.,* 18, 653, 1984.

78. **Coste, M.,** Etude sur la mise au point d'une méthode biologique de détermination de la qualité des eaux en milieu fluvial. Etude effectuée par la division analyse des eaux pêche et pisciculture, Centre Technique du Génie Rural et des Eaux et Forêt, Paris, 1974, 78.

79. **Osborne, L. L., Davies, R. W., and Linton, K. J.,** Use of hierarchical diversity indices in lotic community analysis, *J. Appl. Ecol.,* 17(3), 567, 1980.

80. **Pielou, E. C.,** *Ecological Diversity,* John Wiley & Sons, New York, 1975, 168.

81. **Godfrey, P. J.,** Diversity as a measure of benthic macroinvertebrates community response to water pollution, *Hydrobiologia,* 57, 111, 1978.

82. **Patrick R.,** A proposed biological measure of stream conditions based on a survey of the Conestoga basin, Lancaster County, Pennsylvania, *Proc. Natl. Acad. Sci. U.S.A.,* 101, 277, 1949.

83. **Patrick, R., Hohn, M. K., and Wallace, J. H.,** A new method for determining the pattern of the diatom flora, *Not. Nat. Acad. Nat. Sci. Philadelphia,* 259, 12, 1954.

84. **Elouard, J. M. and Jestin, J. M.,** Impact of temephos (Abate) on the non-target invertebrate fauna. A utilisation of correspondence analysis for studying surveillance data collected in the onchocerciasis control programme, *Rev. Hydrobiol. Trop.,* 15, 23, 1982.

85. **Ukeles, R.,** Growth of pure cultures of marine phytoplancton in the presence of toxicants, *J. Appl. Microbiol.,* 10, 532, 1962.

86. **Wurster, C. F.,** DDT reduces photosynthesis by marine plankton, *Science,* 159, 3822, 1474, 1968.
87. **Goulding, K. H. and Ellis, S. W.** The interaction of DDT with two species of freshwater algae, *Environ. Pollut. Ser. A,* p. 271, 1981.
88. **O'Connors, H. B., Wurster, C. F., Powers, C. D., Riggs, D. C., and Rowland, R. G.,** Polychlorinated biphenyls may alter marine trophic pathways by reducing phytoplancton size and production, *Science,* 201, 737, 1978.
89. **Harding, L. W. and Phillips, J. H.,** Polychlorinated biphenyls (PCB) effects on marine phytoplancton photosynthesis and cell division, *Mar. Biol.,* 49, 93, 1978.
90. **Bohm, H. K.,** Sorption und Wirkung von Chlortriazin, und Phenoxyfettsaüren in statischen und kontinuierlichen Kulturen planktisher blau — Kiesel und Grunalgen, Diss., University of Hohenheim, 1977, 128.
91. **Gunkel, G.,** Untersuchungen zur okotoxikologischen Wirkung eines Herbizids in einen aquatischen Modellokosystem, *Arch. Hydrobiol.,* 65, 235, 1983.
92. **Maule, A. and Wright, S. J. L.,** Physiological effects of chlorpropham and chloroaniline on some cyanobacteria and green algae, *Pestic. Biochem. Physiol.,* 19, 196, 1983.
93. **Maule, A. and Wright, S. J. L.,** Herbicide effects on the population growth of some green algae and cyanobacteria *J. Appl. Bacteriol.,* 57(2), 369, 1984.
94. **Goacoulou, J. and Echaubard, M.,** Influence des traitements phytosanitaires sur les biocenoses limniques. III. Le phytoplancton, *Hydrobiologia,* 148, 269, 1987.
95. **Renberg, I., Kellberg, T., and Nilsson, M.,** Effects of acidification on diatom communities as revealed by analysis of lake sediments, Anderson, F. and Olson, B., Eds., *Ecol. Bull.,* 37, 219, 1985.
96. **Bryan, G. W.,** Pollution due to heavy metals and their compounds, in *Marine Ecology,* Vol. 5, Kine, Ed., John Wiley & Sons, New York, 1984, 1289.
97. **Grahn, O., Hultberg, H., and Laudner, L.,** Oligotrophication — a self accelerating process in lakes subjected to excessive supply of acid substances, *Ambio,* 3, 93, 1974.
98. **Kiser, R. W., Donaldson, J. R., and Olson, P. R.,** The effects of Rotenone on zooplancton populations in freshwater lakes, *Trans. Am. Fish. Soc.,* 92, 17, 1963.
99. **Stenson, J. E. and Oscarson, N. G.,** Crustacean zooplankton in the acidified lake Garsjon system, in *Lake Garsjon, an Acidified Forest-Lake,* Anderson F. and Olson, B., Eds., *Ecol. Bull.,* 37, 224, 1985.
100. **Hanson, M.,** The zooplancton in lakes with low pH in the mountain, Inf. *Inst. Freshwater Res. Drottningholm (Sweden),* 5, 17, 1974.
101. **Sprules, W.,** Midsummer crustacean zooplancton communities in acid stressed lakes, *J. Fish. Res. Board Can.,* 32, 389, 1975.
102. **Nyman, H. G., Oscarson, N. G., and Stenson, J. A.,** Impact of invertebrates predators on the zooplancton composition in acid forest lakes, *Ecol. Bull. (Stockholm),* 37, 239, 1985.
103. **Ramade, F., Le Bras, S., Moreteau, J. C., and Echaubard, M.,** Response of limnetic communities of macroinvertebrates to pesticide use in adjacent cultivated areas, in Abstr. 4th Int. Cong. Ecology, University of Syracuse, Syracuse, N.Y., 1986, 280.
104. **Francis, A. J., Olson, D., and Bernatsky, R.,** Effects of acidity on microbial processes in a forest soil, in *Effects of Pollutants at the Ecosystem Level,* Sheehan, P. J. et al., Eds., John Wiley & Sons, Chichester, 1984.
105. **Persson, G. and Broberg, O.,** Nutrients concentration in the acidified Lake Gardsjon, *Ecol. Bull. (Stockholm), 37,* 158, 1985.
106. **Anderson, G.,** Decomposition of Alder leaves in acid lake waters, in *Lake Gardjson, an Acid Forest Lake and Its Catchment,* Anderson F. and Olson, B., Eds., *Ecol. Bull.,* 37, 293, 1985.
107. **Horne, A. J. and Goldman, C. R.,** Suppression of nitrogen fixation by blue-green algae in a eutrophic lake with trace additions of copper, *Science,* 183, 411, 1974.
108. **Pandey, A. K., Srivastava, V., and Tiwan, D. N.,** Toxicity of the herbicide propanil on *Nostoc calcicola, Z. Allg. Mikrobiol.,* 24(6), 376, 1984.
109. **Hurlbert, S. H., Mulla, M. S., and Wilson, H. R,** Effects of an organophosphorous insecticide on the phytoplancton, zooplancton and insect populations of freshwater ponds, *Ecol. Monogr.,* 42, 269, 1972.
110. **Henrikson, L. and Oscarson, G.,** Waterbugs (Corixidae, Hemiptera) in acidified lakes: habitats selection and adaptations, *Ecol. Bull. (Stockholm),* 37, 232, 1985.
111. **Ramade, F.,** *Ecotoxicology,* John Wiley & Sons, Chichester, 1987, 262.
112. **Amiard, C.,** personal communication.

Chapter 9.1

FATE AND BEHAVIOR OF TRACE METALS IN A SHALLOW EUTROPHIC LAKE

W. Salomons

TABLE OF CONTENTS

I. INTRODUCTION

Large amounts of trace metals are transported by various rivers in the direction of the southern North Sea. The major part of the heavy metal load originates from anthropogenic sources. The Rhine River is the most important fluvial source of trace metals for the North Sea. This influence is reflected in the sediments along the coast, in the Wadden Sea area in the northern part of the Netherlands, and in two freshwater bodies fed by the Rhine River (Figure 1). The high metal load of the Rhine River is not of recent origin. In fact, this river has been polluted with trace metals for more than 60 years.[1] Cadmium and zinc concentrations in sediment samples from the Rhine River in 1922 were 4.4 and 1050 μg/g, respectively. The concentrations of cadmium in 1922 were already 15 times the baseline level of 0.25 μg/g.

The history of the metal pollution of the Rhine river, as reflected in its sediments, is shown in Figure 1. Between 1920 and 1958 all metal concentrations increased, while between 1958 and 1980 this rise continued for cadmium only. Striking are the decreases for arsenic (after 1958) and for mercury (after 1973). Part of the metal load enters the IJsselmeer, the largest freshwater lake in Europe.

II. HYDROLOGY AND SEDIMENTOLOGY

The IJsselmeer is an artificial lake which was created in 1932 when a former lagoon (Zuiderzee) was shut off from the North Sea (see Figure 2). The surface area of the lake has gradually decreased as a consequence of reclamation projects. The present surface area is 1230 km^2, including the Ketelmeer (the mouthing area of the IJssel River). The mean depth of the lake is 4.5 m, the volume of the water body is 5.4×10^9 m^3, and the residence time of the water in the lake is about 6 months. The IJssel River (a distributary of the Rhine River) is the main source of water and is largely responsible for the heavy metal and phosphorus input in the lake. The other source is the atmosphere (Table 1)

Concentrations of suspended mater in the lake vary between 5 and 50 mg/l, with a mean value of about 20 to 30 mg/l. The total suspended matter mass in the lake is 108,000 to 162,000 tons. No data are available on the concentrations during stormy periods. However, experiences from other similar areas indicate that concentrations of several hundred milligrams per liter may be expected, with local values up to 1000 mg/l during severe storms. A mean value of 200 mg/l during stormy periods shows that the total amount of sediment in suspension is on the order of millions of tons, which is several times the annual input by the IJssel River.

The annual sediment transport of the IJssel River (mean sediment concentrations in the water about 40 mg/l) is about 330,000 tons/year. Water from the IJssel River has a residence time of about 1 week in the Ketelmeer. During this period part of the suspended load settles. The concentration of the suspended matter in the water leaving the Ketelmeer is about 15 to 20 mg/l. The decrease is caused by sedimentation in the Ketelmeer, which amounts to about 165,000 tons/year. The transport of the remaining suspended matter entering the IJsselmeer is affected by dispersion by turbulence, (erratic) wind-induced drift currents. Furthermore, the erosion during and sedimentation after stormy periods will cause a mixing of older with more recent sediments.

In considering the sediment balance, it should be taken into account that in the IJsselmeer a large internal production of sediment takes place. The high phosphorus load of the lake causes massive blooms of algae and production of organic matter which partly decomposes, but also partly accumulates in the sediments. Model calculations show that about 335,000 tons of algal organic matter is transported to the sediment in the IJsselmeer. In addition, in the lake a large production of calcium carbonate takes place.[2] From these data a production

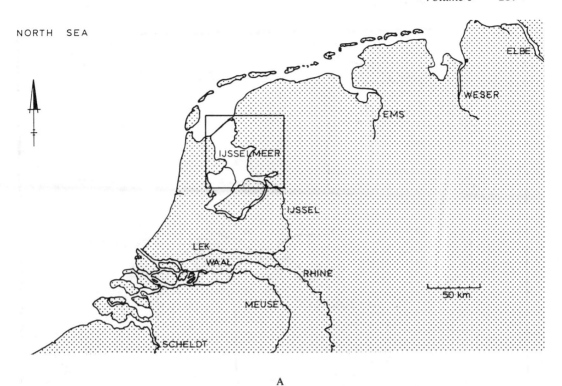

A

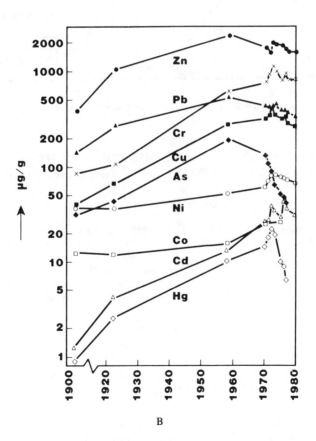

B

FIGURE 1.   (A) Map showing the IJsselmeer; (B) history of metal pollution of the Rhine River as reflected in its sediments in the Netherlands.

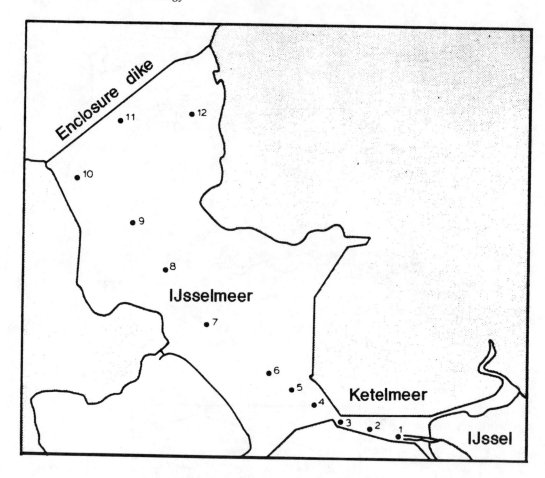

FIGURE 2.   Sampling localities in the IJsselmeer.

Table 1
FLUVIAL AND ATMOSPHERIC INPUT OF SOME
TRACE METALS (TONS/YEAR) IN THE IJSSELMEER

	As	Cd	Cr	Cu	Ni	Zn
Fluvial input	13.5	11.4	112.8	85.2	83.1	688
Atmospheric input	0.45	0.27	0.18	5.0	0.9	52

From Salomons, W., Kerdijk, H., van Driel, W., and Boxma, R., in *Effects of Waste Disposal on Groundwater and Surface Water*, Exeter Symp., International Association of Hydrological Sciences, 1982, 139. With permission.

of 300,000 tons of calcium carbonate per year can be calculated. The internal sediment production exceeds the input of material by the river.

III. METAL CONCENTRATIONS IN PARTICULATES: SEDIMENTS, SUSPENDED MATTER, AND ALGAE

Sediment samples in the IJsselmeer were collected in 1974 and 1977. A total of 12 localities were sampled. At each locality about 10 samples were collected. The mean values for the composition of the sediments sampled in 1977 are presented in Table 2.

Table 2
MEAN VALUES FOR METAL CONCENTRATIONS IN SEDIMENTS
FROM THE IJSSELMEER SAMPLED IN 1977

Locality	%			µg/g						
	% <16 µm	Al	P	Zn	Cu	Cr	Pb	Cd	Ni	Hg
Ketelmeer										
March	26.1	2.63	0.481	1182	173	476	233	24.8	49.4	4.10
	68.5	4.51	0.596	2353	331	637	381	46.3	85.3	7.94
IJsselmeer										
March	20.9	1.84	0.155	573	82	149	95	8.7	30.4	1.58
June	13.1	1.13	0.045	172	19	49	31	2.5	13.4	0.33
	39.0	3.03	0.072	289	23	82	72	2.8	26.0	0.38
	21.1	1.63	0.049	380	32	63	51	1.6	22.6	0.62
	4.7	0.54	0.041	154	10	38	24	1.8	7.8	0.23
	10.1	1.00	0.030	133	12	41	24	1.2	10.6	0.27
	12.8	1.17	0.028	147	16	41	27	1.9	10.8	0.31
	58.7	2.91	0.131	576	47	109	83	3.2	36.1	1.41
	77.6	2.70	0.227	607	58	133	93	4.1	43.0	1.66
	38.5	2.59	0.082	278	39	76	64	1.6	23.3	0.70

Note: Mean values were obtained from about 10 samples from each locality. For the localities see Figure 2.

The data show an extreme variety in metal concentrations. Such a range in metal concentrations is often due to variations in grain size composition. Clay-rich samples with their high surface areas have high metal concentrations, and sandy samples contain low metal concentrations. In order to determine processes affecting metal concentrations in sediments, it is necessary to correct for these differences in grain size composition. Several methods are in use to correct for grain size differences.[3] For this study the correlation between metal concentrations and the percentage of particles with a diameter less than 16 µm was used to correct for grain size differences. In general, linear correlation is observed between these two parameters (Figure 3). A mean regression line was calculated for all samples, and, as an example, the mean values for Cu and Pb from Table 2 together with the regression line are presented in Figure 4. Inspection of this figure shows that the large differences in metal concentrations are mainly due to differences in grain size distribution. Furthermore, scarcely any gradients in metal concentrations are observed in the lake bottom deposits. This shows the intensity of the reworking of the bottom deposits during stormy periods. The only differences which exist are between the Ketelmeer (the mouthing area) and the IJsselmeer. In the Ketelmeer the metal concentrations are much higher and reflect the contemporary pollution of the Rhine River. In the lake, on the other hand, the older deposits are mixed with the more recent material, resulting in lower concentrations. In fact, it has been possible to determine the metal concentrations in sediments from the lake taken about 50 years ago. In Table 3, the metal concentrations in these samples are compared with the baseline for the Netherlands,[4] the sediments from the Ketelmeer, and from the IJsselmeer. Notable is the strong increase of cadmium: an increase from 0.4 to 3 µg/g. The data also show that in 1933 the sampled area was already influenced by the polluted Rhine River, since the concentrations are higher compared with the baseline.

This lack of gradient shows the efficiency of the erosion processes in the lake to level off any regional differences. An extensive mixing of recent suspended matter transported into the lake with older sediments takes place, which results in the low metal concentrations observed in the bottom deposits of the lake. In addition, the internal production of sediment in the lake — organic matter from algae and calcium carbonate — causes a further dilution of the metal concentrations.

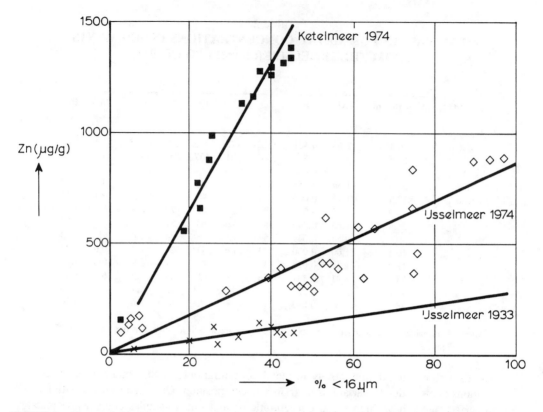

FIGURE 3. The correlation between the percentage <16 μm in sediment samples and the zinc concentrations in the samples. Note the differences between the Ketelmeer and the IJsselmeer and the increase compares with samples taken in 1933.

This intensive mixing of old sediments and new fluvial sediments from the IJssel River is also reflected in the metal concentrations in the suspended matter. Metal concentrations in the IJsselmeer are much lower compared with those in the Ketelmeer (Figure 5).

A strong gradient is observed for aluminum. This element is characteristic for allochthonous mineral particles. This decrease shows the dilution of the riverborne components with autochthonous carbonate and organic matter components. Also, a strong decrease for all other trace metals, with the exception of manganese, is observed. The slight increase in the Ketelmeer (at 5 km) might be due to adsorption processes of metals onto the suspended matter due to pH changes (see Figure 6).

The overall decrease in metal concentrations in the suspended matter cannot be explained by release of metals from the suspended matter, because the dissolved trace metal concentrations also decrease (Table 3). Significant is the decrease in the aluminum content, showing that the inorganic mineral part of the suspended matter decreases. As discussed, a large autochthonous sediment production takes place in the lake, which is higher than the input of suspended matter by the IJssel River. Both carbonates and organic matter contain relatively small amounts of trace metals. Thus, the decrease in metal concentrations can simply be explained by dilution of the suspended matter from the IJssel River with low metal carbonates and organic matter. Metal concentrations in bottom sediments are lower compared with those in the suspended matter. Therefore, the contribution of eroded bottom sediments may further cause a decrease in metal concentrations.

Studies in Swiss lakes[5-7] have shown that in these lakes algae are the most important factor in removing dissolved trace metals from the water column. The determination of metal concentrations in algae is hampered by the relatively large amounts of inorganic

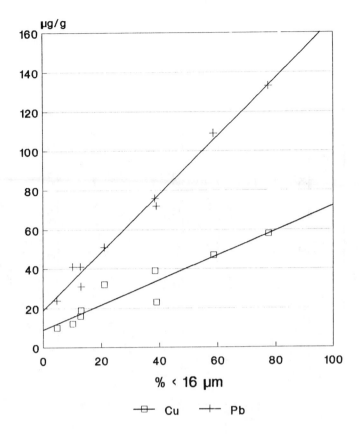

FIGURE 4. The mean concentrations (Table 2) for sediment samples from the IJsselmeer (localities 4 to 12) as a function of the mean percent <16 μm.

material present in the suspended matter, which is fluvial material and eroded bottom sediments. As a result, the algae sampled with a plankton net are invariably contaminated with inorganic suspended particles. These particles with their high metal contents contribute to the metal concentrations in the algae samples. However, it is possible to correct for this admixture. Inorganic suspended particles are characterized by their high aluminum content. Algae samples with high aluminum contents have high metal concentrations, showing the admixture of high metal-containing inorganic particles.[2] The mean composition of trace metals in these samples are presented in Table 4 and compared with other data reported for metal concentrations in algae.

The data are all of the same order of magnitude. It should be noted, however, that the data for the other lakes probably are not corrected for admixtures from inorganic particles, as this is not always clearly stated in the literature.

IV. DISSOLVED METAL CONCENTRATIONS

In the IJsselmeer a continuous change in the pH of the water is observed. Mean values in the river water entering the lake are about 7.4, and in the water leaving the lake, close to 8.5. The pH is subject to seasonal changes, as might be expected from the algal blooms in the lake during the summer period. Mean values for the dissolved trace metals for a number of localities are presented in Table 5.

The data show the removal of a number of trace metals from the surface waters and the increase in pH in the IJsselmeer. A major part of the removal already takes place in the

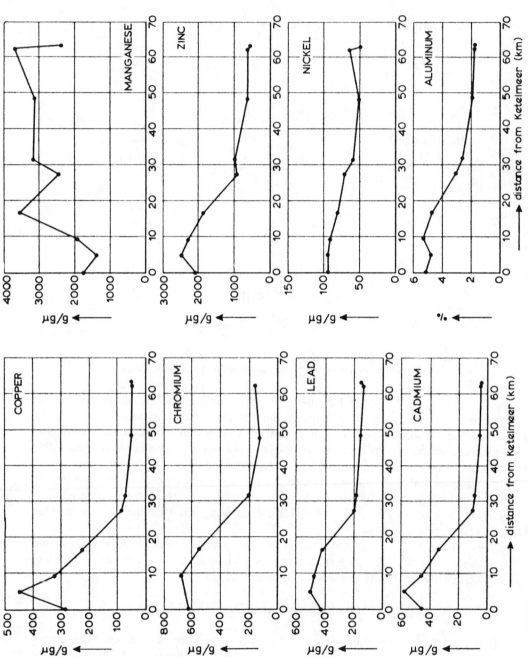

FIGURE 5.   Metal concentrations in the suspended matter (mean of 12 monthly data) as a functon of the distance from the IJssel River.

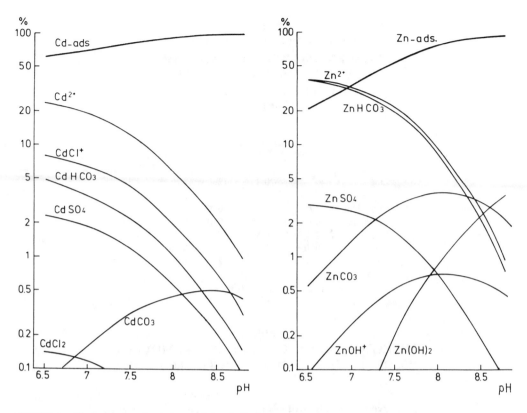

FIGURE 6.   The speciation of cadmium and zinc (including the adsorbed species) in the IJsselmeer. Calculated values using laboratory adsorption data.

Table 3
METAL CONCENTRATIONS (50% <16μm) IN SAMPLES FROM THE KETELMEER AND THE IJSSELMEER (LOCALITIES 4—12)

	%		μg/g						
	P	**Al**	**Ni**	**Zn**	**Cu**	**Cr**	**Pb**	**Cd**	**Hg**
Ketelmeer	0.45	3.64	68.3	1826	253	485	290	34.5	6.0
IJsselmeer	0.16	1.94	29.3	417	39	93	69	3.0	1.1
IJsselmeer 1933			39	133	19	88	39	0.4	
Baseline			29	68	13	72	21	0.25	

Table 4
METAL CONCENTRATIONS IN ALGAE FROM THE IJSSELMEER COMPARED WITH DATA REPORTED FOR OTHER LAKES

Values in μg/g

	Zn	**Cu**	**Cr**	**Pb**	**Cd**	**Ni**	**Al**
IJsselmeer	247 (140)	29 (16)	12 (12)	26 (10)	1.5 (0.8)	25 (8)	0.35 (0.12)
Lake Alpnach[6]	163	38	—	21	5.6		
Bieler See[9]	653	76	—	95	—	—	—
Ploessee[10]	111	60	—	—	—	—	—

Note: Mean values are from 24 samples. Standard deviation is in parentheses.

Table 5
MEAN DISSOLVED METAL
CONCENTRATIONS (μg/l) AND THE pH
AT A NUMBER OF SELECTED
SAMPLING POINTS IN THE
IJSSELMEER

	Locality				
	3 + 4	**6**	**8**	**10—12**	
Zinc	40	20.4	12.4	4.6	4.3
Copper	5.1	2.7	2.5	2.4	1.5
Chromium	1.6	1.0	0.7	0.3	0.2
Cadmium	0.6	0.15	0.10	0.04	0.04
Nickel	7.4	6.8	6.0	6.3	4.5
Arsenic	0.76	0.85	1.1	0.86	0.90
pH	7.4	7.7	8.2	8.4	8.4

Note: Values are average of approximately 12 monthly samples. For the localities see Figure 2.

Table 6
IMPORTANCE OF ALGAE FOR THE REMOVAL OF
DISSOLVED TRACE METALS IN THE IJSSELMEER

	Accumulation dissolved metals	**Removal by algae**	**% Removal by algae**
Zn	469	82	17
Cu	23	10	43
Cr	20	4	20
Cd	4	0.5	13
Ni	12	9	75

Note: Accumulation is measured in tons/year.

southern part of the lake. Very little removal is observed for nickel and arsenic and for copper between localities 3 and 12.

With the metal concentrations in the algae and the net total removal of organic matter by algae, it is possible to compare semiquantitatively the relative importance of organic and inorganic removal processes. The net removal of organic matter by algae to the sediment is about 335,000 tons/year, and, with the metal concentrations in the algae (Table 4), it is possible to calculate the removal of metals (Table 6).

This calculation shows that the removal by algae is important for nickel and for copper, two elements which of all the metals are the least removed from solution. For zinc, cadmium, and chromium, removal by organic processes is not very important. The relative nonimportance of algae for the removal of metals in this lake is caused by the strong pH gradients in the lake. The increase in pH causes a chemical removal through adsorption onto the particulates. In Figure 6 the speciation of cadmium and zinc as a functon of pH including the adsorbed species are shown. This figure, based on adsorption data obtained in the laboratory,[4] shows the strong shift toward the adsorbed species when the pH increases. Very little copper is removed since its adsorpton edge is not in the pH range observed for the IJsselmeer.

The removal of dissolved metals, however, does not show up as increased concentrations

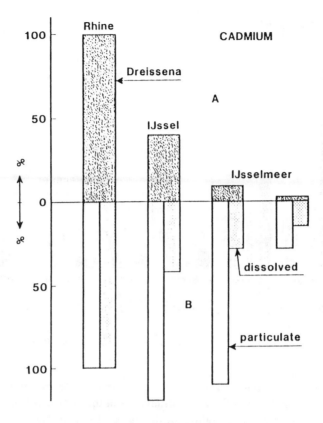

FIGURE 7. Cadmium accumulation by *Dreissena polymorpha*[11] compared with changes in dissolved and particulate metal concentratons in
the Rhine River, its distributary the IJssel River, and in the IJsselmeer.
(From Salomons, W., Kerdijk, H., van Driel, W., and Boxma, R.,
in *Effects of Waste Disposal on Groundwater and Surface Water,* Exeter
Symp., International Association of Hydrological Sciences, 1982, 139.
With permission.)

in the suspended matter. The increase due to adsorption is canceled out by the quantitative,
more important dilution processes (admixtures of low metal-containing bottom sediments
and authigenous phases) described in the former section.

V. INFLUENCE OF THE METAL GRADIENTS IN THE LAKE ON METAL ACCUMULATION BY ORGANISMS

Strong gradients in metal concentrations are observed in the lake. These gradients already
start partly in the river IJssel, due to a small pH increase. The removal of dissolved trace
metals from solution is also reflected in metal uptake by the bivalve *Dreissena polymorpha*
(Figure 7). Between Lobith and Kampen (Rhine River and IJssel River) the pH increases
slightly, affecting the distribution of dissolved cadmium over the dissolved and particulate
phase (e.g., the dissolved metal concentrations decrease, and those in the suspended matter
increase). Although the total cadmium burden of the organism stays constant, the actual
uptake decreases, showing that dissolved trace-metal fractions are more bioavailable than
the particulate forms.

The gradients in metal levels in the lake are also reflected in this bivalve.[8] Strong gradients
are observed for cadmium (which is to a large extent removed from solution in the lake)

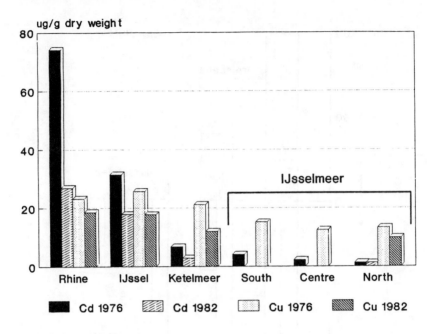

FIGURE 8.   Gradients in copper and cadmium concentrations in the Rhine, IJssel, and IJsselmeer.[11,12]

and smaller gradients for copper. The latter metal is far less removed from solution in the IJsselmeer. Figure 8 also shows the large differences in the cadmium levels between 1976 and 1982. This is probably due to halting of discharge by copper plants along the Rhine.[8]

VI. METAL BUDGET CALCULATIONS FOR THE IJSSELMEER

The data from the monthly surveys of water and suspended matter were used to construct budgets of carbon, calcium, phosphorus, and other metals on a yearly basis.[4] The results (Figure 9) show that the inflow of most dissolved components is higher than the outflow. Only for arsenic is there a net flux from the IJsselmeer to the Wadden Sea area.

The particulate metal and P balances show that, except for Mn, P, and As, there is accumulation in the lake. For P and As, concentrations in the suspended matter leaving the lake are lower than in the suspended matter entering the lake. The net flux from the lake is due to the higher output of suspended matter compared with the input. For manganese both the particulate output and the concentrations in the suspended matter are higher. The particulate and dissolved budgets show that an internal source exists for manganese and arsenic. Man-made inputs, other than the IJssel River, are negligible, so that these can be ruled out.

The internal source which is able to supply manganese and arsenic are the pore waters in the sediments. Average values for the pore waters and for the overlying surface waters are presented in Table 7. The results show that the concentrations of Zn, Cu, Cd, and Ni are on the same order of magnitude as found for the overlying waters. The concentrations of Cr, As, Fe, Mn, and P, on the other hand, are much higher. This difference in concentration results in an upward flux of these metals to the surface waters. The flux was calculated with Fick's law (Table 7).

When the diffusive flux of P, Fe, Mn, and As encounters the more oxic conditions at the sediment/water interface, a partial precipitation will take place. From studies in other lakes, it is known that this will result in an almost complete precipitation for iron. However, the oxidation rate of manganese is much lower, and part may escape to the surface waters. The

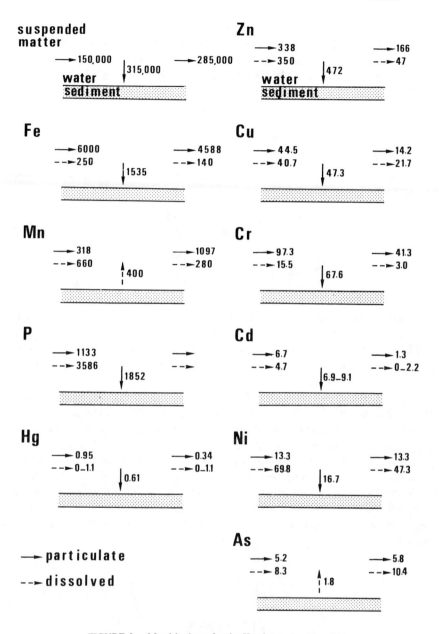

FIGURE 9. Metal budgets for the IJsselmeer (tons/year).

high concentrations of manganese in the suspended matter in the northern part of the lake can be explained by this process. The arsenic apparently stays partly in solution (most likely due to the high pH) and is partly adsorbed by the suspended matter. Although these data are not conclusive since the complicated processes at the sediment/water interface could not be taken into account, they offer a qualitative explanation for the observed phenomena.

VII. SUMMARY OF PROCESSES AFFECTING TRACE METALS IN THE IJSSELMEER

The IJsselmeer acts as a sink for both particulate and dissolved zinc, copper, chromium, nickel, cadmium, and lead. The processes causing their removal from the incoming river

Table 7

DISSOLVED CONCENTRATIONS IN THE PORE WATER OF THE 2-cm TOP LAYER, IN THE OVERLYING WATERS, AND THE FLUX FROM THE SEDIMENT PORE WATERS TO THE OVERLYING WATERS IN THE IJSSELMEER

Metal	Conc 2-cm top layer (mg/m³)	Conc overlying surface water (mg/m³)	Flux F (tons/year)
Zn	10	9.7	—
Cu	1.6	2.4	—
Cr	6.3	0.57	8
Cd	1.2	<0.2	1.4
Ni	5.8	6	—
As	21	0.9	28
Fe	21,000	20	33,000
Mn	9,160	50	13,000
P	5,380	145	7,000

water are settling processes (mainly in the Ketelmeer) of riverborne particulates. Dissolved trace metals (zinc, cadmium, chromium) are removed to a large extent by adsorption processes which occur on mixing of Ketelmeer water with the IJsselmeer in the southern part of the lake. A further removal takes place in the remainder of the lake due to uptake by algae and a further pH increase.

The output of arsenic and manganese from the lake is higher than the input. Determination of metal concentrations in the pore waters show that these two metals are highly enriched in the pore waters. Flux calculations show that the pore waters may be a significant source for the arsenic and manganese in the surface waters. The high pH in the lake causes a reprecipitation/adsorption of the released manganese and arsenic on the suspended matter, which subsequently leaves the lake through the sluices in the enclosure dike. The results show that the IJsselmeer, with regard to processes, is "particle-dominated" and that the pH-increase induced by algal blooms is a major factor in the removal of trace metals. The changes in dissolved metal levels are reflected in metal uptake by the bivalve *Dreissena polymorpha*.

REFERENCES

1. **Salomons, W., Kerdijk, H., van Driel, W., and Boxma, R.,** Help! Holland is plated by the Rhine, in *Effects of Waste Disposal on Groundwater and Surface Water,* Exeter Symposium, International Association of Hydrological Sciences, 1982, 139.
2. **Salomons, W. and Mook, W. G.,** Biogeochemical processes effecting metal concentrations in lake sediments (IJsselmeer, The Netherlands), *Sci. Total Environ.,* 16, 217, 1980.
3. **Förstner, U. and Salomons, W.,** Trace metal analysis on polluted sediments. I. Assessment of sources and intensities, *Environ. Technol. Lett.,* 1, 494, 1980.
4. **Salomons, W.,** Baseline for Dutch Sediments, Delft Hydraulics Rep. No. R1790, 1983.
5. **Baccini, P.,** Untersuchungen über den Schwermetallhaushalt in Seen, *Schweiz. Z. Hydrol.,* 38, 121, 1976.
6. **Baccini, P. and Joller, J. T.,** Transport processes of copper and zinc in a high eutrophic and meromictic lake, *Schweiz. Z. Hydrol.,* 43, 176, 1976.
7. **Sigg, L., Sturm, M., Stumm, W., Mart, L., and Nürnberg, H. W.,** Schwermetalle im Bodensee — Mechanismen der Konzentrationsregulierung, *Naturwissenschaften,* 69, 546, 1982.
8. **Salomons, W. and Förstner, U.,** *Metals in the Hydrocycle,* Springer-Verlag, Heidelberg, 1984.

9. **Santschi, P.,** Dissertation, University of Bern, Switzerland, 1975.
10. **Groth, M.,** Untersuchungen über einigen Spurenelementen in Seen, *Arch. Hydrobiol.,* 68, 305, 1971.
11. **Marquenie, J. M.,** The freshwater mollusk *Dreissena polymorpha* as a potential tool for assessing bio-availability of heavy metals in aqueous systems, in *Heavy Metals in the Environment,* CEP Consultants, Edinburgh, 1981, 409.
12. **Hueck-van der Plas, E. H.,** Summary report: heavy metals in aquatic systems, Stud. Inf. TNO, Delft, 1984, 42.

Chapter 9.2

MERCURY IN THE OTTAWA RIVER (CANADA)

A. Kudo

TABLE OF CONTENTS

I. INTRODUCTION

The largest industry in Canada is the pulp and paper industry. The amount of production in recent years has been valued at some 11 billion Canadian dollars, which accounts for 3 to 3.5% of the gross national product. The industry is extremely export oriented and earns about 9 billion Canadian dollars a year, which represents about 10% of the total Canadian exports.

Along the Ottawa River there are several pulp and paper factories using trees grown in the vast forest of the drainage area. This is natural as the river itself can be used as a transportation device (log-driving) to the mills downstream. The river also supplies fresh water to these factories and receives wastewater from the mills. In fact, the pulp and paper industry is one of the largest water consumers. Every ton of paper produced requires between 200 and 600 tons of fresh water. It is, therefore, hardly surprising that most pulp and paper factories are located along rivers, on lakes, and along the seashore.

Mercury has been used in the industry for many decades. One application has been as elemental mercury in anodes for the production of sodium hydroxide (NaOH) and chlorine gas (Cl_2), which are essential chemicals for the pulp and paper industry. Another use for mercury in the pulp and paper mills is as slimicide (mostly phenylmercuricacetate) which prevents bacterial growth in the production processes. Of course, significant amounts of these mercury compounds have been either spilled or discharged during the process and were eventually released to the river along with organic waste from the mills.

In the 1970s, fish living in the Ottawa River frequently contained more than the legal limit of 0.5 ppm of mercury, and the bottom sediments of various parts in the Ottawa River were contaminated with mercury at levels well over 1 ppm. However, there was (and still is) a considerable lack of understanding concerning the behavior of mercury in the aquatic environment.

In order to acquire a better understanding of the distribution, transport, and transformation of mercury, a major research project was created in 1972 at the National Research Council of Canada. The aim of the project was to determine methods to control mercury in the environment, safely and economically. The project was a multidisciplinary investigation in collaboration with the University of Ottawa. An investigation period of 5 years and funding of $2 million (including salaries) were allocated to the project, including 12 principal scientists, engineers, and professors.

A section of the Ottawa River was selected for a detailed study as a model for the natural aquatic environment.

This chapter describes the behavior of mercury, its distribution, transport, and transformation in a section of the Ottawa River. After 5 years of intensive field and laboratory study, the overall dynamics of mercury in the river were understood with quantitative data. This understanding of mercury behavior in the aquatic environment will help to produce the economic control methods for other pollutants.*

II. OTTAWA RIVER AND STUDY SECTION

The Ottawa River is approximately 1113 km long, and the average annual flow rate is 1700 m^3/s (about the same size as the Rhine River in Europe). The river originates in the drainage area of Lake Timiskaming (in northeastern Ontario) and eventually joins the St. Lawrence River at Montreal (Figure 1). Its drainage area is 148,000 km^2 (57,000 m^2), and the total precipitation in the area averages 864 mm/year. The snowfall contributes about 25% (or 212 cm/year) of the total precipitation. Therefore, the entire drainage area provides

---

* This chapter summarizes results previously published elsewhere.

FIGURE 1. Location of Ottawa River.

(in theory) a mean flow rate of 4000 m³/s. This means that 42.5% of the precipitation eventually leaves the region via the Ottawa River, and, consequently, the surface runoff is fairly high. Formerly, the ratio of maximum to minimum recorded flow rate was as high as 12:1, but the construction of various dams in recent years has reduced this ratio significantly (at present, 4:1).

The river has been used extensively to transport timber (log-driving) from upstream forests to pulp and paper mills. Furthermore, it has been receiving wastes in the form of wood fragments (fiber, chips, and bark) from adjacent pulp and paper industries for nearly 100 years. Extensive organic deposits are found in the river (up to 4 m in depth), and, through the normal transport processes, these materials have become an integral part of the total sediment complex. This feature (organic-rich sediments) is an important characteristic of the Ottawa River sediments.

Though the river has been used for recreation and water supply, it has never supported an important commercial fishery. However, a warning was issued in 1970 discouraging consumption of any fish caught downstream from the city of Ottawa, since analysis indicated that mercury concentrations in various fish were sometimes above the legal limit of 0.5 ppm.

For the detailed study on mercury, its distribution, transport, and transformation (Figure 2), a 4.88-km (3-m) reach of the river, 1.5 km wide, was selected. This section began 1.6 km downstream from the city of Ottawa, the Canadian capital, and was chosen especially for the complicated form and varied environments it represented. In recent years, the mean flow rate has varied from approximately 1000 m^3/s (winter flow) to 4000 m^3/s (spring flood) with the maximum mean velocity, corresponding to the latter condition, approaching 270 cm/s. Bulk transport of bed sediments was therefore expected to be significant during the brief spring flood period. An observation in June 1974 (after spring runoff had tapered down) indicated that the bed-load was 450 tons/d. In this case two different methods were applied for the determination of the bed-load transport rate. Some of the environmental data which influenced general characteristics in the study section of the Ottawa River are shown in Table 1. On average, the study section was covered by ice for a period of 4 months (December to April), and the frost-free period for plant growth was about 3 months in this area. In other words, this area was hot in the short summer (extreme maximum *average* air temperature for the past 75 years is 33.5°C) and cold in the long winter (extreme minimum *average* air temperature for the past 75 years is −29.5°C).

III. COMPONENTS OF THE RIVER AND THEIR MASS MOVEMENT

Six components of the Ottawa River were examined in detail. The components comprised three physical ones (water, suspended solids, and bed sediments) and three biological ones (higher aquatic plants, benthic invertebrates, and fish). Of course, there were other important components such as algae, plankton, and bacteria; however, the resources were concentrated only on these six components.

Table 2 shows the mass of major components and their movement (replacement) in the study section. The volume of water in the section was 23 million tons and was replaced 7.8 times per day.

The calculation of the amount of bed sediments, 0.28 million tons, was based on 4-cm top layer of bed sediments. In fact, the average active depth of bed sediments was estimated at 7.1 cm, as described in the latter part of this report. Three types of bed sediments were identified in the study section. Much of the total surface of the bed sediments (70%) was covered by sand (median diameter from 150 to 600 μm), 27% by wood-chip sediments (a mixture of wood chips, bark, fiber, and silts), and 3% by clay (cohesive sediments). Detailed information of distribution and characteristics of bed sediments as well as their interaction with mercury has been published elsewhere.[1]

The mass of suspended solids was obtained by filtering Ottawa River water using 0.45-μm Millipore filter papers. The suspended solids concentration varied drastically with geographical locations, seasons, and sometimes the depth of water. The concentration was as high as 150 ppm at a spring flood period in the main stream and as low as 1 ppm or even less in the winter period. The suspended solids concentration, 16.1 mg/l, was obtained considering the flow rate and geographical locations throughout the year.

Aquatic plants (total 55 species) grew over 5% of the study section area with the productivity of the plants estimated at 500 tons/year (wet).[18] Another 700 tons of plants were considered to be a carryover to the next year through winter: roots and hard stems. Therefore, the average total biomass for the plants was estimated as 1200 tons (wet) in the summer (Table 2). Of course, the amount of biomass varied from year to year. For example, the mass in 1973 was 4.3 times greater than that in 1972.

The amount of plants was obtained collecting all plants by digging sediments of a unit surface area of 0.5 m^2. A few dozen locations in the study section were sampled in this way every year. The production rates of benthic invertebrates and fish were less well defined, though amounts living in the study section were estimated at 180 tons (wet) for invertebrates (which included 23 species) and 13 tons (wet) for fish (with 22 species).[10-12] The number

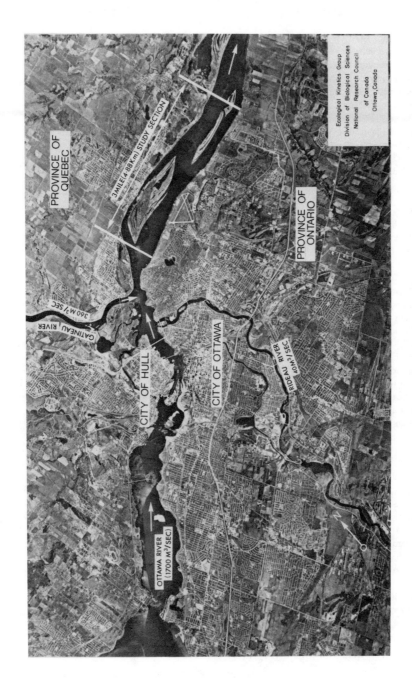

FIGURE 2.   Study section of Ottawa River.

Table 1
ENVIRONMENTAL FACTORS AT STUDY SECTION OF OTTAWA RIVER (1972—1977)

Factors	Mean	Minimum	Maximum
Atmospheric			
Air temperature (highest monthly mean, °C)	20.5[b]	19.9	22.4
Air temperature (lowest monthly mean, °C)	−15.6[b]	−12.7	−11.5
Bright sunshine (h/year)	2015[b]	1813	2112
Total precipitation (mm/year)	864[b]	792	1175
Total snowfall (m/year)	2.12[b]	1.9	3.16
Water (unit ppm, unless otherwise noted)			
Temperature, °C	13	0.5	23
Color, Hazen units	47	40	58
Turbidity J. T. U.	6.7	3.0	10
pH	7.2	6.9	7.5
Oxygen consumed ($KMnO_4$)	7.9	2.4	10.7
Phenolphthalein alkalinity	0	0	0
Total alkalinity	27	22	34
Residue on evaporation at 105°C	67	57	86
Loss on ignition at 550°C	24	20	30
Specific conductance, ($\mu\Omega$ at 25°C)	83	71	96
Total hardness, as $CaCO_3$	30	22	38
Noncarbonate hardness as $CaCO_3$	12	9	13
Calcium (Ca)	11.1	7.5	14.6
Magnesium (Mg)	1.2	0.3	2.6
Iron(Fe) total (dissolved)	0.30 (0.03)	0.04 (0.01)	0.80 (0.06)
Aluminum (Al)	0.03	0	0.06
Manganese (Mn) total	0.01	0	0.03
Copper (Cu) total	0.004	0	0.011
Zinc (Zn)	0.012	0.003	0.034
Sodium (Na)	1.9	1.5	2.6
Potassium (K)	0.76	0.60	1.00
Ammonia (NH_3)	0.04	0	0.20
Sulfate (SO_4)	12.2	10.7	15.6
Chloride (Cl)	1.5	1.0	2.4
Phosphate (PO_4) total	0.07	0.04	0.18
Nitrate (NO_3)	0.42	0.10	0.53
Silica (SiO_2)	4.3	2.3	5.2
Fluoride (F)	0.10	0.15	0.15
Standard plate counts (ml)	1727	40	3000
Coliform colonies (100 ml)	474	8	2500
Fecal coliform colonies (100 ml)	45	0	200

[a] Agriculture Research Institute, Ottawa.
[b] Average of 1890—1965.
[c] Lemieux Island Purification Plant, Ottawa, during 1972.

of species given here comprised only those identified, and the actual number was probably larger.

For the estimation of invertebrate biomass, the same method was applied as for plants. Sediment to a depth of 15 cm was collected from a unit surface area, and the sediment was screened with mesh to isolate all invertebrates. The collected invertebrates were then identified as to species, weighed for mass, and eventually analyzed for mercury.

The most difficult task was to obtain the mass of fish present in the study section at the

Table 2
AVERAGE MASS OF MAJOR COMPONENTS AND
THEIR MOVEMENTS IN THE STUDY SECTION OF
THE OTTAWA RIVER

Components	Total mass (average) (g)	Discharge or production rate (g/year)
Water	2.3×10^{13}	5.3×10^{16}
Suspended solids	3.7×18^8 (16.1 mg/l)	8.6×10^{11}
Bed sediments	2.8×10^{11} (D)	2.0×10^{11}
Higher aquatic plants (55 spp.)	1.2×10^9 (w)[b]	5.0×10^8
Benthic invertebrates (23 spp.)	1.8×10^8 (w)	Unknown
Fish (22 spp.)	1.3×10^7 (w)	Unknown

[a] D = dry weight.
[b] W = wet (fresh) weight.

sampling moment. Three approaches were attempted: (1) catching 5000 fish (mostly bull-head) and releasing them with identification tags, (2) netting fish in selected areas, and (3) calculating the mass of fish from the primary biomass production rate. Because methods 1 and 2 gave little quantitative information (only three tagged fish out of 5000 released were netted within 1 week of intensive netting), the mass of fish, 13 tons, was obtained using method 3.

IV. MERCURY DISTRIBUTION

During 1972 to 1975, a field survey was conducted for detailed mercury analysis of representative samples of the six components of the river system. Some of the sampling was conducted regularly throughout the year, breaking the 70-cm-thick ice cover. Total mercury including inorganic mercury was measured by flameless atomic absorption. A detailed determination of various chemical forms of mercury was conducted using gas chromatography and thin layer chromatography (TLC). Digestion methods and extraction methods for various samples have been reported elsewhere.[18]

Table 3 summarizes the reported results (average) of mercury concentrations from a relatively large number of samples for each of the major components of the study section.

The mercury concentration of filtered water was 0.013 ppb in 1975 to 1976. This average concentration was determined by using 250-ml water samples. Considering the detection limits and background value of mercury in various analytical chemicals used, this value of 0.013 ppb was considered to be less accurate. The more accurate measurements were conducted in 1978 using 2-l samples and the dithizone-benzene extraction technique. The 1978 value for the water was 0.0066 ppb.

The average mercury content of suspended solids was 1140 ppb in 1975 to 1976. Concentration was 87,700 times higher than the concentration of mercury in the same weight of water. The fine particles of suspended solids, indeed, accumulated a considerable amount of mercury from the surrounding water. The average value of 1140 ppb was obtained from a set of 104 samples collected at various locations and at various times of the year. An interesting point concerning suspended solids was that the concentration of these solids in the river water changed drastically (from 150 ppm to less than 1 ppm), while the mercury

Table 3
AVERAGE MERCURY CONCENTRATION AND DISTRIBUTION IN THE STUDY SECTION OF THE OTTAWA RIVER

Components	Total mercury concentration (ppb)	Organic mercury (%)	Mercury existing (g)	Fraction in study section (%)
Filtered water[a] (*n* = 104)	0.013	<1	300	1.3
Suspended solids[a] (*n* = 104)	1140 (d)[b]	<1	420	1.8
Bed sediments (*n* = 1153)	80.6 (d)	3.8	22,600	96.7
Higher plants (*n* = 472)	14.2 (w)[c]	20	17.1 ⎫	
Benthic invertebrates (*n* = 141)	223 (w)	40	40.1 ⎬	0.2
Fish (*n* = 535)	162 (w)	85	2.1 ⎭	
Total in the study section			23,379	100.0

Note: n = number of samples.

[a]   1975—1976 average.
[b]   d = dry weight.
[c]   w = wet weight.

concentration in the suspended solids remained relatively uniform throughout the year and at various locations.

The bed sediments had a relatively low average mercury concentration of 80.6 ppb. This low average was due to the fact that a large area of Ottawa River bottom was covered by course and medium sands. This type of sediment contained less mercury.

The average for higher aquatic plants, benthic invertebrates, and fish, respectively, was 14.2, 223, and 162 ppb of mercury on a wet-weight basis. As the variation due to species, age, sex, season, and anatomical location was considerable, a large number of samples were required to present meaningful average values.

Table 3 shows that bed sediments contained approximately 97% of the total mercury distributed in the study section. Higher plants, benthic invertebrates, and fish shared a negligible amount (0.2%) of the mercury present in the study section, although these components contained high concentrations of mercury.

A high organic mercury fraction was found only in the biological components. Thus, although the distribution of organic mercury would have been of prime importance for short-term biological toxicity studies, studies of the bed sediment-related pollutant dynamics were essential for any understanding of the long-term behavior of mercury in the river system.

Mercury concentrations for all components of the study section had been decreasing since the start of the measurement in 1972. Water concentration decreased from 0.038 ppb in 1972 to 0.013 ppb in 1975 to 1976 and 0.0066 ppb in 1978. The concentrations in bed sediments decreased within 2 years to a level of less than half of the value in 1972. The decrease might have been related to the stoppage of mercury discharge from the pulp and paper industry into the Ottawa River since 1969. There was, however, no means of distinguishing whether the mercury remaining in the sedments originated from industrial output more than 7 years earlier or from the weathering of natural mercury-containing bedrock.

V. GEOGRAPHICAL TRANSPORT OF MERCURY

Questions to be answered in the Ottawa River Project were *what* amount of mercury was coming in and out of the study section, *how* was it transported — by water, by suspended

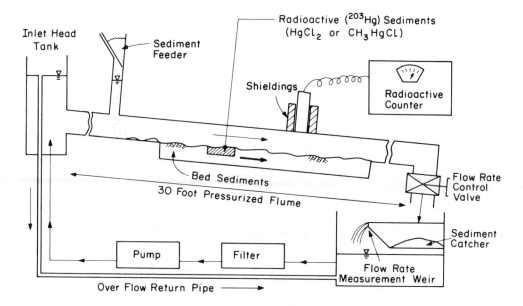

FIGURE 3.   Pressurized laboratory flume for mercury transport by moving bed sediments.

solids, by bed-sediment movements, or by biological activities including movements of fish — and what was the relative importance of these means of geographical transport. In order to obtain the answers, an intensive field and laboratory investigaton was initiated. Because of the small contribution of the biomass, the major effort was focused on the three physical components, i.e., filtered water, suspended solids, and bed sediments.

The most difficult part of the estimation of the mercury transport was the role of bed-sediment movements. The main reason for the difficulty was that the mechanism of bed-sediment transport itself is not fully understood.[19] Furthermore, the Task Committee of the American Society of Civil Engineers concluded after an intensive literature review that nothing is hydraulically known about cohesive sediment movements,[20] which play a critical role in mercury transport associated with bed-sediment movements. Several well-known formulas for bed-sediment transport have been proposed, but their applicability is limited to bed sediments for which particle diameters are larger than 60 to 80 μm. In the study section of the Ottawa River, a portion of bed sediments under 38 μm contained 50 times more mercury than particles of 200 to 300 μm.[7] In short, the bed-sediment transport formulas were *not* directly applicable to the estimation of mercury transport (they may be applicable to *sediment* transport itself). A further limitation was imposed on observations in the field of bed-sediment movements in the study section of the Ottawa River at the time when the movement was the most active, i.e., during the period of spring flood. During this period, not only vast quantities of turbulent water, but ice of various sizes and logs prohibited the operation of small boats in the middle of the stream, and cold water (close to freezing point) made scuba-diving extremely dangerous. There was no bridge within or close to the study section of the Ottawa River.

This combination of difficulties led to the construction of a 30-ft-long hydraulic flume inside a laboratory. The flume was specially designed to observe *mercury* transport by bed-sediment movements, *not* to observe bed-sediment transport itself (Figure 3). A radioactive tracer for various forms of mercury compounds was used to determine quantitative information of mercury transport associated with bed-sediment movements. Various hydraulic conditions as well as various types of bed sediments were used to simulate conditions found in the study section of the Ottawa River.

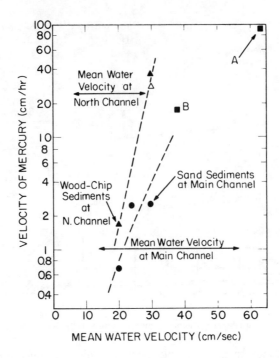

FIGURE 4.   Velocity of mercury associated with bed sediment movements. (Open triangle is for methylmercury, and closed triangles and circles
are for mercuric chloride. A = ^{192}Ir with North Loup River sediments;[36]
B = ^{198}Au with sieved sediments.[37] (From Kudo, A., Townsend, D. R.,
and Miller, D. R., *Prog. Water Technol.*, 9, 923, 1978. With permission.)

Figure 4 shows the results of the flume study. The velocity of mercury associated with
moving bed sediments is expressed (cm/h) in the vertical scale and the mean water velocity
(cm/s) in the horizontal scale. There were three phases concerning mercury movements
associated with moving bed sediments and related to water velocity: (1) no mercury movement
phase, (2) slow mercury movement phase (as compared to the water movement), and (3)
fast mercury movement phase (same speed as the water velocity).

In phase 1, mercury attached to bed sediments did not move at all, because the bed
sediments did not move at all until the mean water velocity reached a certain level. Until
this velocity was reached, there was no observed mercury *desorption* from the bed sediments
into the flowing water. For wood-chip and sand sediments, mean water velocities up to and
including 15 cm/s belonged to phase 1.

In phase 2, mercury gradually moved with bed-sediment movements. The relationship
between mean water velocity and mercury velocity is expressed in Figure 4. Wood-chip
sediments moved three to four times faster than did the sand sediment. This phase was at
15 to 30 cm/s of water velocity for wood-chip sediments and at 15 to 35 cm/s of water
velocity for sand sediments.

Mercury attached to wood-chip and sand sediments moved at the same speed as did the
water with velocities faster than 30 and 35 cm/s, respectively. At this velocity all bed
sediments were suspended by the force of water movement.

After the flume study and the field observatons, the amount of mercury transported through
the study section was estimated. The amount of 1687 kg of mercury had been transported
every year through the study section of the Ottawa River (Table 4). Most mercury transported
was either a soluble form or attached to suspended solids. Filtered water (soluble mercury)
contributed 41% (689 kg/year) of all mercury transported, and suspended solids accounted

Table 4
MERCURY TRANSPORT AT THE STUDY SECTION OF THE
OTTAWA RIVER (AVERAGE)

Components	Mass transported (ton/year)	Mercury transported (kg/year)	Fraction in Hg transported (%)
Water (filtered)	5.3×10^{10}	689	41
Suspended solids	8.6×10^5 (D)[a]	982	58
Bed sediments	2.0×10^5 (D)	16.1	1
Higher plants (dead)	500 (w)[b]	0.007 ⎫	
Benthic invertebrates	Unknown	<0.040[c] ⎬	0.003
Fish	Unknown	<0.002 ⎭	
Total mercury transported		1687	100.0

[a] D = dry weight.
[b] W = wet weight.
[c] Unknown, assuming all invertebrates and fish existing in the study section are replaced every year.

for 58% (982 kg), in spite of a low mercury concentration in filtered water (13 ng/l) and in suspended solids (18.3 ng per suspended solids in 1 l of river water [ng/SS-l]).

The amount of mercury transported by filtered water and suspended solids (total 1671 kg/year) was 72 times more than all mercury present in the study section (23.4 kg). This large quantity of mercury being transported through the study section raised some questions: (1) Where did it come from and where did it go? (2) Was the study section accumulating or discharging mercury, or holding an equilibrium? Only 1% of the 1671 kg mercury being transported every year (16.7 kg) was equivalent to 72% of all mercury existing in the study section. At the time, there was no way to detect a 1% difference on mercury concentratons in waters collected between far upstream and far downstream (4.9-km distance) of the study section.

To overcome this analytical difficulty, water samples were taken at 50 and 500 km upstream from the study section, to magnify the difference of mercury concentrations in water between the study sections and upstream. The analytical results showed there was no major difference in mercury concentrations of filtered water and suspended solids between the study section and at 50 and 500 km upstream. This indicated that there might be a kind of equilibrium of mercury among various components of the Ottawa River throughout the 500 km upstream from the study section. However, this indication did not solve the question of the origin of the mercury. The most plausible explanation at present is that the mercury came from natural weathering, because there were no industries further upstream from the 500 km.[21] Reported natural background concentrations in water ranged from 5 to 50 ng/l.[22] The mercury concentration in Ottawa River water fell well within this range. The Rhine River (Netherlands), for example, is about 600 ng/l.[23] All mercury eventually goes to the Atlantic Ocean.

An interesting point is that mercury contents in suspended solids were relatively constant regardless of changes in suspended solid parameters. For example, suspended solid concentrations ranged from 1.42 to 55.39 mg/l on the Ontario shore of the study section during 1975 to 1976. However, mercury concentrations attached to suspended solids ranged only from 0.45 ppm (dry) to 2.94 ppm (dry).

Furthermore, average mercury concentration of suspended solids at the three sampling locations (two within the study section and one at 50 km upstream from the study section) were very similar at 1.08, 1.19, and 1.45 ppm (dry). Of course, the constituents of suspended solids could have been totally different from those in the spring flood period to those in the winter when the river was covered by 70-cm-thick ice. However, this difference did not change drastically the relative mercury concentration attached to suspended solids. The total amount of suspended solids transported downstream was 0.86 million tons/year (average) on a dry weight basis.

Reported mercury ratio in suspended solids to the whole water (filtered water plus suspended solids) varied from 92% in the Thames River in England to 20% in the LaHave River in eastern Canada.[24,25] Between them were 60% in the Rhine River (West Germany), 56% in the Rhine River (Netherlands), and 52% in the Colombia River in western U.S.[23,26,27] The ratio for the study section was 58%. The results (92%) for the Thames were calculated from analysis for the whole water (filtered plus suspended solids) and for suspended solids and *not* based on two independent analyses of filtered water and suspended solids.

Surprisingly, the role of bed-sediment movements in the mercury transport was very small, only 1%. Every year, the amount of 16.1 kg of mercury was transported downstream attached to sediment particles being moved, whereas water and suspended solids carried 689 and 982 kg of mercury in the study secton of the Ottawa River. The main reason for the insignificant role of bed-sediment movements on mercury transport was considered to be that the Ottawa River had been developed fully for hydroelectric dams. In other words, the energy slope of the river was small (1.5×10^{-5} even at spring flood), and most of the bed sediment materials were physically stopped by the dams, except some of the suspended materials. Furthermore, the dams regulated water flow rate considerably. Therefore, this observation of the small contribution of bed-sediment movements on mercury transport in the study section cannot be applicable to other rivers where hydroelectric dams are not fully developed.

The amount of 16.1 kg of mercury being transported by bed-sediment movements was significant in comparison to the amount of mercury existing in bed sediments of the study section (22.6 kg). In other words, over 71% of the mercury present in the bed sediments moved out every year by bed-sediment movements from the section, and the same amount of new mercury attached to bed-sediment particles came in. The flume study showed little desorption of mercury from bed-sediment particles being moved[4] and that on average the velocity of mercury (*not* sediment itself) associated with bed sediments was 2.4 km/year (27 cm/h). This velocity was similar to the results obtained by a set of flume experiments simulating hydraulic conditions found in the study section of the Ottawa River. Of course, the velocity of mercury attached to the bed sediments at the middle of the stream was far greater than that at the shore. Also the water velocity at the north channel (CIP channel) was much slower than that at the mainstream. Furthermore, the velocity of mercury attached to the bed sediment was a function of types and particle size of bed sediments. With this information, the average active depth of bed sediments, which contributed to mercury transport, was also estimated as 7.1 cm, for the bed sediments of the study section. Again, the active depth could be 200 cm for the sand dune at the middle of the stream (especially during a period of spring flood) and less than 1.0 cm for the clay (cohesive) sediments along the Ontario shore. The difference in the mercury velocity between inorganic- and methyl-mercury attached to bed sediments being moved was insignificant when considering other factors such as type of sediments, as shown in Figure 4.

The role of the biomass on the mercury transport was totally insignificant, only 0.003% (0.049 kg/year) in the study section of the Ottawa River.

VI. METHYLMERCURY

As described previously, most mercury (up to 97%) found in aquatic systems was associated with bed sediments. Fish, invertebrates, and plants contained only 0.2% of the total, though they contained a higher ratio of organic mercury. Reported field surveys showed that the proportion of organic mercury in bed sediments ranged from 0.1 to 2.1%.[28-32] Jacobs and Kenney placed river sediments mixed with inorganic mercury in the Wisconsin and Fox Rivers.[33] After 12 weeks exposure in the natural environment, the concentration of methylmercury in the sediments was only 3%. The results also showed that this concentration of methylmercury was attained within 4 weeks after sediments were placed in the bottom of the rivers. This indicated that within 4 weeks the processes of methylation and demethylation of mercury were in equilibrium under these environmental conditions. However, no experimental demonstration was conducted to show the degradation of methylmercury in sediments of these Wisconsin and Fox rivers experiments.

The most popular methylation study was conducted by Jensen and Jernelöv in 1968.[34] After this report numerous investigations were performed on the aspect of methylmercury production by microorganisms in bed sediments. Very little attention, however, was paid to the proportion of methylmercury to the total mercury present in bed sediments. Spangler et al. observed degradation of methylmercury in mixed cultures from sediments.[35] The degradation reached up to 50% within 5 d. They also found methylmercury degradation in sediments. However, detailed information about the sediments is not available and results fluctuated widely.

What is *not* known about methylmercury in aquatic systems is the total amount of methylmercury existing in all components of the systems, including fish, invertebrates, and plants as well as various types of bed sediments. In order to estimate the amount, an experiment was designed to demonstrate methylmercury equilibrium concentration levels in the Ottawa River sediments. With the equilibrium concentrations, it was possible to estimate the amount of methylmercury in various types of bed sediment in the study section, where detailed information of total mercury concentrations (total mercury estimation 22.6 kg) and types of sediment were available.

Two directions (methylmercury production and degradation) toward an equilibrium level were observed during a period of 50 d under identical environmental conditions. Three types of fresh sediments from the study section of the Ottawa River were spiked with either radioactive mercuric chloride or methylmercuric chloride at a level of 1.00 ppm. The locations of the three sediments were (A) mainstream sediments, (B) north channel sediments, and (C) wood-chip sediments which were located in the north channel but just downstream from the pulp and paper company. The sediments were called type A, B, and C sediment, respectively. More detailed information about these sediments has been reported elsewhere.[1]

As no isotopic exchanges were observed between organic and inorganic mercury under natural environmental conditions and in these mercury concentration ranges,[10] it was concluded that increase or decrease of radioactive methylmercury in sediments was caused by methylation or demethylation, not by simple isotopic exchanges. Organic contents of these sediments (A, B, and C) were 0.35, 0.84, and 59.20%, respectively. The sediments were placed in darkness at a temperature of 20°C and enclosed to prevent a loss of moisture as well as mercury from the sediments. TLC and atomic absorption analysis were used to determine quantitative amounts of various chemical forms of mercury, mainly mercuric and methylmercuric, as well as total mercury in sediments. All samples were analyzed in duplicate.

Surprisingly, methylmercury in cleaner sediments (type A, main channel sediments [sands]) degraded more rapidly into inorganic forms (Figure 4). Within 25 d, 80% of methylmercury was broken into inorganic forms in the sand sediments, whereas during the same period, only 14% of the methylmercury was converted into inorganic forms in the organic sediments (type C). Sediments which were mixed with an inorganic form of mercury increased their contents of methylmercury with an increase of time. For highly organic sediments (type C, wood-chip sediments) the portion of methylmercury was 7.4% and for cleaner sand (type

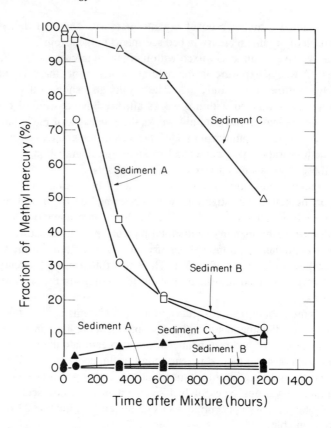

FIGURE 5.   Equilibrium methylmercury concentrations in three types
of bed sediments from the Ottawa River. (From Kudo, A., Akagi, H.,
Mortimer, D. C., and Miller, D. R., *Nature (London)*, 270, 419, 1977.
With permission.)

A, main-channel sediments) the portion was only 0.1% after 25 d. The difference in final
methylmercury concentration between sand (A) and wood-chip (C) sediments was probably
a result of the different rates of formation and decomposition of methylmercury in these
sediments. That is, the wood-chip sediment formed methylmercury more rapidly and also
decomposed methylmercury more slowly than did the sand sediment. During the 50-d
experiments, an average of 1.3% of mercury escaped from the system.

Methylmercury concentrations in each of the three types of sediments converged with
time toward an equilibrium concentration level from both directions, starting with either
mercuric chloride or methylmercuric chloride (Figure 5). From the results, the equilibrium
concentration levels were estimated for sediment types A, B, and C as 0.2, 2.3, and 12.1%,
respectively. The time to attain the equilibrium concentration was also estimated as 3 months
for sediment types A and B and 6 months for sediment type C.

Applying this methylmercuric information to Ottawa River sediments, the amount of
methylmercury present in the study section was calculated to be 1200 g (Table 5). In this
calculation, the 4.9-km section was divided into five regions depending on the type of bed
sediments present in the section of the river.

The amounts of methylmercury in the biomass of the study section were 3.4, 16.0, and
1.8 g for plants, invertebrates, and fish, respectively (Table 5). These biomasses contained
altogether only 21.2 g of methylmercury. In other words, bed sediments contained 57 times
more methylmercury than the biomass. Of course, the degree of accuracy in this calculation
of methylmercury in the sediments was not as high as that for the estimation of methylmercury

Table 5
TOTAL AMOUNTS OF METHYLMERCURY IN
THE STUDY SECTION OF THE OTTAWA
RIVER (AVERAGE)

Components	Amounts of mercury (g)	Fraction (%)
Water (filtered)	3.0[a]	0.2
Suspended solids	4.2[a]	0.3
Bed sediments	1200	97.8
Higher aquatic plants	3.4 ⎫	
Benthic invertebrates	16.0 ⎬	1.7
Fish	1.8 ⎭	
Total methylmercury	1228.4	100.0

[a] Maximum estimation.

in biomass in the study section. However, the calculation clearly showed that the content of methylmercury in the sediment was higher, by at least one or two orders of magnitude, than that in the biomass. This further indicated that a detailed study must be directed toward mercury in bed sediments to understand and control, effectively, mercury pollution in aquatic systems.

VII. CONCLUSION

The following conclusions resulted from the field observations and laboratory studies during the 5-year project:

1. Most of the mercury (96.7% of total mercury and 97.8% of methylmercury) was in bed sediments. Biomass contained an insignificant portion of mercury (0.2% of total mercury and 1.7% of methylmercury). The amount of mercury existing in the study section was about 23.4 kg of total mercury and about 1.23 kg of methylmercury.
2. Suspended sediments (or solids) contributed 58% (982 kg/year) of all mercury transported downstream. Although water (filtered) had a low mercury concentration (13 ng/l, or 0.013 ppb), it accounted for 41% (689 kg/year) of all mercury transported downstream.
3. The role of bed sediment movements on the mercury transport was very small, only 1%, in the study section of the Ottawa River.
4. Methylmercury production and destruction were in equilibrium in sediments without any significant contribution from biological agents (higher aquatic plants, invertebrates, or fish).

Because most results were obtained considering the environmental conditions existing in the 4.9-km study section of the Ottawa River, some caution must be taken before applying these results to other rivers. For example, the study section had a growing season of only 4 months for higher aquatic plants. Obviously, the conditions in tropical rivers, for example, would be entirely different from those in the Ottawa River. Thus, they would have an entirely different ecology: species, growing pattern, productivity, biomass, and so on. Hydraulically, the section of the Ottawa River was developed fully with hydroelectric dams, and the flow rate was well regulated. Therefore, the results cannot be applied to a "wild" river, where

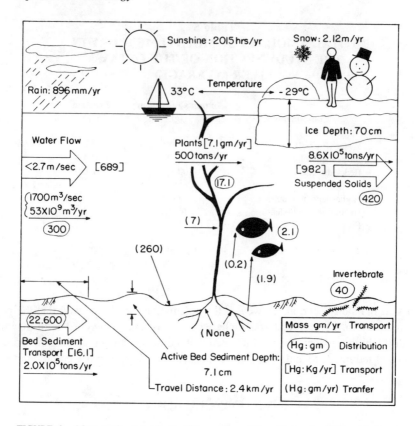

FIGURE 6. Mercury dynamics in the Ottawa River. (From Kudo, A., Miller, D. R., Akagi, H., Mortimer, D. C., Defreitas, A. S. W., and Nagase, H., *Prog. Water Technol.*, 10, 329, 1978. With permission.)

bed sediment transport rates are higher than those in the Ottawa River. In summary, overall mercury dynamics are shown in Figure 6 for the study section of the Ottawa River, including some relevant environmental factors.

ACKNOWLEDGMENT

The author thanks Dr. T. Kauri for his contribution to this chapter.

REFERENCES

1. **Kudo, A. and Hart, J. S.,** *J. Environ. Qual.*, 3, 273, 1974.
2. **Kudo, A., Mortimer, D. C., and Hart, J. S.,** *Can. J. Earth Sci.*, 12, 1036, 1975.
3. **Mortimer, D. C. and Kudo, A.,** *J. Environ. Qual.*, 4, 491, 1975.
4. **Kudo, A., Townsend, D. R., Sayeed, H., and Miller, D. R.,** *Environmental Bio-Geochemistry*, Vol. 31, Nriagu, J. O., Ed., Ann Arbor Science Publishers, Ann Arbor, MI, 1976, 499.
5. **Kudo, A.,** *J. Environ. Qual.*, 5, 427, 1976.
6. **Kudo, A., Akagi, H., Mortimer, D. C., and Miller, D. R.,** *Nature (London)*, 270, 419, 1977.
7. **Kudo, A., Townsend, D. R., and Miller, D. R.,** *Prog. Water Technol.*, 9, 923, 1977.
8. **Kudo, A. and Akagi, H.,** *Bioaccumulation*, Sangyo Tosho, Tokyo, 1977.
9. **Kudo, A., Townsend, D. R., and Miller, D. R.,** *J. ASCE Environ. Eng. Dev.*, 103, 605, 1977.
10. **Kudo, A., Akagi, H., Mortimer, D. C., and Miller, D. R.,** *Environ. Sci. Technol.*, 11, 907, 1977.

11. **Kudo, A., Miller, D. R., Akagi, H., Mortimer, D. C., Defreitas, A. S. W., and Nagase, H.,** *Prog. Water Technol.,* 10, 329, 1978.
12. **Mortimer, D. C. and Kudo, A.,** *Proc. Int. Conf. on Transport of Persistent Chemicals in Aquatic Ecosystems,* National Research Council of Canada, Ottawa, 1974.
13. **Kudo, A., Sayeed, H., Townsend, D. R., and Miller, D. R.,** *Proc. Int. Conf. on Transport of Persistent Chemicals in Aquatic Ecosystems,* National Research Council of Canada, Ottawa, 1974.
14. **Kudo, A., Miller, D. R., and Townsend, D. R.,** *Prog. Water Technol.,* 9, 923, 1977.
15. **Kudo, A. and Mortimer, D. C.,** *Environ. Pollut.,* 19, 239, 1979.
16. **Kudo, A., Miller, D. R., and Nagase, H.,** *J. Water Sci. Technol.,* 13, 305, 1981.
17. **Kudo, A., Nagase, H., and Ose, Y.,** *Int. J. Water Res.,* 16, 1011, 1982.
18. Ottawa River Project Group, Ottawa River Project Report (Final), National Research Council of Canada, Ottawa, Canada, 1977.
19. **Yalin, M. S.,** *Mechanics of Sediment Transport,* Pergamon Press, Oxford, 1972, vii.
20. **Masch, F.,** *J. Am. Soc. Civ. Eng.,* HY44, 1017, 1968.
21. **Marier, J. R.,** NRC Assoc. Comm. Rep. for Environmental Quality, No. 13501, National Research Council of Canada, Ottawa, Canada, 1973.
22. **Wedepohl, K: H., Ed.,** *Handbook of Geochemistry,* Vol. 2, Springer-Verlag, Berlin, 1969.
23. **de Groot, A. J., Salomons, W., and Allersma, E.,** *Proc. of Heavy Metals in Aquatic Environment,* Nashville, TN, 1974.
24. **Smith, J. D., Nicholson, R. A., and Moore, P. J.,** *Nature (London),* 232, 393, 1971.
25. **Cranston, R. E. and Buckley, D. E.,** *Environ. Sci. Technol.,* 6, 274, 1972.
26. **Förstner, U. and Müller, G.,** *Schwermetalle in Flüssen und Seen,* Springer-Verlag, Berlin, 1974.
27. **Bothner, M. H. and Carpeter, R.,** Proc. IAEA Conf. on Radioactive Contaminators of the Marine Environment, International Atomic Energy Agency, Vienna, 1978, 73.
28. **Andren, A. W. and Harris, R. C.,** *Nature (London),* 245, 256, 1973.
29. **Batti, R., Magnaval, R., and Lanzola, E.,** *Chemosphere,* 4, 13, 1975.
30. **Langbottom, J., Pressman, R., and Lichtenburg, J.,** *J. Assoc. Off. Anal. Chem.,* 56, 1297, 1973.
31. **Olson, B. H. and Cooper, R. C.,** *Nature (London),* 252, 682, 1974.
32. **Eganhouse, R. P.,** Southern California Coastal Waste Research Project Report, El Segundo, CA, 1976, 83.
33. **Jacobs, J. W. and Kenney, D. R.,** *J. Environ. Qual.,* 3, 121, 1974.
34. **Jenson, S. and Jernelöv, A.,** *Nature (London),* 223, 753, 1969.
35. **Spangler, W. J., et al.,** *Science,* 180, 192, 1973.
36. **Hubbell, D. W. and Sayre, W. W.,** *J. Am. Soc. Civ. Eng.,* HY3, 39, 1964.
37. **Yang, C. T. and Sayre, W. W.,** *J. Am. Soc. Civ. Eng.,* HY2, 265, 1971.

Chapter 9.3

FACTORS AFFECTING SOURCES AND FATE OF PERSISTENT TOXIC ORGANIC CHEMICALS: EXAMPLES FROM THE LAURENTIAN GREAT LAKES

R. J. Allan

TABLE OF CONTENTS

I. INTRODUCTION

The fate of toxic chemicals in aquatic ecosystems is a vast and complicated subject. Many chemicals introduced from a variety of sources to aquatic ecosystems are short lived. Their degradation takes place by many chemical and biological processes including hydrolysis, photolysis, and bacterial dechlorination. These short-lived organic chemicals may still be highly toxic in the short term. In some cases, aquatic degradation products of chemicals can be more toxic than the original compounds, for example, with the herbicide 2,4-D. Other organic chemicals are persistent in aquatic environments. This persistance may be for days, months, or years. These persistent toxic organic chemicals are those which have caused the most concern in the Great Lakes Basin. They include polychlorinated biphenyls (PCBs); chlorinated benzenes, phenols, and styrenes; chlorinated dioxins and furans; mirex; hexachlorobutadiene; and several organochlorine pesticides.

The Laurentian Great Lakes (Figure 1) are among the largest bodies of fresh water in the world (Table 1).[1] They lie in glacial scour basins and are similar in their physicohemical characteristics. In many ways, large lakes are not that different from much smaller lakes because they interact with the land of their drainage basins; have an interface with the atmosphere; have defined residence times, sedimentation rates, and trophic status; and are subject to many of the same processes of recycling of persistent toxic organic chemicals. Alternatively, as opposed to small lakes, large lakes often have considerable chemical and biological variations in both the horizontal and vertical dimensions. The largest lakes of Canada and elsewhere have in many ways been compared to minioceans. One very important factor affecting toxic chemical fate in large lakes is residence time which, for the Laurentian Great Lakes, is in years and tens of years. Insoluble and persistent toxic organic chemicals are usually lipophilic. In long residence-time lakes, there is plenty of time for such chemicals to enter food webs, bioaccumulate to the highest trophic levels, and be continually recycled through the ecosystem. Another important factor in the great lakes is their low productivity in terms of algal biomass and the important role this plays in the fate of persistent toxic organic chemicals.

The major processes which influence the fate of persistent toxic organic chemicals in large lakes are the same as those in small lakes and those defined and quantified by laboratory testing: processes governed by the physicochemical characteristics of the chemicals, namely, solubility, volatility, hydrophobicity, degree of partitioning (into or onto sediment and other particulates), lipophilicity, and resistance to degradation (by hydrolytic, photochemical, and biochemical routes). Other processes are related to the limological characteristics of the receiving body of water and include suspended particulate concentration and type sedimentation rates, and bioaccumulation pathways. The interconnections between single chemical properties, laboratory tests of chemical and environmental reactions, fate and effects of chemical mixtures in polluted aquatic ecosystems, and ultimately predictive models for contaminant fate are shown schematically in Figure 2.

In the Laurentian Great Lakes, three main limnological characteristics interact to affect the fate of persistent toxic organic chemicals:

1. Low suspended sediment loads
2. Long residence times
3. Low trophic states

Clearly, these are not independent factors. For example, a high inorganic sediment load combined with a high trophic state would lead to rapid sedimentation and burial of persistent toxic organic chemicals with high sediment-water partition coefficients. Such a combination could also lead to rapid bioaccumulation because the (aquatic) sediment-water partition

FIGURE 1. The Laurentian Great Lakes.

Table 1
THE LAURENTIAN GREAT LAKES RANKED
AMONG THE LARGEST LAKES OF THE
WORLD BY AREA

Area	Name	Area (km²)	Area	Name	Area (km²)
1	Caspian	374,000	10	Sap	30,000
2	Superior	82,100	11	Great Slave	28,568
3	Aral	64,500	12	Chad	25,900
4	Victoria	62,940	13	Erie	25,657
5	Huron	59,500	14	Winnipeg	24,389
6	Michigan	57,750	15	Nyasa	22,490
7	Tanganyika	32,000	16	Balkhash	22,000
8	Baikal	31,500	17	Ontario	19,000
9	Great Bear	31,326			

coefficient is related to the octanol-water partition coefficient which, in turn, is correlated with lipophilicity and, hence, degree of bioaccumulation.

The general objective of this paper is to give an overview of these controlling factors and processes as they affect the fate of toxic organic chemicals introduced from various sources to the Great Lakes. There are several ways to organize such an analysis. The approach to be taken here will be to focus on the interaction between ecosystem compartments (Figure 3). The hypothesis is that the critical processes which eventually affect the fate of persistent toxic organic chemicals in large lakes, and elsewhere, are those which operate across interfaces between major ecosystem compartments. These interface interactions encompass the more complex processes which in detail are involved in the transfer of toxic organic chemicals across the interfaces and, hence, between media. The major interfaces are

1. Land-water
2. Air-water
3. Sediment-water
4. Nutrient-water

These four broad groupings are at the highest level of generality. They apply to all lakes, but certain characteristics of large lakes, as noted above, result in particular source-fate relationships for persistent toxic organic chemicals. Better quantification of the role of these

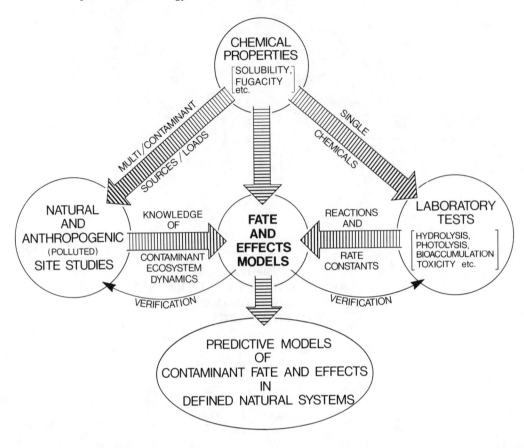

FIGURE 2.   Schematic of the interconnections of field, laboratory, and modeling research to resolve the fate and effects of toxic contaminants in aquatic ecosystems.

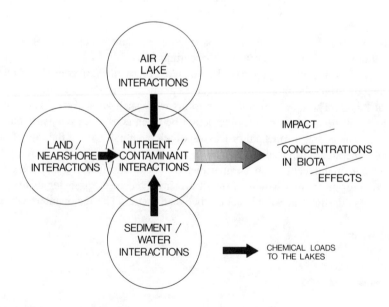

FIGURE 3.   Interactions between the major compartments affecting contaminant fate and effects in aquatic ecosystems.

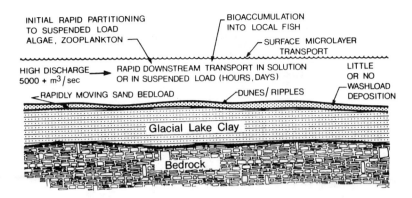

FIGURE 4. Schematic of features and processes operating in the rivers connecting the lower Great Lakes.

interfaces will lead to a better ability to understand, model, and predict fate of toxic organic chemicals in the Great Lakes and elsewhere.

The more detailed objective of this paper will be to discuss the fate of toxic organic chemicals in the context of these critical interfaces and to examine the interactions or movement of chemicals between compartments. Examples from the Great Lakes and their outlet, the St. Lawrence River, will be referred to in the discussion. Particular focus will be on Lake Ontario.

II. LAND-WATER INTERACTIONS

The major source of toxic organic chemicals to the Laurentian Great Lakes is from land-based activities. Chemicals enter the lakes from point sources as industrial, municipal, and urban effluents. They are also derived from nonpoint or diffuse land-based sources, for example, as pesticide residues from agricultural practices.

A. Land-Based Point Sources: The Connecting Channels

Major sites of direct discharge of persistent toxic organic chemicals to the Great Lakes are found on the two connecting channels of the lower Great Lakes and downstream along the St. Lawrence River.[2-4] The two connecting channels are, respectively, the Niagara River between Lakes Erie and Ontario and the Detroit-St. Clair rivers between Lakes Huron and Erie. The shores of these two connecting channels are sites of major petrochemical, chemical, and heavy industry complexes. The industrial-chemical complex on the Niagara River is almost exclusively located along the U.S. shore upstream from Niagara Falls. On the St. Clair River, it is on the Canadian shore between Sarnia and Lake St. Clair. In the U.S., Detroit is situated along the Detroit River between Lake St. Clair and Lake Erie, and Buffalo-Tonawanda is located at the immediate upstream end of the Niagara River. Both of these urban areas are major industrial centers for automobile production, at the former, and steel and chemicals, at the latter. Many major outfalls of toxic chemicals are thus found along the two connecting channels of the lower Great Lakes. The St. Lawrence River is the site of two major petrochemical/chemical complexes, one in the Cornwall area, the other at the eastern end of Montreal. Two large Canadian cities, Montreal and Quebec City, are located on the St. Lawrence River. The most important characteristic of these rivers is their tremendous flow of some 6000 m^3/s. Because of this great flow and short times of travel, the two connecting channels and the St. Lawrence River act essentially as a toxic chemical-collection and downstream-transmission system with little or no temporary removal by burial in bottom sediments (Figure 4).[5] The rivers of the two connecting channels are, thus, a prime point of entry of many toxic organic chemicals to Lake Ontario.[6-10]

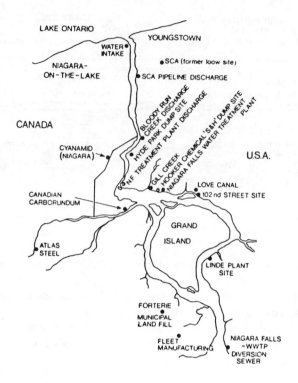

FIGURE 5.   Major effluent outfalls and waste dumps adjacent to the Niagara River.

The St. Clair River is categorized by the International Joint Commission as an area of concern, particularly along the Canadian shoreline where a large petrochemical complex has developed. Numerous industrial discharges, including the largest discharge of treated sewage in the Great Lakes, take place in the Detroit River. The Detroit River is also classified by the International Joint Commission as an area of concern, particularly on the U.S. side downstream from this same River Rouge, where most of the industrial development is located on the U.S. shore. The St. Clair and Detroit rivers and Lake St. Clair, which connects them, supply drinking water to more than 5 million people.

The Niagara River flows from Lake Erie to Lake Ontario and forms the boundary between Canada and the U.S. Steel, chemical, and other industries use it for a variety of purposes. There are several major municipal and industrial outfalls along the Niagara River (Figure 5). The U.S. Environmental Protection Agency has documented 33 major industrial sources of chemical substances to the Niagara River.[12] Fifteen of the major industrial sources collectively discharge 95% of the load of priority pollutants released directly into industrial wastewaters to the Niagara River. In 1977 there were 105 chemical and allied-product establishments in the Buffalo Standard Metropolitan Statistical Area and the City of Niagara Falls, NY. Ranked by sales, 13 of the top 50 U.S. chemical corporations are represented in the Niagara area.

The connecting rivers are, thus, massive transport systems even when concentrations of chemicals are extremely low. At a flow rate of some 6300 m^3/s, a concentration of 1 ppt corresponds to an annual load of some 200 kg. Such concentrations are regularly exceeded in the Niagara River (Figure 6).[13] The transient high concentrations for 1,2,3,4-tetrachlorobenzene (TeCB) and hexachlorobenzene (HCB) can only be due to sudden inputs of these compounds into the river. These can be observed by sampling on a weekly basis at the mouth of the river at Niagara-on-the-Lake. Similar patterns clearly reveal land-based sources

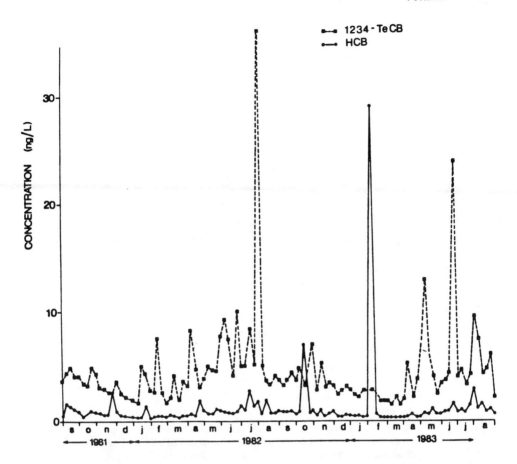

FIGURE 6.   Distribution of α-BHC, total PCBs, 1,2,3,4-TeCB and HCB in whole water at Niagara-on-the-Lake (From Oliver, B. G. and Nichol, K. D., *Sci. Total Environ.*, 39, 66, 1984. With permission.)

along the Niagara River for many other toxic persistent organic chemicals. Direct sampling of water, thus, clearly shows land-based inputs of chemicals along the connecting channels. Another way of demonstrating the impact of land-based sources of persistent toxic organic chemicals along the connecting channels is by analyses of bottom sediments. The concentrations of HCB in lake bottom sediments of Lake St. Clair compared to southern Lake Huron (Figure 7)[14] also clearly reveal major sources of this chemical along the St. Clair River.

B. Land-Based Nonpoint Sources

There are two main nonpoint land-based sources of toxic organic chemicals for the Great Lakes: (1) groundwater and (2) tributaries (other than the connecting channels) to the lakes. Groundwater sources of persistent toxic organic chemicals are thought to be particularly important in terms of their impact on the lower Great Lakes connecting channels because of the suspected leaching of chemicals from landfills and major industrial complexes adjacent to these rivers. The rise in public interest concerning toxic chemical pollution of the lower Great Lakes, started in the late 1970s with revelations about Love Canal, one of the major chemical landfills on the U.S. shore of the Niagara River (Figure 5). Leakage of chemicals to the river is suspected from several large chemical waste dumps,[15] some of which, such as the Hyde Park site, are even larger than Love Canal. Chemicals leach into the Niagara River Gorge face from the Hyde Park landfill and to creeks, such as Bloody Run, which

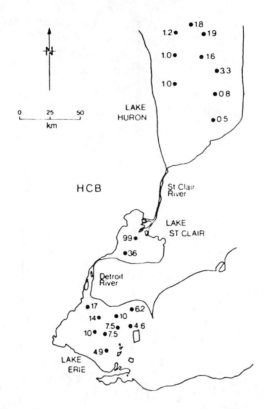

FIGURE 7.   HCB concentrations (ppb) in surficial sediments in the southern end of Lake Huron, Lake St. Clair, and the western end of Lake Erie.
(From Oliver, B. G. and Bourbonniere, R. A., *J. Great Lakes Res.*, 11(3),
369, 1985. With permission.)

drain to the Gorge. However, although concentrations of toxic organic chemicals have been
measured in seeps at the Gorge face (Table 2), loads of specific chemicals have yet to be
quantified. A different situation exists for the St. Clair and Detroit rivers. In the former
case, there is some evidence that contaminated groundwater from shallow aquifers may be
carrying chemicals off the industrial petrochemical complex south of Sarnia to the river.
The Detroit River has chemicals and dredge-spoil dumps in the river proper in the form of
man-made islands, for example, Fighting Island. Quantification of the release of toxic organic
chemicals from these artificial islands into the river is problematic. The assessment of the
role of groundwater as a source of toxic organic chemicals to the Great Lakes is in its
infancy. The resources required to sample groundwater and to derive hydraulic gradients
and heads to construct flow models are at present prohibitive, even for estimation of the
impact of specific chemical landfills.

The second major land-based nonpoint source of toxic organic chemicals to the Great
Lakes is from tributaries (other than the connecting channels) along the shores of the
lakes[16-18] and draining predominantly agricultural areas, at least for the lower Great Lakes
and Lake Michigan. This input has been examined by analyses of water and, especially,
suspended sediments collected at river mouths around the Great Lakes. Residues of persistent
toxic organic chemicals are readily measured, particularly during the spring runoff, in the
suspended sediment loads of these creeks (Table 3).[16]

In summary, land-water interactions are essentially one way, namely, land-based activities
are major sources of persistent toxic organic chemicals to lakes. A much less important but

Table 2
TRACE ORGANIC CHEMICALS IN
GROUNDWATER FROM NIAGARA
GORGE FACE BENEATH THE HYDE
PARK LANDFILL, NIAGARA FALLS,
NY

Chemical	Concentration in μg/l (ppb)
Monochlorobenzene	125
2,4,5-Trichlorophenol	275
Benzene	108
Toluene	295
Trichloroethene	98
Tetrachloroethene	64
Chloroform	132

Note: Values are mean of four seepage face samples.

Adapted from Niagara Gorge/Bloody Run Creek Water Sampling/Analytical Report, New York State Department of Health, August 24, 1984.

Table 3
RESIDUES OF ORGANOCHLORINE
COMPOUNDS IN RIVER MOUTH
SUSPENDED PARTICULATES, NORTH
SHORE OF LAKE ONTARIO (SPRING
MELT)

Stream	Content in dried suspended solids (μg/kg)[a]		
	ΣDDT	HEoD	PCB
Butler Creek	31	0.5	120
Salem River	49	<0.1	70
Colbourne Creek	18	<0.1	500
Shelter Valley Creek	18	1.0	80
Wilmot Creek	7	0.5	130
Oshawa River	97	0.5	120
Lynde Creek	19	3.0	120
Humber River	12	<0.1	230
Etobicoke Creek	11	<0.1	200
Credit River	8	<0.1	120
Oakville Creek	NC[b]	<0.1	1000
Bronte Creek	3	0.5	100

[a] No chlordane or endrin or organophosphate insecticides were identified in suspended solids.
[b] Not calculated.

Adapted from Reference 18.

Table 4
WET DEPOSITION OF
AIRBORNE TRACE
ORGANICS TO LAKE
ONTARIO

Compound	Input[a] (metric t)
Total PAH	1.60
Total PCB	0.48
Total DDT	0.08
α-BHC	0.24
γ-BHC	0.08
HCB	0.03

[a] Because of the range of concentrations at different sites and at different times around the lakes, the values have to be viewed with extreme caution.

Adapted from Reference 19.

partial return of chemicals to the land does occur both by removal and disposal of dredged sediments and from the lakes as a source of toxic organic chemicals to the atmosphere

III. AIR-WATER INTERACTIONS

The air-water interactions which influence toxic organic chemical fate are (1) the atmosphere as a source of toxic organic chemicals to lakes and (2) the lakes as a source of toxic organic chemicals to the atmosphere. The Laurentian Great Lakes have large surface areas, particularly in relation to their immediate, as opposed to upstream, drainage basins. Lake Superior has a particularly small drainage basin in relation to its very large surface area. Lake Ontario also has a small drainage basin if one does not include all of the upstream Great Lakes as part of it. A large surface area can translate into a large direct atmospheric source of toxic organic chemicals in vapor and in wet-fall and dry-fall-particulate phases. There is little buffering of the impact by the drainage basin. Toxic organic chemicals loaded directly into a lake from the atmosphere enter the water in solution or are released into solution from particulates, in response to new water-particle, as opposed to air-particle, equilibria. Once in the water, both in soluble and particulate form, they are immediately available for processing in the lake system.

A. The Atmosphere as a Source

The atmosphere is definitely a source of many toxic organic chemicals to the Great Lakes, although for specific chemicals, there remains considerable debate as to the importance of this source relative to others. Most of the data on atmospheric loads of chemicals to the lakes is derived from analyses of wet precipitation, particularly rain, for toxic organic chemicals such as DDT, PCBs, toxaphene, and γ-BHC (lindane), some of which have now been banned for several years (Table 4).[19] However, adequate samplers necessary to measure toxic organic chemical inputs in all wet, dry, and gaseous phases on a routine and event basis have yet to be deployed in a systematic network in the Great Lakes Basin.

B. The Atmosphere as a Sink

The reverse of the air-to-water pathway is the release of chemicals from water to the atmosphere, a process readily accepted for those chemicals referred to as "volatiles". For

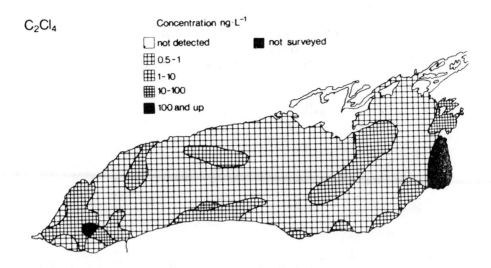

FIGURE 8. Tetrachloroethylene concentrations in Lake Ontario. (From Kaiser, K. L. E., Comba, M. E., and Huneault, H., *J. Great Lakes Res.*, 9(2), 220, 1983. With permission.)

Table 5
THE PROPORTION OF SOME CHLORINATED BENZENE INPUTS TO LAKE ONTARIO FROM THE NIAGARA RIVER LOST BY VARIOUS PROCESSES

Compound	Niagara River loading (kg/year)	Sedimenting (%)	St. Lawrence River (%)	Volatilization[a] (%)
1,2,4-TCB	2400	1	3	96
1,2,3,4-TeCB	760	2	2	96
QCB	240	4	3	93
HCB	120	15	5	80

[a] Calculated by difference.

Adapted from Reference 21.

example, perchloroethylene (Figure 8)[20] is found in measurable concentrations virtually throughout Lake Ontario. This chemical and other volatiles are rapidly lost to the atmosphere, except in ice-covered parts of the Great Lakes during winter months. The fact that there are always measurable concentrations of the chemical is because there are continuous industrial and municipal inputs. Many other more persistent toxic organic chemicals, especially the lower chlorinated isomers of chlorinated benzenes and PCBs are also volatile and are released into the atmosphere. In the case of these chemicals, the contribution to the atmosphere is usually determined as the difference in mass balances. Using this approach, Oliver[21] concluded that greater than 80% of several chlorinated benzenes, including HCB, introduced by the Niagara River to Lake Ontario, were lost by volatilization (Table 5).

Persistent toxic organic chemicals are often hydrophobic. As such they accumulate at water/other compartment interfaces, one of which is the atmosphere. Surface microlayers in water bodies have been shown to acquire much higher concentrations of some toxic organic chemicals than are found in the underlying water (Table 6).[22] The accumulation of such chemicals in microlayers may be an important intermediary step in the release of these chemicals to the atmosphere.

Table 6

**MAXIMUM CONCENTRATIONS OF SOME
CHLORINATED HYDROCARBONS IN THREE
COMPARTMENTS OF THE NIAGARA RIVER**

Chemical	Maximum concentration		
	Microlayer (ng/l)	Subsurface water (ng/l)	Suspended sediment (ng/g)
Total PCBs	57	39	310
Heptachlor epoxide	23	0	0
α-Endosulfan	20	0	14
Lindane (γ-BHC)	0	0.9	0
HCB	0	0	36

Adapted from Reference 22.

In summary, analysis of wet-fall precipitation clearly reveals the presence of several persistent toxic organic chemicals, particularly PCBs and pesticides. The sources of these may be local or in some cases from far outside the Great Lakes Basin. The relative importance of these loadings in a total mass balance awaits a better-established sampling protocol and network and an improved ability to measure dry-fall and gaseous exchange. The reverse process of release into the atmosphere from the water column is far more controversial. Estimates are derived for some chemicals by closing the mass balance, not a very satisfactory procedure, or by laboratory estimates. The interplay of these input-output processes still needs to be resolved in that PCBs collected in wet-fall may simply be a return of PCBs volatilized from a lake over a previous time period.

IV. SEDIMENT-WATER INTERACTIONS

There are three important roles for sediments as they affect the fate of persistent toxic organic chemicals in the Great Lakes:

1. Sorption-desorption
2. Burial
3. Sedimentation-resuspension

In the long term, bottom sediments in the Great Lakes may be the most important interface with the water. The relationship becomes more intimately complex when sediments are in suspension, either through their introduction from tributaries, production *in situ*, or by resuspension from the lake bottom.

A. Sorption/Desorption

In natural waters, the fate of hydrophobic organic chemicals, those compounds with a low solubility, is highly dependent on their sorption to suspended particulates. Sorption is a key process involved in the physical transport of toxic organic contaminants and in their degree of bioavailability and eventual bioaccumulation. Because particulates reduce the pollutant concentration ''in solution'', they buffer the bioavailable aqueous-phase contaminant concentration.

Different chemicals have unique and distinct sorption isotherms for different sediments. However, given a specific concentration of the chemical in water and a specific concentration of suspended sediment, a specific partition coefficient (K_p) can be measured characteristic

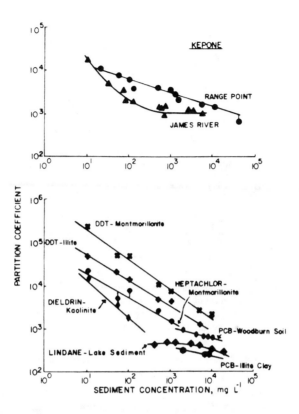

FIGURE 9.   Partition coefficient as a function of the concentration
of paticulates and particulate type. (From O'Connor, D. J. and
Connolly, J. P., *Water Res.*, 14, 1517, 1980. With permission
from Pergamon Journals Ltd., ©1980.)

of both the chemical and of the sediment. The K_p is measured as an equilibrium constant
between the concentration of contaminant in the particulate phase and the concentration in
solution; i.e., at a given sediment concentration, the concentration of the chemical in solution
controls the amount sorbed. This is only true at relatively low sediment concentrations of
up to 20 mg/l. Thus, the first characteristic affecting K_p is the concentration of the chemical
in solution.

The second controlling factor is suspended-particulate concentration. O'Connor and
Connolly[23] showed that there is an inverse relationship between concentration of absorbing
solids and partition coefficient (the ratio of the solid phase concentration to the dissolved
phase concentration in the linear portion of the Langmuir sorption isotherm; see Figure 9.

The third controlling factor is particulate type. The differences between DDT sorption on
illite vs. montmorillonite, and Kepone sorption on James River sediment vs. Range Point
Marsh sediment, are also shown in Figure 9.

The fourth important sediment parameter controlling partitioning sorption is natural organic
content of particulates. Suspended particulates in many aquatic systems are highly organic
and made up of phytoplankton and bacteria. Centrifugation results in a dark organic paste,
which by virtue of bioaccumulation and sorption can contain high concentrations of toxic
organic chemicals. Other suspended particulates, especially resuspended bottom sediments
in the Great Lakes, are made up primarily of inorganic material and usually have an organic
carbon content of some 3%. However, even when low percentages of organic matter are
present, some of this may be in the form of thin coatings on mineral substrates. In fact,
partitioning of toxic organic chemicals onto aquatic particulates may not be so much of a

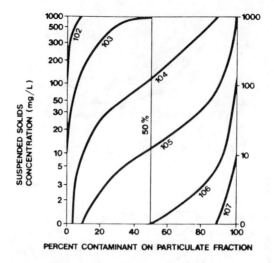

FIGURE 10.   Relationship between concentration of suspended particulates and percent of contaminant transported in the suspended particulate phase (assuming no change in K_p with concentration of particulates and no change in particulate organic content).

sorption process as one of exclusion from water and solubilization of the chemical into an organic (lipid-like) surface layer on particulates. Parks[24] referred to this as hydrophobic bonding. As water solubility increases, the octanol/water partition coefficient (K_{ow}) for a toxic organic chemical decreases[25] or the compound becomes less soluble in lipid-like media. Thus, carbon content (OC) of suspended and bottom particulates is critical to the degree of partitioning. For example, the partitioning of methoxychlor was shown to be linearly related to the organic carbon content of the sediment.[26] Karickhoff[27] defined sediment K_p as equal to $K_{ow} \times OC$, where OC is the fraction of organic carbon in a sediment. Thus, K_p can theoretically be predicted from the K_{ow} for the chemical and the organic carbon content of a sediment.

Last, time affects sorption of toxic organic chemicals in that some 60% of total sorption is usually complete in 2 to 5 min. Equilibrium is established extremely rapidly in relation to the time required for environmental contamination and media sampling. There are, however, two components to the sorption phase:

1.   The rapid component completed in a very short time period
2.   A slower component that often occurs only after a period of minutes but can continue for weeks or longer

Conventional K_ps, as determined in laboratories, are primarily a measurement of the rapid component. In real aquatic ecosystems, however, where exposure times can be very long, the slower sorption component may be important. When sediment contaminants are desorbed, it is usually only the rapid component that becomes quickly bioavailable. However, also with time, in real aquatic ecosystems the slower component may eventually be released, even when the extractant is only water.

For hydrophobic organic chemicals with K_ps of 10^5, in rivers with suspended particulate concentrations of around 10 mg/l (which is just above that of the Niagara River), calculation shows that some 50% of the chemical will be in the particulate phase (Figure 10).[28] If the suspended particulate concentration rises to 100 mg/l, as occurs in many rivers during high-flow periods, then some 90% of the mass of contaminant could be in the particulate phase.

Table 7
PERCENT OF ORGANOCHLORINE
CONTAMINANTS PRESENT IN AQUEOUS (APLE)
PHASE

	Sites		
Chemical	Lake Erie inflow to Niagara River (%)	Niagara-on-the-Lake (%)	Lake Ontario (3 stations) (%)
PCBs	100	22	91
HCB	48	28	100
p,p'-DDT	94	70	100
γ-BHC	100	98	100

Adapted from Reference 11.

In summary, the seven main factors affecting partitioning of toxic organic chemicals onto particulates in aquatic ecosystems are, in decreasing order of importance:

1. Solubility of the chemical
2. Concentration of the chemical
3. Concentration of the sediment
4. Type of sediment, particularly organic content or degree of lipid layering
5. Time (reversible vs. relatively irreversible)
6. Concentration of other chemicals and ions in solution
7. Temperature

Of these factors, only the initial properties of the chemical are defined. All of the others are site related and influenced by various limnological processes. Thus, although laboratory tests do provide some basic understanding that can help to predict aquatic fate, real fate is usually site specific.

The Niagara River has a suspended sediment load of some 10 mg/l on the average. The concentration of suspended particulates in Lake Ontario is far less. Depending on chemical and particulate characteristics as discussed above, different percentages of different toxic organic chemicals are present in the "aqueous" and particulate phases of the river and lake (Table 7).[11] In the Niagara River, the concentrations of toxic organic chemicals in the particulate phase have been determined for the particulates removed by high speed centrifugation. The concentrations in the aqueous phase are determined for the effluent following on-site extraction of 200-l samples in an aqueous phase liquid extractor (APLE).[29,30]

B. Burial

The most important role played in sediments in the Great Lakes is burial of persistent toxic organic chemicals in bottom sediments.[31] For many persistent, hydrophobic, toxic organic chemicals, the largest remaining mass which has not been removed downstream, volatilized/or degraded, is found in the bottom sediments of the lakes. Distribution diagrams for several persistent toxic organic chemicals, such as mirex, PCBs, DDT, and others[32-36] are published for all five of the Laurentian Great Lakes. The results show that the bottom sediments of these lakes are major sinks and sites of eventual burial of toxic organic chemicals. The distribution patterns provide clues to zones or sites of maximum loadings and permit analyses of long-term input trends.

In Lake Ontario, for example, surficial sediment maps of concentrations of specific toxic organic chemicals such as mirex reveal the Niagara River to be a major source.[35] Alternatively, for PCBs (Figure 11), the distribution patterns are more complex and, although the Niagara River is a major source, other sources, particularly the atmosphere, are involved. The patterns and areas of highest concentrations reflect these sources and in-lake processes of sediment resuspension and focusing.

Sediment cores from Lake Ontario have been analyzed to determine changes in historical inputs of organic chemicals to the lake and to find out the rate at which these chemicals are buried.[37] Sediment cores from the Niagara Basin of Lake Ontario at a site where sedimentation rates are high, close to the mouth of the Niagara River, have depth distributions of total chlorobenzenes(CBs) which match U.S. production figures for this chemical (Figure 12).[37] Maximum CB concentrations occurred at depths corresponding to the early 1960s. The same pattern emerges for PCBs. Recent surficial bottom sediments immediately off the Niagara Bar have considerably lower concentrations of these chemicals than was the case in past years. Mirex and octachlorostyrene are also excellent examples of how highly insoluble, hydrophobic, persistent toxic organic chemicals are buried at high sedimentation-rate sites once point sources are removed.[37]

C. Resuspension/Recycling

In Lake Ontario and in other lakes, the major inputs of particulate-associated chemicals, excluding atmospheric sources, include sediments from inflowing rivers, sediments from shoreline erosion, resuspended bottom sediments, and particulates produced in the lakes. Varying amounts of different contaminants entering Lake Ontario are adsorbed to particulates to different degrees for reasons discussed earlier. Soluble contaminants entering the lake may be adsorbed to the suspended particulates derived from nearshore erosion, from bottom sediment resuspension, or by increased biological productivity.

At a 20-m depth in Lake Ontario, higher organic chemical concentrations in particulates were observed at offshore than at nearshore stations, probably owing to bioaccumulation followed by settling of seston.[38] When the downflux of chemicals was calculated using sediment traps, however, the effect of the Niagara River was very apparent. For example, HCB downfluxes were 8 ng/m^2/d for offshore vs. nearshore Niagara River inflow sites. In the open lake, there was considerable variation in contaminant downfluxes, but many of the values for specific chemicals were in the hundreds of ng/m^2/d. Although many organochlorine contaminants were detected in the settling particulates, the most prevalent were the CBs. The concentrations in deeper sedimentation traps were also substantially elevated for PCBs, HCB, and mirex, all of which are strongly partitioned onto bottom sediments.

Resuspension in the Great Lakes produces "nepheloid layers". Particulates in nepheloid layers consist of clay-size particles, biological debris, and aggregated particles. The initial investigation of a nepheloid layer in the Great Lakes was in Lake Michigan.[39,40] The nepheloid layer had a thickness of about 10 m at the base of the offshore slope and gradually thinned farther offshore. In Lake Michigan, internal waves along the thermocline and wind-generated currents were considered the main forces that caused resuspension of bottom sediments. In 1982, a much thicker nepheloid layer than in Lake Michigan was first detected in Lake Ontario.[41,42]

Relative inputs from resuspended bottom sediments, uncontaminated shoreline clays, and river detritus were separated by comparing the ratios of concentrations of various toxic organic chemicals in Niagara River suspended particulates and Lake Ontario surficial sediments, with concentration ratios in particulates collected by sediment traps. Because sediment core analysis has shown that changes have occurred in CB-isomer loadings in recent years, some ratios represent present input, whereas others represent resuspended bottom sediments. The expected resuspended bottom sediment ratios were obtained by collecting cores, gently agitating them to produce a suspended cloud, and collecting and analyzing

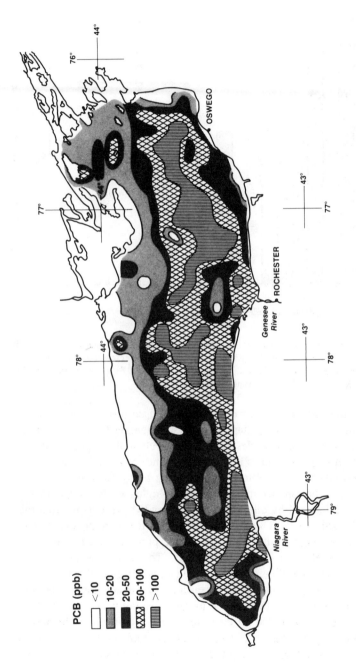

FIGURE 11. Distribution of total PCBs in surficial bottom sediments from Lake Ontario (3-cm depth). (From Thomas, R. L., *J. Great Lakes Res.*, 9(2), 122, 1983. With permission.)

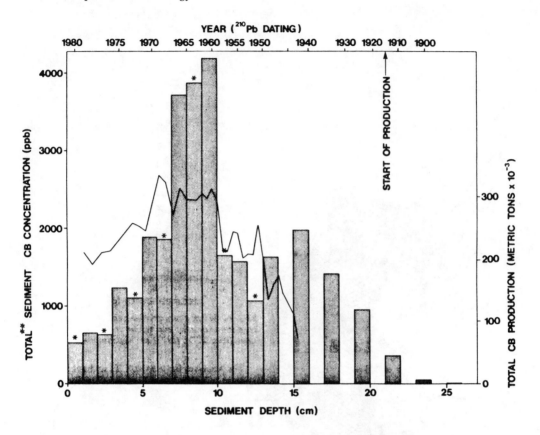

FIGURE 12.   Total concentration of di- through hexa-CBs vs. depth in a sediment core and approximate ^{210}Pb dating. Starred sections were freeze-dried prior to analysis, so data for these sections represent minimum values due to possible volatilization losses. Total U.S. production figures for mono- through hexa-CBs. (From Durham, R. W. and Oliver, B. G., *J. Great Lakes Res.,* 9(2), 165, 1983. With permission.)

this resuspended material. Oliver and Charlton[38] developed and used this technique to determine the relative contribution of organic contaminants in sedimenting (from the epilimnion) and resuspended (from the bottom) particles at different levels in the nepheloid layer. A significant fraction of the settling particulates in the deeper zones near the Niagara River and at all depths in the open lake was found to be resuspended bottom sediments (Table 8).[38] The percent contribution by bottom sediments was much higher in the middle of the lake compared with the area influenced by direct input of suspended particulates from the Niagara River. Mirex was loaded to Lake Ontario by the Niagara River in the 1970s. The production and use of mirex have been limited since 1976, and present-day concentrations on suspended particulates in the Niagara River are very low. The high concentrations of mirex in many of the deeper, offshore, sediment traps are indicators of the continual physical recycling of contaminants by bottom sediment resuspension. Large amounts of bottom sediment-associated contaminants are thus resuspended in the overlying water, and the lake water column is, thus, repeatedly reexposed to these previously sedimented chemicals, increasing the time required for permanent burial and removal from food webs. This is particularly a problem in the Great Lakes areas of concern, where *in situ*-contaminated sediments play a major role in the recycling of contaminants (Figure 13). Remedial action plans to reduce contamination of most of these areas of concern will have to deal with severely contaminated bottom sediments.

Table 8
PERCENT CONTAMINANT
CONTRIBUTION FROM BOTTOM
SEDIMENT RESUSPENSION

Site	Depth (m)	Bottom sediment (%)
Near the Niagara		
River inflow	20	0
	40	0
	60	22
	68	41
Central Niagara Basin		
of Lake Ontario	20	62
	40	57
	80	48
	90	70
	98	52

Adapted from Reference 38.

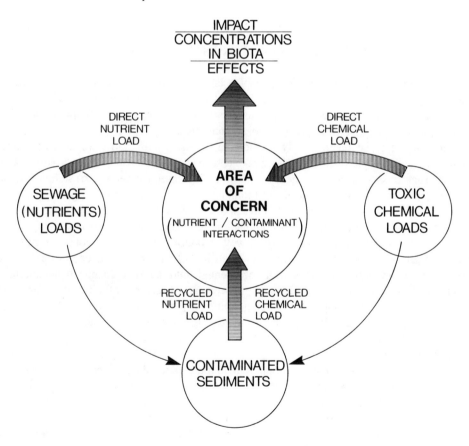

FIGURE 13.   Schematic of the role of internal recycling of nutrients and contaminants in Great Lakes "Areas of Concern".

Table 9

CONCENTRATIONS OF SOME TOXIC ORGANIC CHEMICALS IN SIZE-FRACTIONATED SUSPENDED PARTICULATES

Niagara River suspended solids approximate size range and unfiltered water	Concentration (ng/g dry weight) (ng/l water)					
	HCBD	1,2,4,5-TeCB	1,2,3,4-TeCB	QCB	HCB	PCBs
NOTL <75 μm	NS	NS	NS	NS	NS	NS
NOTL 75—175 μm	9.9	16	38	26	230	560
NOTL 175—300 μm	11	24	58	34	19	580
NOTL 300—500 μm	36	26	86	110	97	1700
NOTL 500—700 μm	9.6	19	56	40	180	1700
NOTL >700 μm*	37	51	300	270	89	3800
Fort Erie >700 μm*	ND	ND	6.0	3.0	5.0	NA
NOTL water	0.6	1.2	2.6	0.8	0.8	11.0
Fort Erie water	ND	ND	0.06	0.05	0.02	NA

Note: * = representative of all size fractions; ND = not detected; NS = no Sample; NA = not analyzed. Data collected during July from the Niagara River at Niagara-on-the-Lake (NOTL) and Fort Erie.

Adapted from Reference 43.

V. NUTRIENT-WATER INTERACTIONS

Nutrient inputs control the trophic state of a lake, and this, in turn, dominates species composition and food webs within the lake. Productivity, primarily a result of nutrient levels, has a strong influence on three major processes affecting toxic, organic chemical fate: (1) sedimentation, (2) degradation, and (3) bioaccumulation.

''Nutrients'' may be taken simply as the role of phosphorus and nitrogen in primary productivity and the role of this in setting lake and river trophic levels, which in turn could influence such processes as toxic chemical burial by increased (or decreased) sedimentation rates; degradation rates of chemicals by biological, photolytic, and other processes; and bioavailability and bioaccumulation. ''Nutrients'' can also be seen as a basic measure of the more complex trophic state, levels both natural and altered (e.g., by fish stocking), and the relative importance of both top-down as well as bottom-up control of contaminant fate. The processes at this interface involve an extension of understanding of eutrophication controls, which are now reasonably well documented, into the toxic chemicals aquatic fate and effects field. The controlling pathways, processes, and factors in the Great Lakes need to be resolved because trophic state is being continually reduced to lower levels while the lakes are being simultaneously impacted by increasing toxic chemical loads.

A. Sedimentation

The effect of nutrients on increasing phytoplankton and zooplankton population densities is well established. In turn, this leads to greater seston concentrations and increased sedimentation rates, especially important in open-water areas of the Great Lakes where shore-derived concentrations of inorganic sediments are low. Sorption of toxic organic chemicals to these trophic level organisms, both when alive and dead, is rapid as has been seen in the Niagara River. Concentrations of toxic organic chemicals in suspended particulates at Niagara-on-the-Lake, the outflow to Lake Ontario, were often much higher than at the start of the river at Fort Erie (Table 9).[43] The larger size fractions were made up primarily of zooplankton, and clearly these had already bioaccumulated or sorbed toxic organic chemicals during the short travel time (<24 h) of the river.

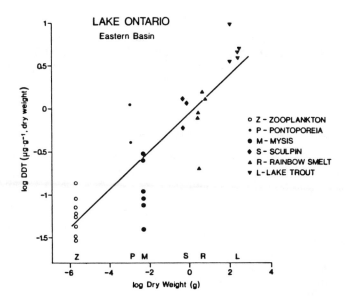

FIGURE 14. DDT concentrations as a function of body size. (From Borgmann, V. and Whittle, D. M., *Can. J. Fish Aquat. Sci.,* 40, 333, 1983.
With permission.)

High nutrient levels produce sediments in which certain biota predominate. The usual
biota in highly organic-rich (eutrophic) sediments are oligochaete worms. These can accumulate high levels of certain toxic organic chemicals relative to the sediments[44] and different
isomers of different toxicity. Hence, they can introduce shifts in isomer ratios between the
sediments and benthic predator organisms. These burrowing worms also contribute to recycling of sedimented organic chemicals by bioturbation of surface layers of sediments to
depths which can represent several years of sedimentation.[45]

B. Degradation

Degradation rates and processes may be strongly influenced by the trophic state. At higher
trophic levels, bacteria become more abundant as do fungi. Both of these are well established
in laboratory testing as major degraders of many toxic organic chemicals. Because of the
many variables involved in biodegradation, this process has, so far, been mainly quantified
by laboratory tests. The use of cyclone fermentor systems allows measurement of both
aerobic and anaerobic degradation rates for specific chemicals. In terms of nutrient-water
interactions, increased nutrient loads increase productivity which in turn increases food
sources for bacteria and fungi. In laboratory tests with chemicals, increased organic nutrient
levels commonly result in more rapid rates of biodegradation for chemicals in both aerobic
and anaerobic conditions.[46,47]

C. Bioaccumulation/Food Webs

Given equal loadings of persistent toxic organic chemicals, the concentrations in biota at
top trophic levels appear to bear an inverse relationship to the trophic status of the water
body. The cause may be the length of the food chain involved or the degree to which a
greater total biological mass can dilute the same chemical load. In Lake Ontario, an oligotrophic lake with a long food chain to top predators, Borgmann and Whittle[48] found a
positive correlation between concentrations of DDT, a typical lipophilic, toxic organic
chemical and body size of various biota (Figure 14).[48] More productive lakes, such as Lake
Erie which is mesotrophic, have higher seston concentrations. They also have more organic

Table 10

**MEAN SIZE AND CONTAMINANT CONCENTRATION
(DRY WEIGHT BASIS) FOR ORGANISMS FROM THE
EASTERN AND WESTERN BASINS OF LAKE ONTARIO
(PPB [NG/G]**

Organism	Body weight (g)	PCBs		Total DDT	
		E. Basin	W. Basin	E. Basin	W. Basin
Zooplankton	2×10^{-6}	200	310	60	80
Amphipods		1,700	1,670	690	480
Mysids		290	390	120	280
Sculpins		3,820	—	970	—
Smelt		2,380	6,750	780	2200
Lake trout		14,100	20.700	4920	5700
Coho salmon		—	9,140	—	3260

Note: Number of samples are 20 and 4 for zooplankton; 8 and 4 for amphipods; 12
and 41 for mysids; and over 50 for all three fish in the Eastern and Western
basins, respectively.

Adapted from Reference 48.

resuspended sediments. Thus, the particulates in suspension have a greater ability to sorb
toxic organic chemicals, consequently buffering their bioconcentration.

To understand fully the role of trophic state in the biological fate of contaminants, estimates
are required of concentrations of specific chemicals at all trophic levels. Borgmann and
Whittle[48] determined concentrations of PCBs and DDT in various trophic level biota from
Lake Ontario (Table 10). Concentrations in the netplankton were similar to those found in
mysids, which spend at least part of the day close to or in contact with bottom sediments
and their resuspended component in the lower nepheloid layer. Clearly, even extremely
insoluble toxic organic chemicals which are found almost exclusively in suspended and
bottom particulates remain bioavailable and pass into food webs by direct uptake by benthic
and planktonic organisms.

Bioaccumulation of toxic organic chemicals in Lake Ontario, as in other lakes, is influenced
by trophic state and the corresponding food web in Lake Ontario. Lake trout from Lake
Ontario have body burdens of persistent toxic organic chemicals, such as mirex and dioxins,
which are also highly partitioned to particulates and rapidly buried in bottom sediments.
Although transfer of organic contaminants to biota directly from suspended particulates has
been demonstrated by Harding and Phillips,[49] the more a fish species relies on benthic
organisms as a food source, the more likely is bioaccumulation of sediment-associated
hydrophobic organic contaminants. For example, bottom sediments and their associated pore
waters are major sources of polycyclic aromatic hydrocarbons to benthic organisms. Reduction in contamination of fish species is slowest for those fish which rely on benthic
organisms or their predators as major food sources.

In Lake Ontario, two ubiquitous fish at the base of the food web are smelt and alewife
(Figure 15). The food sources of smelt (*Osmerus mordax*),[50] based on their main stomach
contents, are 51% amphipods, 28% mysids, and 20% crustacean zooplankton for small smelt
(<11 cm); 40% amphipids, 32% fish, 23% mysids for medium smelt (11 to 16 cm); and
22% amphipods, 21% mysids, 54% sculpins (a benthic organism eater which ingests bottom
sediment with food), and other smaller smelt and alewife, for large smelt (>16 cm).

Alewife (*Alosa pseudoharengus*) stomach contents, unlike those of smelt, do not change
with fish size. The unweighted mean percent composition of alewife stomach contents was

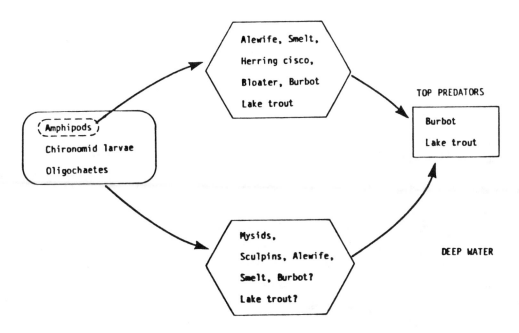

FIGURE 15.   Deep-water Lake Ontario food web.

81% crustacean zooplankton (55% copepods); 13% amphipods (*Pontoporeia affinis* and *Gammarus*); and 4% mysids (*Mysis relicta*). In Lake Ontario, stomach contents of juvenile lake trout (*Salvelinus namaycush*) of 10 to 45 cm in length were analyzed by Elrod.[51] For all seasons, the major forage fish (42% by weight) was the slimy sculpin (*Cottus cognatus*). In April and May, alewife made up 28% of the fish remains in stomachs. Rainbow smelt were the principal forage fish during the July/August period and constituted 25% of the fish remains. Amphipods (*Pontoporeia affinis*) and crustaceans (*M. relicta*) were also found in the trout stomachs. Invertebrates were a major diet of 15-cm lake trout in the April to May period. The stomach contents of coho salmon largely consist of smelt and alewife.[52]

Data on toxic organic chemicals in Lake Ontario biota are very disjointed. Very crude mean values, based on lumping of many diverse values, simplify this data base for a few selected chemicals (Table 11),[43,48,53,54] so that an overall picture of the ranges of concentrations in various organisms and media becomes evident. Gaps in the data base and wide ranges in concentrations are usually related to the many ecosystem variables, critical in predictions of toxic organic chemical fate, but seldom recorded, and most often never measured. These variables include: water nutrient data at the sample site; organic content of bottom sediments; particle-size distributions of sediments and suspended particulates; benthic organism densities; benthic organism species composition; and fish size, age, lipid, and stomach contents. Most toxic organic chemical data exist for the easily collected and analyzed media of bottom sediments and fish. Only scant data are available on intermediate food web biota such as benthic organisms and plankton (both phytoplankton and zooplankton).

Bioaccumulation of toxic organic chemicals in planktonic and benthic biota and in lower food-chain fish (Tables 10 and 11) is eventually reflected in the concentrations of these chemicals found in top predator fish species. In the case of Lake Ontario, the top predators are lake trout and coho salmon (Table 11 and Figure 16). The distinctly lower concentrations of DDT and its main degradation product, chlordane, and PCBs in coho salmon in Lake Erie vs. Lake Ontario (Figure 13), imply a trophic state control of bioaccumulation because the load of chemicals to the lakes is not dissimilar. What is different is the trophic state of

Table 11
SUMMARY AND RANGE OF CONCENTRATIONS OF TOXIC ORGANIC CONTAMINANTS IN LAKE ONTARIO MEDIA (PPT OR NG/KG [L])

Chemical	Raw water	Bottom sediment	Benthos	"Suspended" sediments	Plankton	Fish	Herring gull eggs
Total DDT	0.3—57	25,000—218,000	440,000—1,088,000	40,000	63,000—72,000	620,000—7,700,000	7,700,000—34,000,000
PCBs	5—60	110,000—1,600,000	470,000—9,000,000	600,000—6,000,000	110,000—6,100,000	1,378,000—17,000,000	41,000,000—204,000,000
Mirex	0.1	144,000	41,000—228,000	15,000	ND-12,000	50,000—340,000	1,800,000—6,350,000
CBs	1—54	11,000—4,500,000	NA	574,000	27,000	6,000—370,000	300,000
Dioxins	0.01—0.03	8,000	NA	NA	NA	5—107	44—1,200
Lindane	0.4—11	46,000	NA	1,000—12,000	12,000	2,000—360,000	78,000

Note: These values are only of the crudest nature and are not statistical means. Where only one reference existed, the numbers are means, often of widely ranging values. Where a range is given, several sources of data were involved. ND = not detected; NA = not analyzed.

Compiled from References 43, 48, 53, and 54.

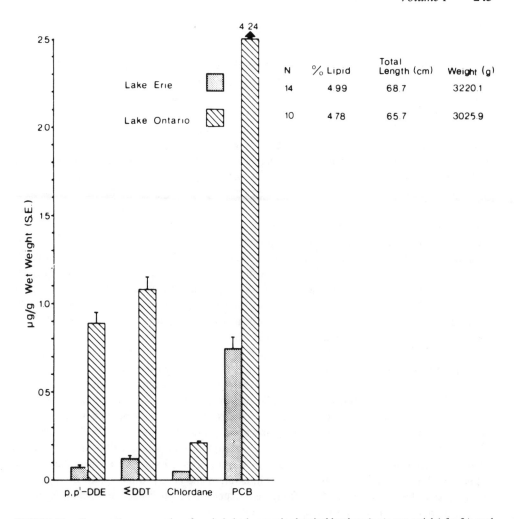

FIGURE 16. Comparative mean values for whole body organic chemical burdens (μg/g wet weight) for 3⁺ aged coho salmon (*Onchorbynchus kisutch*) from Lake Erie and Lake Ontario. (From Whittle, D. M. and Fitzsimons, J. D., *J. Great Lakes Res.*, 9(2), 297, 1983. With permission.)

the two lakes and many of the processes and factors described above are known or thought to be active. For example, the productivity of Lake Erie is higher than in Lake Ontario, resulting in a greater degree of partitioning of hydrophobic toxic organic chemicals to particulates and, thus, reduced bioavailability. Sedimentation is greater at most sites and sedimentation rates higher at most locations in Lake Erie than in Lake Ontario, resulting in more rapid burial of hydrophobic organic chemicals. Bacterial populations in water and sediments are higher and degradation rates may thus also be more rapid in Lake Erie. Because of the larger amount of decaying seston, humic matter should be higher in Lake Erie and photochemical degradation correspondingly enhanced.[56]

Alternatively, in shallower, higher trophic-level lakes such as Lake Erie, resuspension of contaminated sediments is enhanced. Benthic biota populations are more dense with correspondingly greater bioturbation of surface bottom sediments. These processes of resuspension and bioturbation may cause passage of organic chemicals from bottom sediment surfaces directly into the lake food web, or they could result in a dilution of immediate bottom surface-sediment concentrations, by bioturbation, and then an increased partitioning to resuspended-surface sediments.

VI. CONCLUSIONS

Future toxic organic chemical research should focus on the interfaces between the major compartments influencing the fate of toxic organic chemicals in lakes, on the interactions between these compartments, and on the processes which transfer chemicals between them. Interactions between land, air, lake-bottom sediments, lake-trophic state, and toxic organic chemicals all need to be understood and eventually quantified. To do so at specific sites will be costly and labor intensive but is the only valid means of understanding and accurately modeling contaminant fate at these sites. Only on the basis of such knowledge can meaningful and realistic control actions be implemented and remedial action plans developed for in-place pollutants.

Land-water interactions in lakes have been largely limited to measurements of toxic chemical discharges by tributaries, effluent streams, or landfill leachates. The requirements are similar to those studied earlier for nutrients but will be more complex to resolve for organic chemicals. Transmission of chemicals via rivers to the Great Lakes and the controlling factors and processes involved in this transmission, or lack of it, need to be resolved. Included would be improved knowledge of soil loss equations, in-stream degradation pathways, temporary in-stream storage processes, and the role of riverine biota. Of particular importance is the need to develop the concept that erosion and runoff are event based and that the research should focus on events such as snowmelt or storms.

At the atmosphere-water interface, there is a need to more adequately quantify the atmospheric load of specific toxic organic chemicals to the Great Lakes. Adequate networks of samplers based on validated protocols may be required. Better proxy methods of assessing this load, such as by analyses of dated lake-bottom sediment cores, may be an alternative as may be extrapolation from calibrated watersheds, such as has been successfully used in acid-rain loading estimates. The return pathway of chemicals from lakes to the atmosphere also needs to be quantified, perhaps by more accurate mass balances, but more likely by laboratory experiment and modeling combined with direct, over-lake measurements. The latter would involve special equipment and innovative and costly in-lake platforms. To date, the emphasis has been on wetfall. Better estimates of gaseous exchange, and especially dryfall particulate inputs, are needed.

Sediment-water interactions are complex and in lakes can be all pervasive. This area of research has been a focus of activity for many years but seldom from a process approach, and even less on toxic organic chemicals. Many questions about the role of sediments or, in the broader sense, particulates in the fate of toxic organic chemicals in aquatic systems remain. If certain procedures were followed, our knowledge of the role of sediments in toxic chemical fate would be greatly enhanced. For example,

1. When suspended particulates are collected and analyzed for organic contaminants, the quantity of suspended solids should be measured along with their organic carbon content so that partitioning can be better estimated and aqueous phase concentrations better predicted.
2. At polluted sites, spatial and temporal variations in suspended particulate concentrations should be determined; for example, in nepheloid layers in the Great Lakes.
3. Sedimentation and resuspension rates for particulates should be measured at polluted sites.
4. Surface areas, morphology, and surface characteristics of particulates require further quantification in terms of sorption processes relevant to contaminant bioavailability.
5. The colloidal fraction of the "aqueous" phase needs to be characterized better with respect to its role in contaminant transport and sedimentation.
6. Transfer rates of toxic organic chemicals into and out of bottom sediments by chemical, biological, and physical processes need to be better quantified.

7. The effects of diagenetic changes on toxic organic chemical bioavailability require quantification.

The fourth interface or area of interaction is also complex. It has been studied little, and, at best, indirectly, by recognizing apparent differences in bioaccumulation in lakes of similar toxic organic chemical loading but different trophic state. A great deal of the required research is simply the need to have a detailed knowledge of the role of direct transmission of toxic organic chemicals from the lake components of water and sediments to biota and then, the role of food webs in transferring these chemicals or their degradation products to higher trophic levels in lakes of different trophic state. Data bases thus need to be extended to include many intermediate food web biota. More sites will need to be studied, and sampling frequencies will have to increase. A more flexible, toxic organic-chemical oriented sampling of aquatic media should aim at providing not only a monitor of conditions, but a more readily interpretable database of bioavailabilty and bioaccumulation processes. In summary, there are requirements for:

1. More data at specific polluted field sites if we are to understand pathways and produce realistic predictive models of contaminant fate
2. More information on the nature of the biotic and abiotic components of suspended loads
3. More concentration data on lower trophic-level biota, different types of benthic organisms, and phyto- and zooplankton species
4. More information on food webs to ensure enough concentration data for all media and biota relevant to exposure to higher organisms and humans
5. Measurement of degradation rates and biological transfer rates in both nonpolluted and polluted aquatic ecosystems of different trophic states
6. More frequent testing of laboratory measurements of critical processes such as sorption, photolysis, hydrolysis, and degradation at a variety of sites of different trophic states; i.e., field vs. laboratory comparisons made for a range of organic chemicals to encompass those expected to be virtually insoluble, moderately soluble, and very soluble

Research to resolve the processes and interactions at the interfaces in aquatic ecosystems is needed to understand how toxic organic chemicals are loaded to and naturally cleansed from different trophic state ecosystems, or become concentrated in media and result in exposure to, and, hence, risk to, aquatic biota and humans. Models for predicting the environmental fate of toxic metals and organic chemicals are often based on laboratory research which provides information on rates of reactions. Although these models can predict compartmentalization on a gross scale, they often fall afoul of the complexities of natural ecosystems. They usually cannot predict concentrations in specific media, and such concentrations are required for estimation of real risk. Predicting the fate of toxic organic chemicals in a specific aquatic ecosystem requires knowledge of the following: all inputs (loadings from the land, point and nonpoint sources, the atmosphere, and bottom sediments), initial conditions (concentrations of all relevant media at specific times), process pathways and rates (of transport, volatilization, photolysis, hydrolysis, degradation, sedimentation, burial, resuspension, and bioaccumulation).

Toxic metal research was well established by the late 1970s. Only in the 1980s have organizations applied sufficiently sophisticated laboratory and field research techniques to toxic organic-chemical fate in aquatic ecosystems. We have now begun to comprehend the complexity of the processes affecting toxic organic-chemical fate in real ecosystems. As in any field of research, there have been lone individuals ahead of their time. We do have considerable information on concentrations in various aquatic media and on distributions and transport phases in specific systems. The Great Lakes in total are now one of the best-documented sites worldwide for toxic organic chemicals. Yet even here there are major gaps in our knowledge of the controls of contaminant fate, especially of processes and their rates. The ability to predict the aquatic fate and effects of toxic organic chemicals is a critical requirement because there are still many sources of such chemicals to aquatic ecosystems, and there is an expanding need to dispose of toxic chemical wastes. More extensive use of toxic chemicals in agriculture and forestry, industrial and urban development, and extension of these pollutant problems to the Third World has resulted in a global concern about the fate of these materials.

REFERENCES

1. **Beeton, A. M.,** The world's great lakes, *J. Great Lakes Res.,* 10(2), 106, 1984.
2. **Allan, R. J., Mudroch, A., and Munawar, M., Eds.,** The Niagara River/Lake Ontario pollution problem, *J. Great Lakes Res.,* 9(2), 109, 1983.
3. **Chau, Y. K., Maguire, R. J., Wong, P. T. S., and Sanderson, M. E., Eds.,** Detroit River-St. Clair River. Special issue, *J. Great Lakes Res.,* 11(3), 191, 1985.
4. **Lawrence, J., Ed.,** St. Clair river pollution. Special issue, *Water Pollut. Res. J. Can.,* 2, 283, 1986.
5. **Allan, R. J.,** The limnological units of the lower Great Lakes-St. Lawrence River corridor and their role in the source and aquatic fate of toxic contaminants, *Water Pollut. Res. J. Can.,* 21(2), 168, 1986.
6. **Warry, N. D. and Chan, C. H.,** Organic contaminants in the suspended sediments of the Niagara River, Inland Waters Directorage, Water Quality Branch, Ontario Region, Burlington, Ontario, 1981.
7. **Warry, N. D. and Chan, C. H.,** Organic contaminants in the suspended sediments of the Niagara River, *J. Great Lakes Res.,* 7, 394, 1981.
8. **Kuntz, K. W. and Warry, N. D.,** Chlorinated organic contaminants in water and suspended sediments of the lower Niagara River, *J. Great Lakes Res.,* 9(2), 241, 1983.
9. **Couillard, D.,** BPC et pesticides organochlorines dans le système Saint-Laurent, *Can. Water Res. J.,* 8(2), 32, 1983.
10. **Kuntz, K. W.,** Toxic contaminants in the Niagara River, 1975-1982, Tech. Bull. No. 134, Water Quality Branch, Inland Waters Directorate, Ontario Region, Burlington, Ontario, 1984, 47.
11. **McCrea, R. C., Fisher, J. D., and Kuntz, K. W.,** Distribution of organochlorine pesticides and PCB's between aqueous and suspended sediment phases in the lower Great Lakes region, *Water Pollut. Res. J. Can.,* 20(1), 57, 1985.
12. **Vincent, J. and Franzen. A.,** An overview of environmental pollution in National Enforcement Investigation Center, U.S. Environmental Protection Agency, Denver, CO, 1982.
13. **Oliver, B. G. and Nicol, K. D.,** Chlorinated contaminants in the Niagara River, 1981—1983, *Sci. Total Environ.,* 39, 57, 1984.
14. **Oliver, B. G. and Bourbonniere, R. A.,** Chlorinated contaminants in surficial sediments of Lakes Huron, St. Clair, and Erie: implications regarding sources along the St. Clair and Detroit Rivers, *J. Great Lakes Res.,* 11(3), 366, 1985.
15. **Elder, V. A., Proctor, B. L., and Hites, R. A.,** Organic compounds found near dump sites in Niagara Falls, New York, *Environ. Sci. Technol.,* 15, 1237, 1981.
16. **Frank, R.,** Pesticides and PCB in the Grand and Saugeen River basins, *J. Great Lakes Res.,* 7, 440, 1981.
17. **Frank, R., Sirons, G. J., Thomas, R. L., and McMillan, K.,** Triazine residues in suspended solids (1974—1976) and water (1977) from the mouths of Canadian streams flowing into the Great Lakes, *J. Great Lakes Res.,* 5, 131, 1979.
18. **Frank, R., Thomas, R. L., Holdrinet, H., McMillan, R. K., Fraun, H. E., and Dawson, R.,** Organochlorine residues in suspended solids collected from the mouths of Canadian streams flowing into the Great Lakes, 1974—1977, *J. Great Lakes Res.,* 7, 363, 1981.
19. **Eisenreich, S. J., Looney, B. B., and Thornton, J. D.,** Assessment of airborne organic contaminants in the Great Lakes ecosystem, a report to the Science Advisory Board, Canada-U.S. International Joint Commission, Appendix A, 1980.
20. **Kaiser, K. L. E., Comba, M. E., and Huneault, H.,** Volatile halocarbon contaminants in the Niagara River and Lake Ontario, *J. Great Lakes Res.,* 9(2), 212, 1983.
21. **Oliver, B. G.,** Distribution and pathways of some chlorinated benzenes in the Niagara River and Lake Ontario, *Water Pollut. Res. J. Can.,* 19(1), 47, 1984.

22. **Maguire, R. J., Kuntz, K. W., and Hale, E. J.,** Chlorinated hydrocarbons in the surface microlayer of the Niagara River, *J. Great Lakes Res.,* 9(2), 281, 1983.
23. **O'Connor, D. J. and Connolly, J. P.,** The effect of concentration of absorbing solids on the partition coefficient, *Water Res.,* 14, 1517, 1980.
24. **Parks, G. A.,** Adsorption in the marine environment, in *Chemical Oceanography,* Vol. 1, Academic Press, New York, 1975, 241.
25. **Chiou, C. T., Freed, V. H., Schmedding, D. W., and Kohnert, R. L.,** Partition coefficient and bioaccumulation of selected organic chemicals, *Environ. Sci. Technol.,* 11, 475, 1977.
26. **Karickhoff, S. W., Brown, D. S., and Scott, T. A.,** Sorption of hydrophobic pollutants on natural sediments. *Water Res.,* 13, 241, 1979.
27. **Karickhoff, S. W.,** Semiempirical estimation of sorption of hydrophobic pollutants on natural sediments and soils, *Chemosphere,* 10, 833, 1981.
28. **Allan, R. J.,** The role of particulate matter in the fate of contaminants in aquatic ecosystems, Inland Waters Directorate Sci. Ser. No. 142, 1986.
29. **McCrea, R. C.,** Development of an aqueous phase liquid-liquid extractor (APLE), Interim Rep., Water Quality Branch, Inland Waters Directorate, Ontario Region, Burlington, Ontario, 1983.
30. **Oliver, B. G. and Nicol, K. D.,** Field testing of a large volume liquid-liquid extraction device for halogenated organics in natural waters, *Int. J. Environ. Anal. Chem.,* 25, 275, 1986.
31. **Allan, R. J.,** The role of particulate matter in the transport and burial of contaminants in aquatic ecosystems, in *The Role of Particulate Matter in the Transport and Fate of Pollutants,* Hart, B. T., Ed., Pub. Water Studies Centre, C. I. T., Melbourne, Australia, 1985, 1.
32. **Frank, R., Thomas, R. L., Holdrinet, M., Kemp, A. W. L., and Braun, H. E.,** Organochlorine insecticides and PCB's in sediments of Lake St. Clair (1970 and 1974) and Lake Erie (1971), *Sci. Total Environ.,* 8, 205, 1977.
33. **Frank, R., Thomas, R. L., Holdrinet, M., Kemp, A. W. L., Braun, H. E., and Dawson, R.,** Organochlorine insecticides and PCB in the sediments of Lake Michigan (1975), *J. Great Lakes Res.,* 7, 42, 1981.
35. **Thomas, R. L.,** Lake Ontario sediments as indicators of the Niagara River as a primary source of contaminants, *J. Great Lakes Res.,* 9(2), 118, 1983.
36. **Kaminsky, R., Kaiser, K. L. E., and Hites, R. A.,** Fates of organic compounds from Niagara Falls dumpsites in Lake Ontario, *J. Great Lakes Res.,* 9(2), 183, 1983.
37. **Durham, R. W. and Oliver, B. G.,** History of Lake Ontario contamination from the Niagara River by radiodating and chlorinated hydrocarbon analysis, *J. Great Lakes Res.,* 9(2), 160, 1983.
38. **Oliver, B. G. and Charlton, M. N.,** Chlorinated organic contaminants on settling particulates in the Niagara River vicinity of Lake Ontario, *Environ. Sci. Technol.,* 18, 903, 1984.
39. **Chambers, R. L., Eadie, B. J., and Rao, D. B.,** Role of nepheloid layers in cycling particulate matter, *Coastal Oceanogr. Climatol. News,* 2(2), 32, 1980.
40. **Chambers, R. W. and Eadie, B. J.,** Nepheloid and suspended particulate matter in southeastern Lake Michigan, *Sedimentology,* 28, 439, 1981.
41. **Sandilands, R. G. and Mudroch, A.,** Nepheloid layer in Lake Ontario, *J. Great Lakes Res.,* 9(2), 190, 1983.
42. **Charlton, M. N.,** Downflux of sediment, organic matter and phosphorus in the Niagara River area of Lake Ontario, *J. Great Lakes Res.,* 9(2), 201, 1983.
43. **Fox, M. E., Carey, J. H., and Oliver, B. G.,** Compartmental distribution of organochlorine contaminants in the Niagara River and the western basin of Lake Ontario, *J. Great Lakes Res.,* 9, 287, 1983.
44. **Oliver, B. G.,** Uptake of chlorinated contaminants from anthropogenically contaminated sediments by oligochaete worms, *Can. J. Fish. Aquat. Sci.,* 41(6), 878, 1984.
45. **Eadie, B. J., Robbins, J. A., Landrum, P. F., Rice, C. P., Simmons, M. S., McCormick, M. J., Eisenreich, S. J., Bell, G. L., Pickett, R. L., Johansen, P., Rossman, R., Hawley, N., and Voice, T.,** The cycling of toxic organics in the Great Lakes: a three-year status report, NOAA Tech. Memorandum ERL GERL-45, National Oceanic and Atmospheric Association, Washington, D. C., 1983.
46. **Liu, D., Strachan, W. M. J., Thomson, K., and Kwasniewska, K.,** Determination of the biodegradability of organic compounds, *Environ. Sci. Technol.,* 15, 788, 1981.
47. **Liu, D., Carey, J., and Thomson, K.,** Fulvic-acid-enhanced biodegradation of aquatic contaminants, *Bull. Environ. Contam. Toxicol.,* 31, 203, 1983.
48. **Borgmann, U. and Whittle, D. M.,** Particle-size-conversion efficiency and contaminant concentrations in Lake Ontario biota, *Can. J. Fish. Aquat. Sci.,* 40, 328, 1983.
49. **Harding, L. W., Jr. and Phillips, S. H., Jr.,** Polychlorinated biphenyls: transfer from microparticulates to marine phytoplankton and the effects on photosynthesis, *Science,* 202, 1189, 1978.
50. **Heberger, R. F. and House, R.,** Food of rainbow smelt and alewives in Lake Ontario, 1972, 17th Conf. Great Lakes Res. (abstr.), Hamilton, Ontario, 1974, 83.

51. **Elrod, J. H.,** Seasonal food of juvenile lake trout in U.S. waters of Lake Ontario, *J. Great Lakes Res.,* 9(3), 396, 1983.

52. **Norstrom, R. J., Hallet, D. J., and Sonstegard, R. A.,** Coho salmon *(Oncorhynchus kisutch)* and herring gulls *(Laus argentatus)* as indicators of organochlorine contamination in Lake Ontario, *J. Fish. Res. Board Can.,* 35, 1401, 1978.

53. **Strachan, W. M. J. and Edwards, C. J.,** Organic pollutants in Lake Ontario, in *Toxic Contaminants in Lakes,* Nriagu, J. and Simmons, M., Eds., Wiley Interscience, New York, 1984, 239.

54. **Weseloh, C. D. V.,** A report of levels of selected contaminants in eggs of herring gulls from the Niagara River 1979—1982, *Can. Wildl. Serv. Rep. Ser.,* 1983.

55. **Whittle, D. M. and Fitzsimons, J. D.,** The influence of the Niagara River on contaminant burdens of Lake Ontario biota, *J. Great Lakes Res.,* 9(2), 295, 1983.

56. **Baxter, R. M. and Carey, J. H.,** Evidence for the photochemical generation of superoxide ion in humic waters, *Nature (London),* 306, 575, 1983.

Chapter 10.1

THE "ENCLOSURE" METHOD: CONCEPTS, TECHNOLOGY, AND SOME EXAMPLES OF EXPERIMENTS WITH TRACE METALS

O. Ravera

TABLE OF CONTENTS

I. INTRODUCTION

The pollution hazard has a different significance when it is applied to human health rather than to the survival of other organisms. The pollution hazard for man is evaluated on the actual (or potential) damage produced at the individual and population levels. For other organisms the hazard is exclusively evaluated at the population and community levels. Consequently, environmental pollution is considered noxious only if the population viability is reduced. Therefore, the final aim of studies on ecosystem pollution is to obtain knowledge of the effects on populations and communities. It is evident that protecting populations also protects resources such as air, soil, and water. Research on the concentration and distribution in the environment of pollutants, their associated metabolites, and their physicochemical forms is very interesting. On the other hand, the results from these studies may be utilized for the protection of the environment only if they are finalized to the knowledge of the actual (or potential) effects on the community. From this point of view, a close connection between research on pollutants and that on their biological effects seems to be necessary to establish relationships between the pollution level and the consequent damage to the biota. In several cases there is, unfortunately, no collaboration between chemist and biologist, and the research of the latter is often limited to laboratory experiments and the description of communities living in polluted environments.

To predict the effects of pollution on an entire community from the results of laboratory experiments is rather aleatory. In fact, the conclusions drawn from these experiments are related to constant and controlled conditions established by the researcher, and, consequently, the complexity of the natural environment with its daily and seasonal variations cannot be simulated. In addition, the experiment is commonly carried out on groups of individuals belonging to those species which have adapted to live under laboratory conditions and is rarely carried out on actual populations. Indeed, the population cannot be identified with a simple group of organisms, because it is a unit with a well-defined demographic and genetic structure, adapted to its ecosystem, and occupying a niche in the community. As a consequence, from information about the pollutant effects on a group of individuals, although all the developmental stages are taken into account, the effects at the population level can only be predicted with great caution. Because predation and competition are obviously absent in laboratory experiments, no information on the effects due to the variations of these relationships caused by the pollutant can be obtained. In addition, chelating substances and suspended matter, which act on the physicochemical forms of several pollutants and on their toxicity, are normally absent from these experiments. On the other hand, laboratory experiments are necessary (1) to evaluate the potential toxicity of single pollutants and their mixtures (2) to compare the relative toxicity of a series of chemicals suspected of being dangerous, and (3) to obtain basic information for legislation on environmental protection. These experiments are also useful for evaluating the reliability of hypotheses drawn from studies on polluted environments.

Research on the effects at levels lower than that of the individual (i.e., the effects on the cell and subcellular structure and functions) is very important because it provides knowledge on the action mechanisms of the pollutant, those of detoxification, and the causes of the damage observed in the single individual. On the other hand, the results of these studies cannot be utilized to predict the effects at population level.

Important results, often utilized to develop mathematical models, have been obtained by experiments carried out with polluted and nonpolluted "balanced microcosms". Unfortunately, this method has some significant disadvantages. For example,

1.   The microcosm size is always relatively small.
2.   The population densities of the tested species must be far greater than those of the natural species to have sufficient material.

3. The ratios between the species are decided by the researcher.
4. Temperature and light patterns are often different from those of the natural environment.

In making conclusions, it is necessary to evaluate the effects of pollution in populations and communites in their natural environment. Unfortunately, if in the natural environment the pollution level is not sufficiently high to alter the community structure dramatically, its effects are difficult to detect. Information on these dangerous effects may be obtained from comparison between communities of polluted environments and those living in similar but nonpolluted environments. Conclusions from these studies must be drawn with great caution because the comparison is based on the assumption that similar environments must have similar communities. However, this is not always true. The effects may be easily evaluated in those environments studied for an adequate time before and after being polluted, but the number of these environments is rather limited. The effects at community level may also be evaluated in those environments which were not studied before they were polluted, if the pollutant loading is quantitatively constant in time and multiannual studies are carried out. Indeed, one may presume that the continuous accumulation of pollutants in the environment progressively eliminates the sensitive species and enhances the population density of the more resistant species.

Very useful information has been obtained from studies on natural experimentally polluted environments. One may cite the experiments carried out in several lakes of the Precambrian Canadian shield. Some lakes were enriched with nutrients, others contaminated with various pollutants, and others acidified.[1,2] These lakes were studied for several years before and after the treatment. It was possible to perform these experiments because of the large number of lakes in the area, the low density of the human population resident in the watershed, the well-equipped scientific institutions, and favorable legislation. In areas where conditions are not so favorable, natural water bodies cannot be experimentally contaminated.

One of the best methods for evaluating the effects of pollutants on populations and communities is the enclosure technique, which represents a compromise between laboratory experimentation and investigations in natural environments.

II. ENCLOSURE TECHNIQUE*

The enclosure technique consists of isolating (generally by clear and flexible sheets of plastic material) two phases (water and sediment) representative of the entire ecosystem or one phase of it (for example, the water column without the underlying sediments). For pollution studies an enclosure is contaminated with a given toxic and another kept as control. The effects of pollution may be evaluated by comparing the variations of selected physical, chemical, and biological parameters within the contaminated enclosure and those observed in the control. The first scientists to employ enclosures were Strickland and Terhune,[3] McAllister et al.,[4] and Antia et al.[5] in marine environments, and Lund[6] and Goldman[7] in freshwater ecosystems. Originally, this technique was used to test ecological hypotheses and to follow some ecological processes. It was later employed to study pollution problems. Important results have been obtained with the enclosure technique not only from studies on the direct and indirect effects produced by organic and inorganic pollutants in populations and communities, but also from research on eutrophication, radioactive contamination, and some fundamental ecological processes such as the production and degradation of the organic material. In addition, with this technique the distribution, fate, and speciation of the pollutant

---

* In this paper, only the "enclosure" ("microcosm" or "microecosystem"), consisting of an isolated part of a natural ecosystem, is considered; the "microcosms" developed in the laboratory or outdoors as well as turbidostats, phytotrons, outdoor channels, artificial ponds, balanced aquaria, land-based tanks, and sediment cores are not taken into account.

may be studied under seminatural conditions. These are the reasons why the "enclosure" method has been more and more widely used in recent decades and why there have been improvements in research carried out in fresh water, brackish water, and marine environments. The utility and diffusion of the enclosure method are obvious from the series of papers presented at some meetings; for example, Microcosm in Ecological Research (Augusta, GA, November 8 to 10, 1978) and the 20th Meet. of the "Plankton Ecology Group" (Societas Internationalis Limnologiae) held in Trondheim, Norway (August 23 to 28, 1982).

The main advantages of the enclosure method are the following:

1. The initial physical, chemical, and biological characteristics inside the enclosure are those of the natural ecosystem in which the experiment is carried out.
2. Light and temperature are similar inside and outside the enclosure, because their variations closely follow those of the environment in which the enclosure is anchored.
3. Initially, the ratios between the various species are the natural ones and the intra- and interspecific relationships are not altered.
4. The enclosed ecosystem is, at least for a certain time, self-sustaining and always contains more trophic levels.
5. The enclosed ecosystem may be manipulated, for example, aeration, addition of pollutants and nutrients, reduction of zooplankton, introduction of fish.
6. The environmental conditions before and after the addition of pollutant as well as the amount of pollutant and the modalities with which it has been added to the enclosure are known.
7. The availability of the noncontaminated enclosure is kept as control.

Unfortunately, in using enclosures there are also disadvantages concerning technical aspects, more precisely: (1) the "wall effect", i.e., the colonization of fouling on the enclosure wall as well as the elimination of the exchanges of water and nutrients with the external environment and the reduction of the turbulence and (2) the artificial form of the enclosure which is always different from that of a natural water body.

Various advantages and disadvantages of this method have been discussed by Zeitzschel,[8] Steele and Menzel,[9] Davies and Gamble,[10] and Draggan.[11] Because the disadvantages are generally judged to be less than the advantages, this technique is considered useful in evaluating the effects produced by noxious substances (for example, strong acids, trace metals) in communities under seminatural conditions.

Although enclosures of various types have been used, commonly an enclosure consists of a cylinder open at the top to allow the transmission of light and heat and the exchange of gases with the air. The bottom may be closed by a plastic disk, and, in this case, the enclosure generally is floating and may have a length equal to or smaller than the depth to which it is anchored. If the cylinder is open at the bottom, it is buried in the mud, and it isolates a water column from the lake surface to the sediments. This type of enclosure, simulating a complete ecosystem, allows the study of the processes at the water/sediments level in addition to those at water/air interfaces. To maintain the cylinder in the vertical position, inflated collars, resin-coated elastomers, polyurethane-foam cylinders,[12] or other floating materials are fixed to the upper part of the enclosure. To prevent water exchange between the natural environment and that of the enclosure, the top of the latter must be extended for some decimeters above the lake surface. Some authors fill the enclosure with water collected from the environment in which the experiment is carried out. Because the chemical and biological characteristics of a water body vary with the depth, the water used to fill the enclosure must be composed of a series of samples of equal volume collected from the water surface progressively downward to the depth corresponding to that at which the bottom of the enclosure lies. For example, an enclosure 7 m long must be filled with at least seven discrete samples collected from the surface to a depth of 7 m or better with a continuous sampling obtained by lowering a hose connected to a pump at a constant rate.[13] In some experiments the whole water column is enclosed without modifying the vertical distribution of the chemical and biological characteristics. When it is possible, it is evident that this method is preferable to others already mentioned. To obtain identical starting conditions, enclosures made for the same experiment must be filled at the same time. Effects on phyto- and zooplankton of the nutrient input from bird guano have been demonstrated by some authors.[14,15] Consequently, when there are many birds in the area, bird scarers must be fixed to the top of the enclosure to prevent fertilization of the enclosed water column with their excrement. To prevent the entrance of both the droppings of the birds and atmospheric precipitation, the enclosure may be covered with a plexiglass plate, raised with respect to the upper part of the enclosure.[16]

The size and shape of the enclosure must be chosen in relation to the aims of the research. The diameter may vary from some decimeters to some tens of meters and the length from about 50 cm to several meters. The size of the enclosure should be decided according to the density of the enclosed populations and the size of the individuals of these populations. Indeed, small enclosures may be used for microorganisms and phytoplankton studies, whereas medium-size enclosures should be adopted for zooplankters (except macrozooplankton) and small benthic organisms with high population density. Large enclosures must be used for macroplankton, necton, and large benthic organisms. Indeed, Harte et al.[17] reported that the minimum volume of water for 1 g of fish is normally 2500 l. This must be considered in planning enclosure experiments with fish, because if the ratio between fish biomass and water volume is too high, the fish behavior may be modified and the predation pressure on zooplankton becomes far greater than that which is normal, with consequences also for the phytoplankton populations and nutrient concentration in the water. The size of the enclosures must also be chosen in relation to the season, the trophic level of the environment, and the duration of the experiment. In fact, the microecosystem enclosed tends to become different from the environment from which it has been isolated. This differentiation increases with increased temperature and productivity. From these considerations, large enclosures seem to be preferred. On the other hand, these enclosures are very expensive, and more costly equipment is necessary for their settlement and recovery. In addition, large enclosures make the sampling more difficult, particularly for that of zooplankton. According to Stephenson et al.,[18] in small enclosures (20 m³) no important patterned distribution of zooplankters has been observed, whereas in large (1000 m³) and medium (125 m³) enclosures an edge zone with respect to zooplankton distribution of about 1 m from the wall was evident. Conversely, Williams et al.[19] observed in three 1300-m³ enclosures (Controlled Experimental Ecosystems, or CEEs) that the horizontal distribution of zooplankton was not significantly influenced by the location of the sampling position (center, edge zone). Uehlinger et al.[20] found for both chemical parameters and plankton standing stock that the range of variation between four enclosures (volume about 77 m³) with one sampling site in the center was not significantly different from that of five sampling sites distributed within the surface (about 7 m²) of one enclosure. The researcher must decide if it is more convenient to use one or two large enclosures or several enclosures of medium (or small) size. If there is no particular reason to the contrary, it seems preferable to use a sufficient number of medium (or small) enclosures to plan a statistically more correct experiment.[21] The number of enclosures is, obviously, related to the number of treatments chosen, for example, the number of pollutant concentrations or the number of pollutants to be tested.

At the beginning of the experiment, the physical and chemical characteristics of the water columns inside a series of enclosures are similar, but the number of organisms living in the different water columns is very rarely the same, because of their heterogeneous spatial distribution. The initial differences in population density may increase rapidly and, thus, have a different influence on the physical and chemical characteristics of the water column and on the relationships between the different species. This is particularly true for the populations with high production capacity such as bacteria and algae. For example, an increase in phytoplankton population produces an increase of pH values and oxygen concentration. The alterations of the physical environment, in turn, influence the populations. Because of the variability of the enclosures, each experiment should be repeated at least two or three times. For example, to evaluate the effects of one concentration of a pollutant six enclosures must be used: three contaminated and three as control. The differentiation of the enclosed ecosystem seems to be less rapid in larger than in smaller enclosures; for example, Takahashi et al.,[22] with an experiment which lasted for 30 d, demonstrated that the biota behaved in a very similar manner in four enclosures of 70 m³ each.

The ratio between the diameter of the enclosure and its length is important because it controls the water movement inside the enclosure, which in turn influences the planktonic populations, particularly phytoplankton, and the distribution of solutes and particulate material. For example, the wind action becomes more efficient as the ratio diameter/length of the enclosure increases.

Enclosures have been constructed of various materials, for example, rubber, polyethylene, polyvinyl chloride (PVC), and, in a few cases, of Perspex or aluminum. Enclosures made of flexible material are to be preferred to those of rigid materials because the lake water movement is in part transferred through the flexible wall, increasing the turbulence in the enclosed water column.[23] According to Davies and Gamble,[10] PVC is preferred to polyethylene because it is more resistant and nontoxic. The toxicity of the material used must be tested. For example, rubber may release some toxic substances, and Gächter[24] isolated the enclosed water column from the rubber wall by polyethylene foil to prevent zinc contamination. In certain types of PVC we have measured a small release of cadmium. To prevent contamination of phthalates released from a PVC wall, Davies et al.[25] exposed water-filled enclosures to weather for 2 to 3 d before starting the experiment. The transparency of the enclosure wall must be very high to permit the maximum of light transmission, except in experiments designed to study the decomposition of organic matter.

The wall of the enclosure reduces water movement inside the enclosure, abolishes any exchange with the lake, and enhances colonization by fouling organisms on the external and internal surface of the enclosure. When the fouling growth rate is very rapid, it may reduce the transparency of the enclosure significantly. In addition, the fouling may adsorb some of the pollutant, decreasing the concentration of the toxic material in the water. The metabolism of the algae and animals constituting the fouling influences the chemical characteristics of the water column, acting on the oxygen and nutrient concentrations. The influence of the fouling on the characteristics of the water column is one of the causes of the differentiation of the enclosed ecosystem from the external environment. The fouling growth rate, obviously, depends on the temperature, light, and the abundance of nutrient. The fouling colonization has only one advantage, i.e., the effects of the pollutant on it can be studied. The fouling influence may be reduced by increasing the volume of the water column in relation to its lateral surface, i.e., by increasing the diameter of the enclosure and reducing its length. Some authors[26] control the fouling growth by rotary brushes.

After a certain time from the beginning of the experiment (depending principally on the enclosure size, temperature, and trophic state of the environment), some characteristics of the enclosed water column are found to be more or less modified. This is due to the fact that the enclosed water column does not receive nutrients from the external environment and the reduced turbulence inside the enclosure prevents the transfer of the nutrients from the sediments into the water. The oxygen concentration decreases first in the lower part of the enclosure and then in the entire water column. In addition, a progressive decrease of the phosphate, nitrate, and chlorophyll concentrations is evident as well as an increase of ammonia. As a consequence, the experiment must be ended before the microecosystem shows this collapse, if there is no other reason for its continuation. It is a good rule to measure the same parameters with the same frequency in the control and contaminated enclosures as well as in the water body in which the experiment is carried out. Comparison of the results obtained from the control enclosure and the natural environment indicates when the differences are so great that the experiment should be terminated. The turbulence inside the enclosure may be increased by artificial mixing obtained, for example, with the air lift apparatus described by Reynolds et al.[27] With this apparatus the authors obtained in an enclosure an artificial deepening of the epilimnion during the summer period of thermal stratification and studied its effect on the dynamics of phytoplankton populations. Other authors[3,26] have obtained turbulence inside the enclosure with various stirring devices. To bubble air (or oxygen) into the enclosure has the advantage of both increasing the concentration of the gas and producing turbulence. To imitate the normal replacement of nutrients, some authors[13,22] added them to the enclosure (phosphates, nitrates, and silicates). The artificial water movement combined with oxygenation and the addition of nutrients may delay the collapse of the ecosystem. These manipulations must be done carefully to prevent an excessive input of nutrients and/or oxygen with the consequence that the enclosed ecosystem is so greatly changed that it is too different from the natural one.

III. SOME EXAMPLES OF ENCLOSURES

Some recent and less recent examples have been chosen to give a general view of the variability of the enclosure techniques and the various practical and technical problems faced by this method.

Two very large enclosures (diameter = 45.5 m; length = 16 m; volume = 18,000 m³) were described in Lund[6] and used by Lack and Lund[23] to study the dynamics of phytoplankton of Blelham Tarn, U.K. (Figure 1), in relation to nutrient addition. The same enclosures have been used by Smyly[28] to evaluate the influence of planktivorous fish on zooplankton. The cylindrical enclosures were made of butyl rubber, open at both ends, and the lower end was buried in the sediments by means of a weight. To maintain the buoyancy of the upper part of the enclosure, three inflatable collars were connected to the cylinder, and the uppermost one projected 15 cm above the upper part of the tube to which the bird scarers were fixed. The sampling was carried out by a small rowing boat that could enter and leave the enclosure. Because of the great weight of the enclosures (4 tons each), they were transported to the lake by mechanical handling. The physical conditions have been carefully studied within these enclosures.[23] For example, the temperature variations in the enclosures were similar to those in the lake, and at the surface the temperature differences very seldom exceeded 1°C. The turbulence was, obviously, greater in the lake than in the tubes. The MELIMEX (Metal Limnological Experiment) research project was similar to the marine CEPEX (Controlled Ecosystem Pollution Experiment) project, which consisted of the cooperation of several institutions from the U.S., U.K., and Canada to carry out experiments in seminatural conditions. MELIMEX consisted of a series of studies carried out in a very eutrophic Swiss lake (Baldeggersee) with three large enclosures (diameter = 12 m; length = 12.5m): one as control and two contaminated.[24] The cylinders were constructed from rubber foil isolated from the enclosed water column by a polyethylene film. The base, weighted with heavy chains, was embedded in the sediments, and the floating collar was filled with plastic bags containing styrofoam balls. A wooden platform allowed sampling of the center of the enclosure. The enclosures were settled at a depth of 10 m. The most important characteristic of these experiments was the renewal of enclosed water (Figure 2). The flux (11.5 m³/d) obtained by pumping water into and out of the enclosure was equivalent to the renewal of a natural lake with a mean depth of 10 m. The concentration level of the

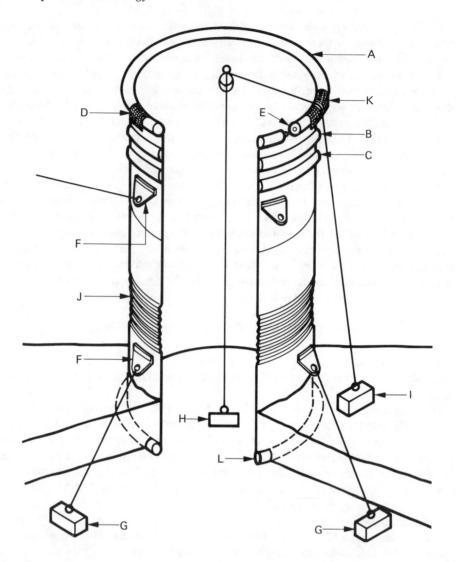

FIGURE 1.   Diagram of a tube with the center cut away to show the component parts, but not to scale. Bird scarers are not shown. A, B, C = the inflation collars. D, E = gaps between two compartments, one compartment end containing the inflation valve; in D, the gap is covered by plastic as it is in the lake and E shows it with the plastic cover removed. F = lugs and eyelets with ropes attaching the upper ring to supports on shore and the lower ring to concrete blocks; G = embedded in the lake deposits. H, I = concrete blocks within and outside the tube, joined by a floating rope, and used, when necessary, to pull a boat in or out of the tube. J = the excess length ("slack") of the tube to allow for rise and fall of lake level. K = the extra sheet of butylite covering the top inflation collar and over which the boat is rowed or pulled. L = the base of the tube through which chain is run in order to sink the tube into the mud. (From Lack, T. J. and Lund, J. W. G., *Freshwater Biol.*, 4, 401, 1974. With permission.)

metals (Hg, Cu, Cd, Zn, and Pb) in the enclosure was maintained by additions of the metals to the inflow. In the enclosures, traps were suspended at 5 and 10 m from the water surface to collect the sedimenting material. To study the processes controlling phosphorus and nitrogen turnover in lake ecosystems, Uehlinger et al.[20] carried out two experiments with enclosures ("limnocorrals") constructed of polyethylene sheets (diameter = 3 m; length = 11 m) and strengthened with aluminum rings every 1.8 m. The enclosures were submerged (by scuba divers) with the bellows closed, and then opened by using a ship's crane. Successively, the enclosures were attached to metal rings supported by buoys. This system was

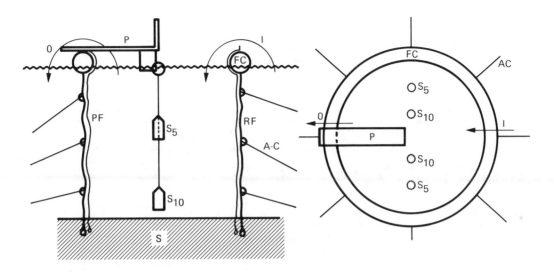

FIGURE 2.   Diagram of limnocorral. I = inflow; O = outflow; P = sampling platform; RF = rubber foil; S_5 and S_{10} = sedimentation trap in 5 and 10 m, respectively; S = sediment; PF = polyethylene foil; FC = floating collar; A-C = anchor-cable. (From Gächter, R., *Schweiz. Z. Hydrol.*, 41, 170, 1979. With permission.)

anchored to the sediment by chains. Platforms fixed to the supporting rings permitted the sampling (Figure 3). The CEEs consisted of large-volume (1300 m³) plastic cylinders which were adapted to study the direct and indirect effects produced by pollutants in a complete community, fish included.[29] These enclosures consisted of an upper cylinder (diameter = 9.5 m; length = 16.2 m) holding about 87% of the total volume, terminating in a 7.3-m-long funnel containing the remaining 13%.[13] For example, Koeller and Wallace[30] studied the effects of mercury on the chum salmon (*Oncorhynchus keta*) with 72-d long experiments, and Koeller and Parsons[31] studied the effects produced by copper on the same species. With the same enclosures, Beers et al.[13] evaluated the alterations of marine zooplankton biomass and structures, particularly the copepod production, induced by low concentrations of mercury (1 to 10 µg/l). Thomas et al.[32] tested the same concentrations of mercury on phytoplankton dynamics. Thomas et al.[33] studied the effects of copper on standing crops and phytoplankton productivity with enclosures, similar to those described by Menzel and Case,[29] but smaller, i.e., their volume was about 68 m³. Studying the degradation rate and the effects on phyto- and zooplankton of naphthalenes was the aim of research carried out by Lee and Anderson[34] with CEPEX enclosures (diameter = 2 m; length = 15 m). The persistence and the effects of two pollutants (methyl parathion and Linuron) on the zooplankton, macroinvertebrates, and flora of some ponds were the subjects of research by Stephenson and Kane,[21] carried out with small cylindrical enclosures (volume = 1 m³). The enclosures were made of transparent polyethylene with the top protruding above the water surface and held by a metal frame and its lower part attached to a fiberglass base. The rim of the base was pushed into the sediment. These experiments, which were repeated three times and which lasted for 6 weeks, were planned in a randomized block design. The results obtained have been compared with those from laboratory toxicity tests and bioassays of water samples coming from the enclosure water column. Gamble et al.[35] studied the effects of natural stresses and copper contamination in marine planktonic communities isolated from totally enclosed cylinders (diameter = 3 m; length = 17 m). These enclosures, with some modification, were used by Davies and Gamble[10] to evaluate the combined effects of mercury (10 µg/l) and fertilization on planktonic populations. The enclosures, made of nylon-mesh reinforced PVC (diameter = 3 m; length = 17 m; volume = 5 m³), had a conical top and

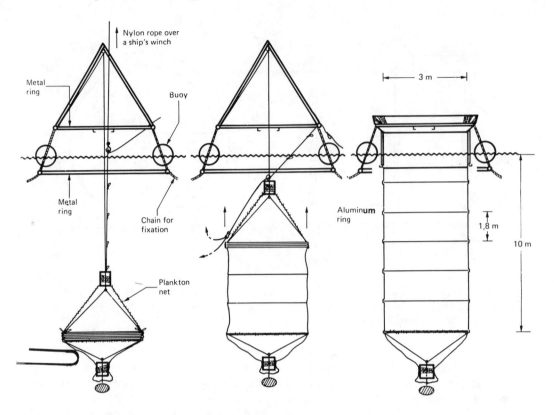

FIGURE 3. Scheme of limnocorral. (From Uehlinger, V., Bossard, P., Bloesch, J., Bürgi, H. R., and Bührer, H., *Verh. Int. Ver. Limnol.*, 22, 163, 1984. With permission.)

bottom and were completely closed and submerged; i.e., they were isolated from the atmosphere as well as from the sediments. The light levels inside the enclosures were 45% lower than those measured in the open water. The sedimented material was concentrated by a cone narrowing to 2 cm and collected by a pump housed on the support raft (Figure 4). Water samples were collected by a pump through a black PVC pipe, and zooplankton was collected with a 68-μm mesh net of 45-cm mouth diameter hauled vertically from 15 m. Before the contamination, the enclosures — filled by a pump — were left *in situ* to settle for about a week. After this period, the experiment was continued for 20 d. The enclosure used by Kuiper[36] consisted of a bag (diameter = 0.75 m; volume = 1.4 m³) composed of two layers: the inner layer made of polyethylene, which is biologically inert, and an outer layer of polyamide, which is very resistant to mechanical stress. The bags, suspended on aluminum frames, were shielded from bird droppings and atmospheric deposition by Perspex disks. The flotation of the enclosures was maintained by PVC buoys (Figure 5). The enclosures were filled with seawater pumped and filtered through a 2-mm-opening size net. With these enclosures, Kuiper carried out experiments on the effects of mercury,[36] and Kuiper and Hanstveit studied the effects of some organic micropollutants[37,38] on phyto- and zooplankton. The influence of nutrients, pH, and CO_2 on the shift from blue-green algae to green algae was studied by Shapiro[39] with a series of polyethylene enclosures (diameter = 1 m; length = 1.5 m) suspended from a raft in a small and shallow lake. The effects of acidification on *Bosmina longirostris* (Cladocera) populations were evaluated by Havens and De Costa[40] using cylindrical enclosures (diameter = 1 m; length = 2 m) suspended from a wooden raft and sealed at the bottom of the sediments. They had a volume of about 2 m³, and their opening was at 0.5 m above the water surface to prevent exchange with lake water.

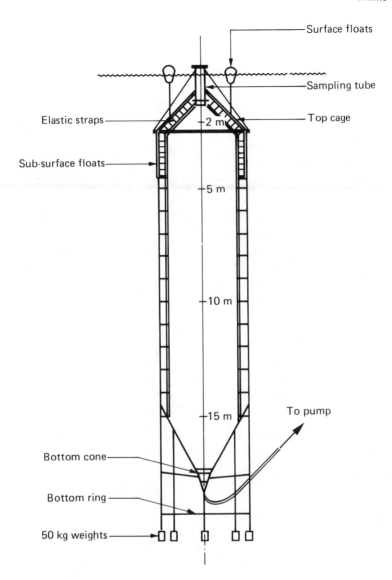

FIGURE 4.   Enclosure used in the 1976 Lake Ewe mercury experiment. (From Davies, J. M. and Gamble, J. C., *Philos. Trans. R. Soc. London Ser. B.*, 286, 527, 1979. With permission.)

Small polyethylene bags (1.57 m³) were adopted by Henry[41] as enclosures to study the role of zooplankton in phosphorus cycling. Their upper end was attached to a 1-m diameter aluminum hoop, and the bottom end was a disk, heat-sealed and secured with tape. The enclosures were supported by a wooden platform. Solomon et al.[42] used a series of ten enclosures to follow the distribution and pathway of a pesticide (Permethrin) in the water column and sediments. The enclosures were made of flexible nylon-reinforced PVC plastic (5 × 5 × 5 m) with a volume from 98 to 120 m³ depending on the depth at which they were settled. The upper part was open, and the bottom was closed by the sediments to which they were anchored. The same enclosures were used by Kaushik et al.[43] to study the effects of Permethrin on zooplankton communities. Large enclosures (10-m diameter, 130-m³ volume) of polyethylene closed at the bottom by the sediments were used by Salki et al.[44] to study the effects of selenium on the interrelations between fish, zooplankton, and phyto-

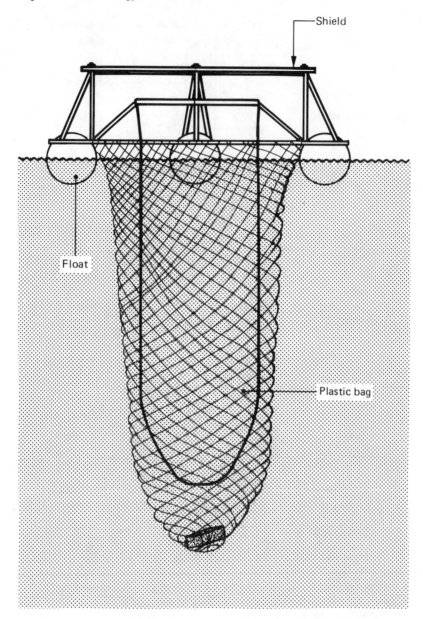

FIGURE 5.   Diagram of a suspended plastic bag. (From Kuiper, J., *Helgol. Wiss. Meer-esunters*, 30, 653, 1977. With permission.)

plankton. To evaluate the effects of some synthetic pollutants (Gamosol, Sefoil, Corexid, Nonylphenol), Lacaze[45] employed cylindrical enclosures of transparent polyethylene (diameter = 60 cm; length = 2 m; volume = about 560 l). The cylindrical form was maintained by three rings of PVC: the first at the top attached to floats of polyurethane foam, the second at 1 m from the open top, and the third buried in the sediments. Goldman[7] employed 10-m-long polyethylene cylinders of two different diameters (1.16 and 0.58 m) to study nutrient-limiting factors of lake phytoplankton. The tubes were held open by wire hoops spaced at 1-m intervals and attached to the outside of the tube with adhesive polyethylene tape. A free-floating, transparent plastic sphere (diameter about 6 m) was used by McAllister et al.[4] to study the kinetics of the phytoplankton. The buoyancy of the sphere maintaining its center at 5.5 m beneath the sea surface was controlled by regulating an enclosed air volume. Antia et al.[5] repeated this experiment, which was prolonged to 100 d, to measure the decay process of phytoplankton. To do this, 28 d after filling, the sphere was "blacked out" by strips of opaque black plastic over the top half and was left in this condition until the end of the experiment. The amount of light which entered by backscattering into the lower point of the enclosure was so small that it could be considered insufficient to promote any plant growth. For reasons of aquaculture practice, Devik[46] studied the possibility of making an open-bottom greenhouse system to obtain *in situ* cultures with relatively constant structure. To this end he carried out experiments with enclosures to ascertain whether the elimination of water horizontal exchange from the surface down reduces the vertical transport and the consequent persistence of fertilizers in the surface layer. The enclosures (diameter = 6 m; length = 12 m), made of floating cylinders open at the two bases, were settled in a bay with a depth ranging from 15 to 20 m. The enclosure wall was made of split-fiber nylon, reinforced in the upper and lower parts by tarpaulin cloth. The wall was closed by a slip-tubing arrangement that ran along the vertical roping of the wall. The neutral buoyancy was obtained by compensating the weights with foam sheeting. The enclosure was surrounded by a collar of polyethylene tubing carrying davits which secured a freeboard of 25 to 30 cm of the wall. In this microecosystem, temperature and salinity patterns demonstrated the elimination of the horizontal exchanges and the decrease of the vertical ones. The enclosed water column was then stabilized by a density gradient. Consequently, the nutrients added (nitrate and phosphate) to the upper layer were sufficiently permanent to allow their complete utilization by phytoplankton before they were dispersed. Blinn et al.[47] studied, with short time experiments (2 to 24 h), the effects of mercury on the primary productivity of phytoplankton with plastic chambers (surface base = 1.7 m^2; length = 3.8 m) of transparent polyethylene. The four sides and the bottom corners of the parallelepiped enclosure were heat sealed, and the upper edge was open for atmospheric exchange. The chambers were filled with about 6 m^3 lake water by towing the enclosure slowly. The chambers, suspended from a wooden frame supported on a styrofoam float, protruded 30 cm above the lake surface and were weighted by suspending weights from the bottom. To separate an area which was large enough to simulate lake conditions, but small enough to carry out experiments with safe and inexpensive tracers at easily measurable concentrations, Hesslein and Quay[48] used an enclosure with a plastic interlocking zipper allowing closure at the triangular area. The enclosure, made of a vinyl sheet, 14.6 m long and 6 m deep, was suspended at the top by surface floats and anchored at three corners to the sediments by lines. McQueen and Lean[49] obtained good results from an experiment with two cylindrical enclosures (diameter = 8 m; length = 14 m; volume = 550 m^3) of polyvinyl-coated nylon to monitor the effects of hypolimnetic aeration. The enclosures had flotation collars and were weighted into the sediments with chains sewn into the lower hems. One enclosure was kept as a control while the other was equipped with a hypolimnetic aerator. The nutrient recycling in a community of decomposing macrophytes (*Myriophyllum spicatum*) was studied by Kistritz[50] with four aluminum cylinders (diameter = 0.80 m; length = 1 m) settled in a shallow reservoir. To prevent the influence of light and precipitation, each enclosure was covered by a lid. Two cylinders open at the bottom were pushed about 15 cm into the sediments, and two were closed at the bottom by an aluminum disk. One cylinder was open at the bottom. One was closed and contained *Myriophyllum*. One open cylinder and one closed cylinder were filled with reservoir water without macrophytes. A special enclosure which was suitable for use in marine areas with strong wind or considerable sea tide was the "plankton tower" described by von Bodungen et al.[51] This bottle-shaped enclosure was composed of a 2-m-diameter cylinder extended from the sediment up to a depth of 2.5 m connected by a conical transition segment to a cylinder 3 m long of 0.8-m diameter which protruded 1.5 m over the sea surface. The volume of the enclosure was about 30 m.^{3} The enclosure was supported by a steel framework which was also utilized for sampling and for supporting some automatic recording instruments. The framework was settled at 11-m depth in the Kiel Bight (Baltic Sea).

IV. EXAMPLES OF TRACE METAL EXPERIMENTS

From 1976 our research program has been focused on:

1.   The direct and indirect effects of trace metals (e.g., copper, cadmium, mercury, and aluminum) on planktonic community
2.   The consequences of these effects on the physical and chemical characteristics of the water
3.   The distribution and fate of the metal added to the enclosed water column

The experiments were carried out with plastic cylinders settled in a very eutrophic and shallow lake of northern Italy: Lake Comabbio (45° 45' 47" N; 8° 41' 32" E). The surface of this lake is 3.5 km^2, its maximum depth is 7.7 m, and its mean depth is 4.4 m. A detailed description of Lake Comabbio was given in References 52 and 53. As an example of these experiments, the method and the most important results reported in three of our papers are summarized.

Cylindrical enclosures made of nylon-mesh reinforced, flexible, and transparent PVC were used. The size of the enclosures (diameter = 1 m; length = 3 m) was so chosen that they were not too expensive, and transport, installation, and sampling were easy. The cylinders could be folded like an accordion for easy transport to the lake. In addition, each experiment could be repeated two or three times. The relatively small size of the enclosures reduced the duration of the experiment to a maximum of 20 to 22 d. Two types of enclosures were employed: one with two open ends and another with an open upper end and a closed lower one. The cylindrical shape of the tube was maintained by rings of rigid PVC fixed by adhesive tape to the upper and lower circumferences. The tube was kept vertical by three closed and empty polyethylene 10-l bottles, each attached to the lower part of the upper ring. To exclude waves the upper ring extended nearly 40 cm above the lake surface. Three lead weights were attached to the external surface of the lower ring. If the bottom base was open, the lower ring was embedded in the sediments for about 40 cm. The lower ring of the tube closed at the bottom was replaced by a PVC disk (fixed by tape to the lower circumference of the tube) with a hole (diameter = 10 cm) at the center. This hole could be closed by a weighted PVC disk when the tube was filled with lake water. The open enclosure was filled by folding it like an accordion onto the embedded lower ring and hauling it to the lake surface. In both types of enclosures the whole water column was enclosed. The open tubes, as well as the closed tubes, were anchored by nylon ropes kept stretched by lead weights buried in the sediments. The open tube enclosed a water column with the underlying sediments, whereas the closed tube isolated the water column from the sediments (Figure 6).

Water, phytoplankton, zooplankton, and sediments were sampled before and after the addition of the metal with a mean frequency of 2 d. The same sampling program was applied to the control enclosures and the lake. In addition to the metal, the following parameters were measured: temperature, water transparency, conductivity, pH, dissolved oxygen, chlorophyll - a and pheopigments, total phosphorus, P-PO$_4$, P-particulate, N-NO$_3$, N-NO$_2$, N-NH$_4$, and N-particulate. Phyto- and zooplankton organisms of each sample were identified and counted. Discrete samples of water, phytoplankton, and seston were collected by a Ruttner bottle of 1-l volume. Integrated water samples were collected through a plastic pipe (diameter = 2.5 cm; length = 3 m) which could be closed at the upper end, and sediment

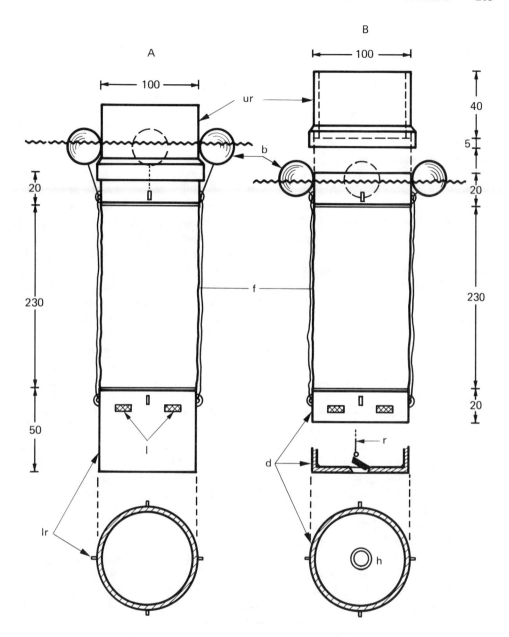

FIGURE 6. Diagram of the enclosure. (A) In contact with sediment; (B) isolated from the sediment. ur = upper ring; b = floating bottle; f = flexible PVC; lr = lower ring; d = PVC ring; h = hole; r = nylon rope; l = lead weight. The measures are given in mm. (From Ravera, O. and Annoni, D., *Atti 3rd Congr. Associazione Italiana Oceanologia e Limnologia,* de Bernardi, R., Ed., Istituto Italiano di Idrobiologia, Pallanza, Italy, 1980, 409. With permission.)

samples were collected through a core sampler consisting of a simple Perspex tube (diameter = 2.5 cm; length = 3 m). Zooplankton was collected by a Perspex sampler developed by P. H. Kerrison (description not yet published) which permitted the collection of a quantitative integrated sample and the simultaneous filtration of the zooplankters on a nylon net of 60-μm mesh. Integrated samples of zooplankton were also collected with a pump and filtered on a 60-μ-mesh size net. To simulate an accident, the metal solution was added once at the beginning of the experiment to obtain initial concentrations ranging from 10 to 100 μg/l for copper (chloride) and cadmium (chloride), and of 6 mg/l for aluminum (sulfate) in the enclosures. The metal solution was mixed into the water column taking care to prevent resedimentation. In the experiments carried out during recent years (data not yet published), strips (5 cm × 3 cm) of the same material of the enclosure wall were suspended in the enclosed water column to measure the metal uptake by the fouling per unit surface. From this information the total amount of metal subtracted from the water column by the fouling of the wall was estimated. Because the lake is eutrophic this amount was always considerable, although it varied with the experimental conditions (e.g., light, temperature, metal concentration in the water column).

The results obtained from three experiments are summarized here. One with copper was carried out from July 21 to August 10, 1976,[54] one with cadmium from September 30 to October 11, 1979,[55] and another with aluminum from October 7 to October 29, 1980.[56]

During the experiments the physical, chemical, and biological characteristic of the water column of the control enclosure were rather similar to those of the lake. Most of the copper and cadmium was absorbed by plankton, suspended particles, the wall of the tube, and by the sediment for the enclosures open at the lower end. There is evidence that the amount of metal adsorbed by the sediments was in relation to the metal concentration in the water column. For example, after 11 d from the beginning of the experiment, the cadmium concentration of the surface sediment was 1 µg/g dry weight in the enclosure contaminated with 10 µg Cd per liter, 1.18 µg/g dry weight, in that with 50 µg Cd per liter, and 3.50 µg/g dry weight in the enclosure with 100 µg Cd per liter. Sediments were enriched with metal by direct adsorption from the water as well as from the sedimentation of suspended material and dead plankters contaminated by the same metal. As an effect of the enrichment of the sediments, the metal concentration in the water column decreased.

Concentration of 10 µg/l of cadmium had a considerable effect on the plankton, whereas the same concentration of copper had only a small and temporary influence. Phyto- and zooplankton were affected by concentrations of copper and cadmium of 50 µg/l and higher. At this concentration the phyto- and zooplankton population density and biomass decreased more or less according to the taxa. For example, diatoms were less influenced by copper and cadmium than Cyanophyta and Pyrrophyta, and cladocerans were more sensitive than copepods and rotifers to both metals. The sediments, decreasing the metal concentration in the water column, reduced the biological effects in the latter. No recovery was observed in the cadmium experiment because of its short duration (11 d), whereas that with copper recovered at least in part.* From the 12th d after copper contamination, the population density of *Microcystis aeruginosa* recovered progressively until the end of the experiment (22nd day), except in the enclosure with 100 µg Cu per liter and isolated from the sediments. The recovery pattern in the enclosures isolated from the sediment, with 50 µg/l, was very similar to that observed in the enclosure with 100 µg/l and in contact with the sediment (Figure 7). At the end of the experiment a complete recovery of the population density of *Anabaena* was noted only at the lower copper concentration (10 µg/l) in both types of enclosure. Because of the dominance in number of *Microcystis* (90% of the total phytoplankton), the chlorophyll concentration recovery followed the pattern of this species. The oxygen concentration and pH value recoveries were in agreement with that of *Microcystis*. This is easily understandable if the phytoplankton influence on these factors is taken into account.

Cadmium and copper were considered because of their high toxicity, whereas the experiment with aluminum was carried out because several research studies in the natural environment deal with the influence of aluminum on the physical and chemical characteristics of water, and relatively few deal with the effects in fish, without considering phyto- and

---

* In successive experiments with a longer duration, the ecosystem recovery for cadmium has also been recorded.

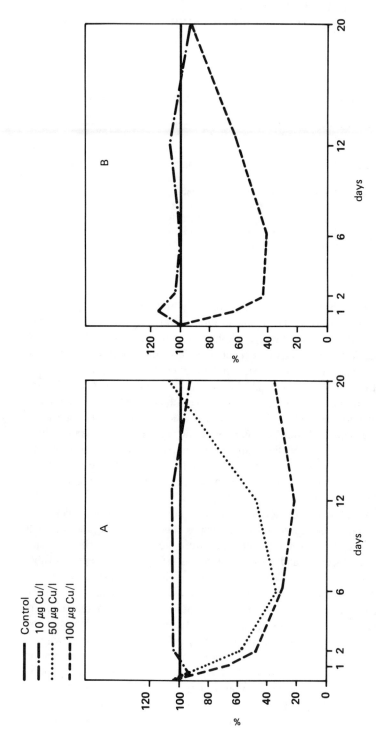

FIGURE 7. Variations of *Microcystis aeruginosa* population exposed at different concentrations of Cu (CuCl$_2$). The values are expressed as percent of control (100). (A) Enclosure isolated from the sediments; (B) enclosure in contact with the sediments. (From Ravera, O. and Annoni, D., *Atti 3rd Congr. Associazione Italiana Oceanologia e Limnologia*, de Bernardi, R., Ed., Istituto Italiano di Idrobiologia, Pallanza, Italy, 1980, 415. With permission.)

Table 1
NUMBER OF ZOOPLANKTERS
PER LITER PRESENT IN
DIFFERENT WATER COLUMNS

	Cylinder		
Taxa	Z_0	Z_1	**%Δ**
Cladocerans	79	31	60.8
Cyclopids	60	30	50.0
Diaptomids	31	20	35.5
Rotifers	44	12	72.7
Total	214	93	56.5

Note: Z_0 represents the control column and Z_1 the cylinder treated with aluminum. Results were measured after 24 h. %Δ is the percent of decrease in the number of zooplankters.

Modified from Reference 56.

zooplankton. In addition, laboratory experiments are generally limited to fish, *Daphnia* and some monospecific populations of algae. The concentration used (6 mg Al^{3+} per liter) was lower than that commonly employed to reduce phosphate in eutrophic lakes, and the low toxicity of aluminum to plankton could thus be expected. A few hours after the addition of aluminum sulfate the water transparency increased, and the water color changed from yellow-brown to pale blue. On the second day the total phosphorus concentration decreased from 100 to 60 μg/l and that of phosphates ($P-PO_4$) from 85 to less than 10 μg/l. These concentrations remained constant until the end of the experiment (22nd day). At the same date, the pH value in the contaminated enclosure in contact with the sediment decreased less than that observed in the enclosure isolated from the sediment, if compared with the corresponding controls. At the same time, a dramatic decrease of the population density and biomass of phyto- and zooplankton was observed. Samples of aluminum floccules were collected from the sediments, and the planktonic organisms trapped by them were examined. Most of the zooplankters were still alive and tried to escape from the floccules. The enclosure experiment was repeated on a small scale in the laboratory with Perspex tubes employing the identical aluminum concentration (6 mg/l) and water and plankton from the same lake. The floccules trapped about 70% of the cladocerans and rotifers and about 40% of the copepods (Table 1). This difference was probably due to the capacity of the organism to escape by swimming from the sedimentating floccules. As a consequence, it seems that the decrease of phyto- and zooplankton in the water column was principally due to the sedimentation of the plankters by metal flocculation and not to pH lowering and/or aluminum toxicity (at least, at the concentration used in this experiment). The decrease of phytoplankton biomass, as witness also the chlorophyll concentration lowering, caused the decrease of the oxygen concentration and, at least partly, of the pH value. In addition, the decrease of phosphate, phosphorus, and nitrogen particulate was also due to the aluminum flocculation as well as to the transparency increase and the color change. From the 13th day after aluminum addition, Cyanophyceae recovered in the open enclosures, but not in those isolated from the sediment. Also for zooplankton the recovery was less evident in the latter enclosures than the recovery in those in contact with the sediment. A few days after the contamination, the recovery of the water transparency, pH, electrical conductivity, and phosphorus particulate was observed

— more clearly in the open enclosures than in those isolated from the sediments. Almost all the aluminum added to the enclosure flocculated and sedimented in a few hours after contamination, leaving only 3% in the water column. The flocculation was facilitated by the high pH of the water (8.7 to 9.0), abundance of phosphate (86 to 110 μg/l), and suspended matter. The metal decrease in the water column explains why the most important effects on the planktonic community were observed at the beginning of the experiment and why later a more or less complete recovery of the ecosystem was noted.

These experiments demonstrate the useful information that may be obtained by using the enclosure technique. In addition, to carry out enclosure experiments, combined with appropriate laboratory tests, we may enhance our knowledge of the effects of trace metals and facilitate the identification and evaluation of the direct and indirect effects produced by pollutants. From the results obtained from this research, the following conclusions may be drawn:

1. Relatively low concentrations of toxic metals may produce dramatic effects on the community.
2. Because the community metabolism controls the characteristics of the water, a decrease of its biomass, produced by metal, has an indirect and important influence on the physical environment which, in turn, acts on the community.
3. The direct effects of the metal on the community and the indirect effects on the physical environment are evident during the first days after the contamination.
4. The recovery of the ecosystem occurs some days after the greatest decrease of the population density and biomass.
5. The recovery is never complete, at least during our experiment (i.e., in a range of 10 to 20 d); indeed, although the population density and biomass tend to reach the initial levels, the structure of the community is greatly modified because some species are more resistant than others.
6. A consequence of the modification of the community structure is certainly the alteration of the relationships between the species.
7. The influence of the sediment is relevant in decreasing the effects of the metal in the first part of the experiment and in accelerating the recovery of the second.

The results obtained from the experiments described are relative to the metal and concentrations used, the season, and the characteristics of the lake in which the experiments were carried out. Indeed, the physical and chemical characteristics of the lake and the biomass and structure of the community vary with the season. These variations have a great importance on the biological effects produced by pollutants. For example, the effect of a metal on phytoplankton will be more evident in the seasons in which the less-resistant species are more abundant (e.g., Cyanophyta). Because the decrease of the metal in the water column is accelerated by the abundance of sestonic particles for the same metal loading, less evident effects may be expected in an eutrophic water body than in an oligotrophic lake. As a consequence, the same experiment must be repeated in different types of lakes and in different seasons in order to arrive at more realistic and complete information. Some studies with this aim have been completed and others are in progress.

V. DISCUSSION AND CONCLUSIONS

The enclosure method has been employed more and more in recent years, especially to investigate problems connected with pollution. The results obtained are encouraging, and some problems can only be tackled with this technique. On the other hand, there is no general agreement on the validity of the enclosure method. The most substantial criticism

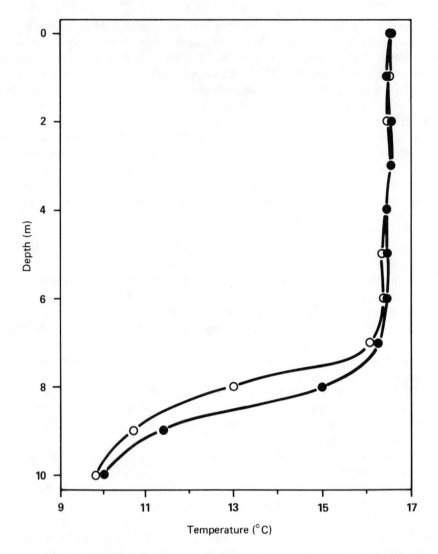

FIGURE 8.   Thermal stratification inside a 10-m-long polyethylene cylinder compared with that of the surrounding Castle Lake water in September 1960. The solid circles represent the temperature inside the cylinder; the open circles, outside. (From Goldman, C. R., *Limnol. Oceanogr.* 7, 99, 1962. With permission.)

concerns the reduction of turbulence in the enclosed water column produced by the wall of the enclosure and, consequently, by a cause that cannot be eliminated.

As an example, the opinions of some authors are reported here. Lack and Lund[23] observed that the phytoplankton of their enclosures was very similar to that of the lake even after 2.5 years of isolation. Kuiper[16] found the same development pattern in planktonic communities isolated in enclosures for 4 to 6 weeks. Goldman[7] demonstrated with temperature measurements inside and outside polyethylene enclosures that thermal stratification is not greatly altered in the enclosed water column (Figure 8). Hesslein and Quay[48] found identical temperature profiles inside and outside the enclosures. Boyce,[57] with a theoretical model, simulated the heat transfer from a stratified lake to a water column isolated by a thin curtain. He concluded that the similarity of the temperature profiles inside and outside the enclosure was the effect of the horizontal heat transfer and not due to the similarity of the vertical transfer processes in the open lake and in the enclosed water column. Kemmerer[58] stated that the enclosure method is far superior to laboratory tests and "the tubes did not measurably affect the physical characteristics of the water". Other authors, including McAllister et al.[4] have affirmed that enclosures give more valid approximations than bottle tests for phytoplankton studies. Conversely, Verduin[59] did not agree with this statement because he noted that Kemmerer[58] " . . . did not measure eddy diffusivity", which, according to Verduin,[59] in the enclosure was reduced to 10% or less of the value in the open lake. In addition, Steele et al.,[60] from research on the vertical mixing rate inside large enclosures (e.g., the enclosures of the CEPEX project), concluded that the eddy diffusivity was reduced by at least an order of magnitude. Gamble et al.[35] and Davies and Gamble[10] demonstrated that, in spite of this reduction, larger diatoms can sustain themselves in the upper layer of the enclosed water column for several weeks. Takahashi et al.,[22] using enclosures about 70 m³ volume, demonstrated, in contrast to the criticism by Verduin,[59] that the enclosed ecosystem behaved in a very similar way to the outside environment. Experiments with cylindrical enclosures (diameter = 0.5 m; length = 10 m), suspended in a lake from a wooden frame, were carried out by Bender and Jordan[61] to compare the primary production measurements in these enclosures with those made in open water. The samples collected from the water column and from the lake were placed in biological oxygen demand bottles (two light and two dark for each depth) and incubated at the respective depths. The differences in the results obtained from the two methods were explained as being due to the reduced diffusion inside the enclosure and, possibly, to the fouling growing on the inner wall of the tube. The authors concluded that the nutrient influence on the primary production cannot be measured with the enclosure method. Bower and McCorkle[62] carried out an experiment with a cylindrical "limnocorral" (diameter = 3.5 m; length = 8 m) anchored to a flotation collar to measure directly the primary production in an enclosed water column labeled with radiocarbon. According to the authors their experiment " . . . was the first opportunity to measure directly the photosynthetic carbon uptake by ¹⁴C technique, and to assess, by direct comparison, the accuracy of a ¹⁴C-bottle method". In addition, they concluded: "The direct measurement of ¹⁴C uptake in limnocorral provided the best available circumstance in which to measure net photosynthesis." The different attitude of Bender and Jordan[61] from that of Bower and McCorkle[62] toward the reliability of the primary production studied with enclosures depended on the differences in the methods used and then on the results obtained. Indeed, the enclosures employed by Bender and Jordan[61] maximized the "wall effect", having a diameter/length ratio equal to 0.5, whereas those used by Bower and McCorkle[62] had a ratio of 2.28 and, therefore, in the latter case the turbulence reduction inside the water column and the amount of fouling were minimized. In addition, the first authors used the "clear and dark bottle" method, whereas Bower and McCorkle employed the ¹⁴C labeled enclosure as an incubation "bottle".

Because of the isolation of the water column, the enclosure is different from the external water body in which it has been settled. As a consequence, the enclosure cannot simulate the natural environment perfectly nor for a long time, i.e., the enclosed ecosystem cannot function as a miniature lake. The disagreement between the authors on this subject is due, at least in part, to the differences in the type of enclosures used, the method adopted, and the environment in which the experiment was carried out. On the other hand, standardization of the enclosure methods is very difficult because the duration of the experiment and the size and shape of the enclosure must be chosen according to the aim of the research and to the environment in which the experiment is carried out. It might be easier and more useful to standardize the size and shape of one (or a few) enclosures for each type of ecosystem. Regarding the experiment with trace metals, or other pollutants, it is evident that the "wall effect" (which consists of the consequences of the water column isolation and the fouling growth) equally influences the control enclosure as well as the contaminated ones. Consequently, the differences between the community of the control and that of the contaminated enclosures may be considered an effect of the pollutant. Therefore, the results obtained may be accepted, although the control characteristics are different from those of the natural environment. In addition, we must consider that the enclosure method is the only method (except the experimentally polluted natural lake) that permits investigations on natural communities living in their experimentally polluted environment. This technique has two main advantages: the amount of pollutant and the way in which it is added are known, and so are the ecological conditions before and after contamination.

REFERENCES

1. **Schindler, D. W.**, Ecological effects of experimental whole lake acidification, in *Atmospheric Sulphur Deposition. Environmental Impact and Health Effects,* Shriner, D. S., Richmond, C. R., and Lindberg, S. E., Eds., Ann Arbor Science Publishers, Ann Arbor, MI, 1980, chap. 44.
2. **Schindler, D. W. and Turner, M. A.**, Biological, chemical and physical responses of lakes to experimental acidification, *Water Air Soil Pollut.,* 18, 259, 1982.
3. **Strickland, J. D. and Terhune, L. D. B.**, The study of "in situ" marine photosynthesis using a large plastic bag, *Limnol. Oceanogr.,* 6, 93, 1961.
4. **McAllister, C. D., Parsons, T. R., Stephens, K., and Strickland, J. D. H.**, Measurements of primary production in coastal sea water using a large volume plastic sphere, *Limnol. Oceanogr.,* 6, 237, 1961.
5. **Antia, N. J., McAllister, C. D., Parsons, T. R., Stephens, K., and Strickland, J. D. H.**, Further measurements of primary production using a large-volume plastic sphere, *Limnol. Oceanogr.,* 8, 166, 1963.
6. **Lund, J. W. G.**, Preliminary observations on the use of large experimental tubes in lakes, *Verh. Int. Ver. Limnol.,* 18, 71, 1972.
7. **Goldman, C. R.**, A method of studying nutrient limiting factors "in situ" in water columns isolated by polyethylene film, *Limnol. Oceanogr.,* 7, 99, 1962.
8. **Zeitzschel, B.**, Controlled environment experiments in pollution studies, Oceanology Int. 78, Tech. Sess. B (Biol. Mar. Tech.), 1978.
9. **Steele, J. H. and Menzel, D.**, The application of plastic enclosures to the study of pelagic marine biota, *Rapp. P. V. Reun. Cons. Int. Explor. Mer,* 1978.
10. **Davies, J. M. and Gamble, J. C.**, Experiments with large enclosed ecosystems, *Philos. Trans. R. Soc. London Ser. B,* 286, 523, 1979.
11. **Draggan, S.**, The microcosm: biological model of the ecosystem, Tech. Rep. Monitoring and Assessment Research Centre, University of London, 1980.
12. **Reynolds, C. S., Harris, G. P., and Gouldney, D. N.**, Comparison of carbon-specific growth rates and rates of cellular increase of phytoplankton in large limnetic enclosures, *J. Plankton Res.,* 7, 791, 1985.
13. **Beers, J. R., Reeve, M. R., and Grice, G. D.**, Controlled ecosystem pollution experiment: effect of mercury on population dynamics and production, *Mar. Sci. Commun.,* 3, 355, 1977.
14. **Manny, B. H., Wetzel, R. G., and Johnson, W. C.**, Annual contribution of carbon, nitrogen and phosphorus by migrant Canada geese to a hardwater lake, *Verh. Int. Ver. Limnol.,* 19, 949, 1975.
15. **Golovkin, A. N. and Garkavaya, G. P.**, Fertilization of water of the Murmansk Coast by bird excreta near various types of colonies, *Biol. Morya,* 5, 49, 1975.
16. **Kuiper, J.**, Development of North Sea coastal plankton communities in separate plastic bags under identical conditions, *Mar. Biol.,* 44, 97, 1977.
17. **Harte, J., Levy, D., Rees, J., and Saegebarth, E.**, Making microcosm an effective assessment tool, in *Proc. Symp. Microcosms in Ecological Research,* Giesy, J. P., Ed., U.S. Department of Energy, CONF-781101, Washington, D.C., 1980.
18. **Stephenson, G. L., Hamilton, P., Kaushik, N. K., Robinson, J. B., and Solomon, K. R.**, Spatial distribution of plankton in enclosures of three sizes, *Can. J. Fish. Aquat. Sci.,* 41, 1048, 1984.
19. **Williams, I. P., Gibson, V. R., and Smith, W. K.**, Horizontal distribution of pumped zooplankton during a controlled ecosystem pollution experiment: implications for sampling strategy in large-volume enclosed water columns, *Mar. Sci. Commun.,* 3, 239, 1977.
20. **Uehlinger, U., Bossard, P., Bloesch, J., Bürgi, H. R., and Bührer, H.**, Ecological experiments in limnocorrals: methodological problems and quantification of the epilimnetic phosphorus and carbon cycles, *Verh. Int. Ver. Limnol.,* 22, 163, 1984.
21. **Stephenson, R. R. and Kane, D. F.**, Persistence and effects of chemicals in small enclosures in ponds, *Arch. Environ. Contam. Toxicol.,* 13, 316, 1984.

22. **Takahashi, M., Thomas, W. T., Siebert, D. L. R., Beers, J., Koeller, P., and Parsons, T. R.,** The replication of biological events in enclosed water columns, *Arch. Hydrobiol.,* 76, 5, 1975.

23. **Lack, T. J. and Lund, J. W. G.,** Observations and experiments on the phytoplankton of Blelham Tarn, English Lake District, *Freshwater Biol.* 4, 399, 1974.

24. **Gächter, R.,** MELIMEX, an experimental heavy metal pollution study: goals experimental design and major findings, *Schweiz. Z. Hydrol.,* 41, 169, 1979.

25. **Davies, J. M., Baird, I. E., Massie, L. C., Hay, S. J., and Ward, A. P.,** Some effects of oil-derived hydrocarbons on a pelagic food web from observations in an enclosed ecosystem, and a consideration of their implications for monitoring, *Rapp. P. V. Reun. Cons. Int. Explor. Mer,* 179, 201, 1980.

26. **Santschi, H. and Lamont-Doherty Geological Observatory of Columbia University,** The MERL mesocosm approach for studying sediment-water interactions and ecotoxicology, in Proc. Symp. Micropollutants in sediment-water system, Delft Hydraulic Laboratories, Delft, Netherlands, 1985.

27. **Reynolds, C. S., Wiseman, S. W., Godfrey, B. M., and Butterwick, C.,** Some effects of artificial mixing on the dynamics of phytoplankton populations in large limnetic enclosures, *J. Plankton Res.,* 5, 203, 1983.

28. **Smyly, W. J. P.,** Further observations in limnetic zooplankton in a small lake and two enclosures containing fish, *Freshwater Biol.,* 8, 491, 1978.

29. **Menzel, D. W. and Case, J.,** Concept and design: controlled ecosystem pollution experiment, *Bull. Mar. Sci.,* 27, 1, 1977.

30. **Koeller, P. A. and Wallace, G. T.,** Controlled ecosystem pollution experiment: effect of mercury on enclosed water columns. V. Growth of juvenile chum salmon *(Oncorhynchus keta),* Mar. Sci. Commun., 3, 395, 1977.

31. **Koeller, P. and Parsons, T. R.,** The growth of young salmonids *(Oncorhynchus keta)*: controlled ecosystem pollution experiment, *Bull. Mar. Sci.,* 27, 114, 1977.

32. **Thomas, W. H., Seibert, D. L. R., and Takahashi, M.,** Controlled ecosystem pollution experiment: effect of mercury on enclosed water columns. III. Phytoplankton population dynamics and production, *Mar. Sci. Commun.,* 3, 331, 1977.

33. **Thomas, W. H., Holm-Hansen, O., Seibert, D. L. R., Azam, F., Hodson, R., and Takahashi, M.,** Effects of copper on phytoplankton standing crop and productivity: controlled ecosystem pollution experiment. *Bull. Mar. Sci.,* 27, 34, 1977.

34. **Lee, R. F. and Anderson, J. W.,** Fate and effect of naphthalenes: controlled ecosystem pollution experiment. *Bull. Mar. Sci.,* 27, 127, 1977.

35. **Gamble, J. C., Davies, J. M., and Steele, J. H.,** Loch Ewe bag experiment, 1974, *Bull. Mar. Sci.,* 27, 146, 1977.

36. **Kuiper, J.,** An experimental approach in studying the influence of mercury on a North Sea coastal plankton community, *Helgol. Wiss. Meeresunters.,* 30, 652, 1977.

37. **Kuiper, J. and Hanstveit, A. O.,** Fate and effects of 4-chlorophenol and 2,4-dichlorophenol in marine plankton communities in experimental enclosures, *Ecotoxicol. Environ. Saf.,* 8, 15, 1984.

38. **Kuiper, J. and Hanstveit, A. O.,** Fate and effects of 3,4-dichloroaniline (DCA) in marine plankton communities in experimental enclosures, *Ecotoxicol. Environ. Saf.,* 8, 34, 1984.

39. **Shapiro, J.,** Blue-green algae: why they become dominant, *Science,* 179, 382, 1973.

40. **Havens, K. and De Costa, J.,** The effect of acidification in enclosures on the biomass and population size structure of *Bosmina longirostris, Hydrobiologia,* 122, 153, 1985.

41. **Henry, R. L.,** The impact of zooplankton size structure on phosphorus cycling in field enclosures, *Hydrobiologia,* 120, 3, 1985.

42. **Solomon, K. R., Yoo, T. Y., Lean, D., Kaushik, N. K., Day, K. E., and Stepehnson, G. L.,** Dissipation of permethrin in limnocorrals, *Can. J. Fish. Aquat. Sci.,* 42, 70, 1985.

43. **Kaushik, N. K., Stephenson, G. L., Solomon, K. R., and Day, K. E.,** Impact of permethrin on zooplankton communities in limnocorrals, *Can. J. Fish. Aquat. Sci.,* 42, 77, 1975.

44. **Salki, A., Turner, M., Patalas, K., Rudd, J., and Findlay, D.,** The influence of fish-zooplankton-phytoplankton interacting on the results of selenium toxicity experiments within large enclosures. *Can. J. Fish. Aquat. Sci.,* 42, 1132, 1985.

45. **Lacaze, J. C.,** Utilisation du'un dispositif expérimental simple pour l'étude de la pollution des eaux "in situ". Effets comparés de trois agents émulsionants antipétrole, *Tethys,* 3, 705, 1971.

46. **Devik, O.,** Marine greenhouse systems: the entrainment of large water masses, *Environ. Sci. Res.,* 8, 113, 1976.

47. **Blinn, D. W., Tompkins, T., and Zaleski, L.,** Mercury inhibition on primary productivity using large volume plastic chambers "in situ", *J. Phycol.,* 13, 38, 1977.

48. **Hesslein, R. and Quay, P.,** Vertical eddy diffusion in the thermocline of a small stratified lake, *J. Fish. Res. Board Can.,* 30, 1491, 1973.

49. **McQueen, D. J. and Lean, D. R. S.,** Hypolimnetic aeration and dissolved gas concentrations: enclosure experiments, *Water Res.,* 17, 1781, 1983.

50. **Kistritz, R. U.,** Recycling of nutrients in an enclosed aquatic community of decomposing macrophytes *(Myriophyllum spicatum)*, *Oikos,* 30, 561, 1978.
51. **von Bodungen, B., von Brockel, K., Smetacek, J., and Zeitzschel, B.,** The plankton tower. I. A structure to study water/sediment interactions in enclosed water columns, *Mar. Biol.,* 34, 369, 1976.
52. **Annoni, D. and Ravera, O.,** Ricerche condotte sul lago di Comabbio (Provincia di Varese, Italia Settentrionale) dal maggio 1976 al maggio 1977, Rapporto CCE, Eur 5890i, 1977.
53. **Annoni, D. and Ravera, O.,** 1977, L'influence d'une longue période d'isothermie sur les caractéristiques des eaux d'un lac peu profond: Lac de Comabbio, *Eaux Ind.,* 19, 52, 1977.
54. **Ravera, O.,** Effetti ecologici del rame studiati per mezzo di un ecosistema sperimentale, in *Atti 3° Congr. Associazione Italiana Oceanologia e Limnologia,* de Bernardi, R., Ed., Istituto Italiano di Idrobiologia, Pallanza, Italy, 1980, 417.
55. **Kerrison, P. H., Sprocati, A. R., Ravera, O., and Amantini, L.,** Effects of cadmium on an aquatic community using artificial enclosures, *Environ. Technol. Lett.,* 1, 169, 1980.
56. **Zarini, S., Annoni, D., and Ravera, O.,** Effects produced by aluminum in freshwater communities studied by "enclosure" method, *Environ. Technol. Lett.,* 4, 247, 1983.
57. **Boyce, F. M.,** Mixing within experimental enclosures: a cautionary note on the limnocorral, *J. Fish. Res. Board Can.,* 31, 1400, 1974.
58. **Kemmerer, A. J.,** A method to determine fertilization requirements of a small sport fishing lake, *Trans. Am. Fish. Soc.,* 97, 425, 1968.
59. **Verduin, J.,** Critique of research methods involving plastic bags in aquatic environments, *Trans. Am. Fish. Soc.,* 98, 335, 1969.
60. **Steele, J. H., Farmer, D. M., and Henderson, E. W.,** Circulation and temperature structure in large marine enclosures, *J. Fish. Res. Board Can.,* 34, 1095, 1977.
61. **Bender, M. E. and Jordan, R. A.,** Plastic enclosure versus open lake productivity measurements, *Trans. Am. Fish. Soc.,* 99, 607, 1970.
62. **Bower, P. and McCorkle, D.,** Gas exchange, photosynthetic uptake, and carbon budget for a radiocarbon addition to a small enclosure in a stratified lake, *Can. J. Fish. Aquat. Sci.,* 37, 464, 1980.

Chapter 10.2

OUTDOOR PONDS: THEIR USE TO EVALUATE THE HAZARDS OF ORGANIC CHEMICALS IN AQUATIC ENVIRONMENTS

N. O. Crossland and D. Bennett

TABLE OF CONTENTS

I. INTRODUCTION

Outdoor exprimental ponds, if properly constructed and managed, may be regarded as realistic replicas of larger, natural aquatic ecosystems. They become naturally colonized by bacteria, fungi, algae, macrophytes, and invertebrates, and they should be capable of supporting small populations of fish. Thus, they contain all, or nearly all, of the components of larger, real-world systems and, like their natural counterparts, their structure and function is regulated by interactions between their component parts and the external environment. Therefore, the outdoor experimental pond provides a more realistic experimental unit than is possible to achieve in the laboratory. In this respect it is analogous to the agronomist's small field plot, forming a bridge between the laboratory and the natural environment.

In ecotoxicology the most important advantage of a pond study, compared with laboratory or microcosm studies, is that it is possible to study effects of contaminants on populations of various species *under conditions of real-world exposure*. On the basis of data obtained in the laboratory there is generally more uncertainty involved in predicting chemical fate than in predicting biological effects. Differences between toxicity in the laboratory and biological effects in the field can often be attributed to the action of the environment on the chemical. Therefore, it is important to carry out fate studies in parallel with biological studies. Other advantages of pond studies are that it is possible (1) to obtain data for species that are not easily maintained in the laboratory; (2) to study indirect effects, such as predator-prey interactions, effects on dissolved oxygen (DO) concentrations, and algal blooms; and (3) to study the rate of recovery of populations from the effects of stress.

II. CONSTRUCTION AND MANAGEMENT OF OUTDOOR PONDS

Outdoor ponds of various shapes and sizes have been used in aquatic ecotoxicology. In general, ponds with a volume less than about 10 m^3 have not proved to be satisfactory for experimental work. Very small ponds are inherently very unstable ecosystems which are subject to rapid successional changes in their flora and fauna. These changes will lead to wide variation in the species composition of a series of experimental units, thus rendering them unsuitable for experimental purposes. For example, Caspers and Hamburger[1] found that variation between small (2 m^3) artificial ponds was so great that they were unable to detect the effects of relatively toxic wastewater against a background of high biological variation.

Larger ponds can be managed in such a way that they will develop into more stable, balanced ecosystems. Hall et al.[2] have described a set of 50 experimental ponds built at Cornell University, Ithaca, NY. The volume of each pond was 650 m^3, the maximum depth was 1.3 m, and the surface area was 700 m^2. Ponds of this size are typically used for rearing fish, and they closely simulate natural freshwater environments in some agricultural areas. Experiments were carried out by Hall et al.[2] using replicated treatments of ponds to study effects of inorganic nutrients on aquatic communities. Outdoor ponds used at the U.S. Fish and Wildlife Laboratory, Columbia, MO were built to similar specifications and have been used to evaluate the fate and effects of herbicides and insecticides on freshwater communities.[3-5]

Ponds of an intermediate size, i.e., with a volume of about 10 to 50 m^3, have been used to study fate and effects of chemicals and oil. Ponds of this size are big enough to provide a habitat for a variety of organisms including small fish populations. Giddings et al.[6] have described a system of 15 ponds of 15-m^3 volume, each 1 m deep with a 3.5 $\times$ 3.5-m bottom and a 5 $\times$ 5-m perimeter, built at the Oak Ridge National Laboratory, TN. They were excavated in clay soil and lined with reinforced plastic. Sediment from a fish pond was transferred into each experimental pond, covering the bottom (but not the sloping sides) to a depth of 15 cm. Water from the fish pond was pumped into each experimental pond to a depth of 80 to 90 cm. One week later a volume of 8 l of *Elodea canadensis* from a natural pond was transferred into each experimental pond. A natural assemblage of pond organisms (bacteria, fungi, algae, zooplankton, insects, and benthic microfauna) was introduced with the water, sediment, and *E. canadensis.*

A set of 12 experimental ponds of 40 m³ has been used to study the fate and effects of various organic chemicals.[7] Each pond is 5 m wide, 10 m long, and has a maximum depth of 1 m (Plate 2). The 10-m dividing walls are vertical and made from concrete. The shorter, 5-m sides are sloping, and they consist, as do the bottom of the ponds, of a mixture of clay, alluvial silt, and organic debris. Clay, removed during excavation, was used to replace more porous substrates, such as gravel intrusions, thus forming a relatively impermeable bottom layer. The ponds are interconnected by pipes, and, when not in use for experiments, the water is occasionally circulated between the ponds to promote the uniform distribution of materials and organisms. When used for experiments, complete separation of ponds is achieved by closing a series of valves situated in the interconnecting pipes.

After filling the newly dug ponds with water from a nearby stream, partially rotted bales of hay were placed in them to provide an initial quantity of organic matter and nutrients. Within a few weeks large numbers of zooplankton were present, and within a few months a wide variety of aquatic insects were seen. The Diptera were early colonizers, particularly Chaoboridae and Chironomidae. After 1 year 19 families of invertebrates were present, but the diversity of the macroinvertebrate community was less than that in neighboring, more mature ponds. During the second year of their development the diversity of the plant and animal communities increased. The dominance of planktonic, unicellular algae was succeeded by the development of filamentous algae and vascular plants. Development of this macrophytic vegetation was associated with the colonization and development of large populations of mayfly larvae and other aquatic insects that are dependent on these plants for food and shelter. By the end of the second year of their development the pond communities contained most, if not all, of the important groups of macroinvertebrates present in more mature ponds and were then considered to be suitable for experimental work.

III. EXPERIMENTAL DESIGN

In order to evaluate the hazard of an organic chemical to aquatic environments, it is necessary to evaluate its chemical fate as well as its effects on aquatic organisms. Before undertaking a pond experiment a detailed review of physicochemical properties and toxicological data should be carried out to identify those areas of uncertainty that are most important to the hazard assessment. If possible, some attempt should be made to predict the chemical fate, e.g., using mathematical modeling techniques. If this can be done, the objective of the pond study might be to validate the predictions, or, if not, the objective might be to obtain some essential piece of information required for modeling. Similarly, it is useful to review the available toxicity data as a basis for choosing dosages for the pond study. Consideration should be given as to whether the species used in toxicity tests are representative of those occurring in the pond. If not, it may be worth carrying out some further laboratory tests before undertaking the pond study.

Having defined the objective as clearly as possible, the need for replication must be considered. Experimental pond facilities are not widely available, and, therefore, it may be difficult to obtain adequate replication of pond ecosystems. However, some degree of system replication is essential if the objective is to study the relation between chemical treatment and biological effects. Careful management of pond systems can do much to reduce variability between ponds, but, however carefully this is done, differences will still exist. The effect of any chemical treatment has to be measured against the natural variation of populations of organisms between ponds. It is important to realize that it is this variation between ponds that matters when carrying out statistical analyses to investigate effects of treatments. Inadequate replication of systems cannot be offset by taking larger numbers of replicate samples within the systems. More intensive sampling of this sort may be useful in obtaining more accurate measurement of the differences between systems, but it may not help to demonstrate that observed differences are a result of chemical treatment. The number of replicates needed for a particular experiment will depend on the species to be studied, the nature of the biological responses to be measured, and the way these vary between ponds. Thus, some pretreatment data should be obtained to assess this variability as a guide to the required degree of replication. The greater the biological variation, the more replicates will be needed to demonstrate the effects of chemical treatment.

IV. APPLICATION OF CHEMICALS

To avoid extraneous contamination of ponds during treatment, inert materials such as stainless steel and glass should be used for the construction of application equipment. Reservoirs, or other nondisposable parts of spraying equipment, should not be made from polymers because such materials can readily absorb some organic chemicals and, if this occurs, it can be very difficult to decontaminate the equipment.

A pressurized hand-held sprayer may be used either to spray chemicals onto the water surface (Section VIII.A) or to distribute them beneath the surface (Sections VIII.B and IX). The equipment illustrated in Plate 3 consists of a compressed air cylinder, a spray reservoir, and a lance attached to a boom and nozzles. Compressed air is fed to the spray reservoir through flexible nylon tubing. The spray reservoir is a 10-l stainless-steel cylinder fitted with a safety valve, carrying handle, and a removable head secured by bayonet fittings. The removable head incorporates an inlet for compressed air and an outlet for spray liquid. The pressurized liquid is fed to four fan spray nozzles, mounted 500 mm apart, via flexible nylon tubing.

Use of this type of equipment will help to ensure that when chemicals are applied beneath the water surface they will be distributed uniformly into the water column. Even so, it is unlikely that homogeneous distribution can be achieved immediately after treatment. However, thermal mixing in shallow ponds (i.e., up to 1 m deep) will generally lead to homogeneous distribution in the water column within a period of 24 h.

V. MONITORING ENVIRONMENTAL VARIABLES

Since it is not possible to exercise control over environmental variables such as pH, temperature, DO concentrations, wind speed, and sunlight, it is important to measure them in sufficient detail to evaluate their effects on experimental results. However, it is not necessary to measure all of such variables in the context of any one particular study. From a consideration of the physicochemical properties for a particular chemical, it should be possible to evaluate which process or processes, i.e., volatilization, sorption, phototransformation, hydrolysis, or biodegradation, will be important in determining its fate. Having established which are the important fate processes, environmental monitoring should then be oriented towards environmental variables that may affect the relevant fate processes; e.g., wind speed will affect the rate of evaporation of volatile substances; intensity and duration of sunlight will affect the rate of phototransformation of aromatic compounds.

PLATE 2. Outdoor ponds used to evaluate fate and effects of organic chemicals in aquatic environments.

PLATE 3.   Pressurized application equipment suitable for applying chemicals to experimental ponds.

Some environmental variables, particularly those that affect the health of species, populations, and ecosystem, should be monitored routinely, whatever the nature of the treatment. These include pH, temperature, and dissoved oxygen. Many natural waters are strongly buffered and, in this case, there may be little variation in pH values between ponds, although algal blooms can cause considerable variation even in these waters. A weekly measurement of pH in each pond will probably suffice for most experimental purposes. In ponds that are about 1 m deep, thermal stratification often occurs during the hours of daylight but after sunset thermal mixing takes place. The water temperature is therefore much less variable than the air temperature. For this reason it is probably sufficient to measure water temperature on only one occasion each day. However, in order to monitor temporal changes, the measurements must always be carried out at the same time of the day.

DO concentration is an important limiting factor that deserves special attention. The solubility of oxygen in water is only about 10 mg/l at 15°C, and aquatic animals, especially active predators, may be intolerant of concentrations much below about 5 mg/l. In any pond ecosystem a diurnal cycle of DO will exist, and its amplitude will depend on interactions between diffusion, photosynthesis, and respiration. In most pond ecosystems the diffusion of oxygen between air and water occurs relatively slowly, and, therefore, the amplitude of the DO cycle is largely dependent on plant photosynthesis and respiration. Thus, the DO cycle is a key indicator of the state of health of the pond ecosystem, and, therefore, it is worthwhile to invest in good instrumentation for monitoring. A variety of electronic probes and data loggers have been developed for this purpose. The availability of this modern equipment means that DO cycles can now be monitored in much greater detail than was possible only a few years ago.

VI. BIOLOGICAL SAMPLING AND ANALYSIS OF RESULTS

Toxic chemicals can have both direct and indirect effects on aquatic ecosystems. Effects can occur at various levels of organization, e.g., at the organism, population, community, or ecosystem level. Direct effects are attributable to toxicity to one or more species, and they may be evaluated by studying biological responses at the population level. Responses are often categorized as structural, e.g., abundance and biomass, or functional, e.g., growth, reproduction, productivity. A variety of biological methods are available to study these responses. Several workers have argued for a balanced approach towards evaluating both structure and function.[8,9] For most organic chemicals a cost-effective approach depends on a careful consideration of toxicity data obtained either from the literature or from laboratory toxicity tests. Before undertaking a pond study, acute and, possibly, chronic toxicity data should be available or obtained for representative aquatic species, e.g., an alga, an invertebrate, and a fish. Such data will indicate which groups of organisms are likely to be the most susceptible, and sampling programs can be oriented toward the study of the most susceptible species.

The simplest to measure and often the most appropriate of structural responses is a potential effect on population density. Chemicals that are toxic but nonpersistent, such as many of the new synthetic insecticides, may be expected to have considerable but transient effects on population densities of susceptible species when contamination of water bodies occurs at doses used in agricultural spraying. It has been shown that aquatic systems can recover relatively quickly from the effects of such contamination, and, therefore, detailed study of effects on ecosystem function may not be appropriate. Examples of the evaluation of structural responses in relation to the acute toxicity of organic chemicals are given in Section VIII.

Population densities of zooplankton species can be sampled using a plexiglass cylinder long enough to reach the bottom of a pond and fitted with a closing mechanism at its lower end. Sampling with a device of this kind ensures that the whole of the water column is sampled, thus avoiding the problem of patchiness in vertical distribution. The number of samples removed from each pond should be sufficient to estimate the population density with reasonable precision, but it should be remembered that the variation of density between ponds, rather than the precision of estimates within ponds, will ultimately determine sensitivity in detecting effects. It has been found that five zooplankton samples, each sample consisting of about 6 l of pond water, yielded estimates with an acceptable degree of precision.[10] Samples should be preserved in a sucrose-formaldehyde solution for later examination. In the laboratory individuals should be counted and identified under a dissecting microscope. Whole samples or subsamples should be counted, depending on zooplankton density. The numbers of the dominant species in the zooplankton samples should be transformed to logarithms, and the transformed data may be used to calculate the logarithmic mean density of each species for each treatment. A two-way analysis of variance, followed by a Dunnett's two-tailed test, may be used to assess statistical significance of differences between population densities in treated ponds compared with control ponds.

The free-swimming (nekton) invertebrates in pond ecosystems generally contain a number of species, (e.g., various species of Ephemeroptera, Odonata, Hemiptera, Coleoptera) that are relatively sensitive to toxic chemicals. They may be sampled with the aid of a net having mesh openings of 1 mm. The net should be long enough to reach to the bottom of a pond, and it should be drawn across a transect of the pond as quickly as possible to minimize the chances of fast-swimming invertebrates escaping the net. Alternatively, the net may be mounted on a sledge, as described by Hurlbert et al.,[11] to assist in pulling the net quickly across the pond. Benthic invertebrate communities may be sampled using a variety of grab or airlift samplers. Both free-swimming and benthic samples should be processed in the laboratory, and the results may be analyzed in the same way as for zooplankton. It has been found that from three to five samples per pond, depending on the size of the sample, yielded estimates of population density of the dominant species with an acceptable degree of precision.[10]

The data obtained from zooplankton and invertebrate samples may also be used to calculate various diversity indexes. These may be useful for describing overall structural changes in the invertebrate communities.

Small populations of fish may be introduced into experimental ponds. In the absence of fish some of the insect predators such as *Notonecta* and *Chaoborus* can become very abundant, leading to an overall reduction in the diversity of the invertebrate community. The density of the fish population should be maintained at a relatively low level, much lower than in most natural waters, to permit the development of high densities of invertebrate populations. Various species may be used to assess acute toxicity. They may be confined to cages within the ponds or released and then recaptured by using nets or by electrofishing.

Functional responses such as growth, reproduction, and productivity are often regarded as more indicative of the health and integrity of the ecosystem than structural responses. They are more important in the context of chronic toxicity than acute toxicity since populations, communities, and ecosystems can usually recover quickly from short-lived effects on functions. Furthermore, such responses are often more sensitive than structural responses, and, therefore, they may provide useful indicators of the limits of chemical contamination that can be tolerated without harmful effects. At present time there is no general agreement on what kind of functional responses should be measured. Sheehan[9] has reviewed numerous possibilities including morphological, physiological, behavioral, and biochemical responses. In Section IX.B examples of the measurement of some functional responses are given in relation to the chronic toxicity of organic chemicals.

VII. SAMPLING AND ANALYSIS FOR CHEMICALS

In order to study the fate of chemicals, four kinds of samples may be required from ponds, namely, surface water, subsurface water, bottom sediment, and biota.

For chemicals having a low water solubility, concentrations in the surface film may be very much greater than in the subsurface water (Section VIII.A). In this case it may be desirable to obtain samples of the surface film quite separately from samples of the subsurface water. An appropriate method of sampling the surface film has been described for petroleum hydrocarbons in the marine environment.[12] This involves the use of fine-mesh stainless-steel disks. Water from the surface film is drawn into the mesh by capillary action.

Subsurface water samples may be obtained using a sealed glass bottle fitted with a mechanical device such that it can be opened and closed underwater at the required depth. The bottle should be lifted to the surface and wiped free of superficial water before transferring its contents to a collecting jar. To obtain a representative sample of water from a treated pond about ten such subsamples may be taken and combined in a collecting jar.

Sediment cores may be obtained using a device such as that described by de Heer.[13] This consists of a sample holder, 8-cm I.D., with a sharp, stainless-steel cutting edge on the lower end and a closing mechanism on top. To obtain a sample the holder, with both top and bottom ends open, is pushed into the sediment. Next, the top of the holder is tightly closed by turning a milled knob situated above the water at the top of a steel rod. This creates a partial vacuum inside the holder above the sediment and so helps to retain the sample intact during removal. The holder is slowly raised to the surface, and its lower end is stoppered while still underwater. The holder and the sediment sample are immediately transferred to a thermally insulated box containing solid carbon dioxide. In the laboratory the frozen sample is removed by pouring warm water over the holder. The frozen core of sediment can then be sectioned into thin layers using a microtome.

Samples of the biota for residue analysis may be removed from the ponds using the methods described in Section VI.

Sample preservation is necessary to arrest the processes of degradation which would otherwise remove the chemical from the sample. For samples of sediment and biota, transport from the field to the laboratory in a thermally insulated box containing solid carbon dioxide is ideal. However, this procedure is not recommended for preservation of water samples, since glass containers are liable to fracture at low temperatures while, if polymeric containers are used, the polymers may interact in various ways with trace contaminants in the water. On the other hand, chemical procedures can be readily applied to the preservation of water samples. They are of two kinds: (1) the addition of a biocide to halt microbiological action and (2) the addition or use of a chemical to extract organic chemicals from the sample matrix. However, the use of chemicals in this way must always involve consideration of their purity, possible effects on the sample and analyte, and what, if any, constraints might be imposed on subsequent analytical procedures.

A description of procedures for the analysis of organic chemicals in environmental samples is beyond the scope of this chapter. The reader is referred to the literature for particular chemicals, to other chapters in this book, and to various handbooks and manuals for sample handling and analysis.[14-17] A summary of procedures used for organic chemicals described in this chapter (Sections VIII and IX) is given in Table 1.

VIII. EXAMPLES OF POND EXPERIMENTS TO ASSESS THE HAZARD OF ACUTE TOXICITY TO THE AQUATIC ENVIRONMENT

The most widely used insecticides today are relatively nonpersistent, biodegradable compounds, such as many organophosphorus and pyrethroid insecticides. The normal agricultural

Table 1
SUMMARIZED ANALYTICAL METHODS

Chemical and sample type	Analytical step		
	Extraction	Purification	Determination
MEP			
Subsurface water	Solvent partition (hexane)	None	glc/NPfid[a]
Sediment	Blend with Na₂SO₄; methanol extraction	Centrifugation	glc/NPfid
Fish	Blend with Na₂SO₄; acetone/hexane extraction	Liquid/liquid partition Florisil column chromatography	glc/NPfid
Cypermethrin			
Subsurface water	Solvent partition (hexane)	None	glc/ecd[b]
Surface film water	Solvent (acetone)	None	glc/ecd
Sediment and fish	Blend with Na₂SO₄; acetone/hexane extraction	Liquid/liquid partition Florisil column chromatography	glc/ecd
Vegetation	Solvent (acetone/hexane)	Florisil column chromatography	glc/ecd
PCP			
Subsurface water	Liquid/solid extraction (SEP-PAK C18)	None	Reverse phase HPLC/UV adsorption
Sediment	Solvent (acetone/formic acid)	pH adjustment, solvent wash	glc/ecd after derivatization
Fish	Solvent (acetone/hexane/formic acid)	pH adjustment, solvent wash	glc/ecd after derivatization
DCA			
Subsurface water	None	None	Reverse phase HPLC/UV adsorption

[a]   Gas-liquid chromatography/nitrogen-phosphorus selective flame ionization detection.
[b]   Gas-liquid chromatography/electron capture detection.

use of these chemicals may present a hazard to the aquatic environment since many of them are acutely toxic to aquatic organisms. However, the risks may not be very great if the amount of contamination is small or if the contaminant is not in a form that is bioavailable. In this section two examples of pond studies are given to illustrate their value in making hazard assessments for chemicals of this kind (cf. Section IX which deals with hazards arising from potential chronic toxicity).

In the first example (Section VIII.A) unreplicated pond experiments are described in which a pyrethroid insecticide was sprayed onto the surface of outdoor ponds to study its subsequent fate and the exposure of aquatic organisms to toxic concentrations in the water. As mentioned in Section III, some degree of replication is essential to demonstrate that biological effects are related to experimental treatments and not to some other factor. However, in these experiments the primary objective was to study distribution and degradation of the pyrethroid in relation to its known toxicity. In this case, the need to collect and analyze relatively large numbers of samples of water, sediment, vegetation, and fish for pyrethroid residues in a single pond was more important than replication of systems.

In the second example (Section VIII.B) a replicated pond experiment is described which was carried out to evaluate the fate and effects of the organophosphorus insecticide methyl parathion (MEP) under field conditions and to compare the results with predictions based on laboratory studies. In this case the residue sampling program was limited to the study of MEP in the water and sediment. Replication of treated and control ponds permitted a detailed study of direct and indirect effects of the treatment.

A. Fate and Effects of a Pyrethroid Insecticide

The insecticide cypermethrin [RIPCORD, NRDC 149, (R,S)-α-cyano-3-phenoxybenzyl (IR, IS, *cis, trans*)-3-(2,2-dichlorovinyl)-2,2-dimethyl cyclopropane-carboxylate] is a synthetic pyrethroid used to control a variety of insect pests. In normal agricultural use recommended dosages are in the range 25 to 100 g/ha. In the laboratory the 96-h LC_{50} for various species of fish is in the range 0.4 to 2.2 μg/l, and for sensitive species of aquatic invertebrates the 24-h LC_{50} is 0.05 to 2.0 μg/l.[18] These data indicate that because of its high toxicity to aquatic organisms, cypermethrin could present a hazard to the aquatic environment if it were used near water. However, such toxicity data are, in themselves, insufficient to make a reliable assessment of hazard because it is not possible to calculate how much of the insecticide that contaminates the water surface will subsequently be available for uptake by aquatic organisms. Pond studies were therefore undertaken to investigate the distribution, degradation, and biological effects of cypermethrin when it was sprayed over the surfaces of experimental outdoor ponds.

Two unreplicated pond experiments were carried out when mature, unlined ponds, 0.7 m deep, were oversprayed with an emulsifiable concentrate formulation of cypermethrin at a dosage of 100 g/ha using a knapsack sprayer fitted with a 2-m boom held about 1 m above the water surface.[19] Residues of cypermethrin in surface water, subsurface water, sediment, aquatic vegetation, and fish were sampled and analyzed using methods described in Section VII. Effects on zooplankton and invertebrates were evaluated using methods described in Section VI. Effects on survival of fish were evaluated using wild populations of fish (in the first experiment) or introduced populations of rudd, *Scardinius erythrophthalmus* (in the second experiment).

Concentrations of cypermethrin recovered from various kinds of pond samples are shown in Figure 1. Very high concentrations were associated with surface vegetation and the surface water film soon after treatment, but only low concentrations (<2.6 μg/l) were found in the subsurface water. It was calculated that if all of the applied cypermethrin had dispersed homogeneously into the water column, the water concentrations soon after treatment would have been in the range 16 to 18 μg/l. In fact, the maximum concentrations found in subsurface water samples were 2.6 μg/l, 48 h after treatment in the first experiment, and 1.4 μg/l, 1 h after treatment in the second experiment. Thus, only 8 to 16% of cypermethrin sprayed onto the surface of the ponds was subsequently found in subsurface water.

There were no mortalities, or other observed effects, among populations of rudd even though the pond water contained apparently lethal concentrations of cypermethrin. In the first pond experiment these fish were exposed to a water concentration of 1.5 to 2.6 μg/l during the period 0 to 96 h after treatment. In the second experiment they were exposed to 0.5 to 14 μg/l for the same period of time. These exposures may be compared with a 96 h LC_{50} value of 0.4 μg/l obtained in the laboratory using filtered, dechlorinated tap water. Survival of fish in pond water containing apparently lethal concentrations of cypermethrin was attributed to its sorption by suspended solids in the pond water. The partition coefficient of cypermethrin is very high (K octanol/water $\cong$ 2.1 × 10^6), and, therefore, it is strongly bound onto organic matter. At the same time its water solubility is very low. Thus, a relatively small amount of suspended matter in the pond water (2 to 4 mg/l) was sufficient to reduce the aqueous concentration of cypermethrin to sublethal levels.

Populations of zooplankters and aquatic insects were severely affected by the cypermethrin treatments. Effects on these populations were consistent with the results of laboratory toxicity

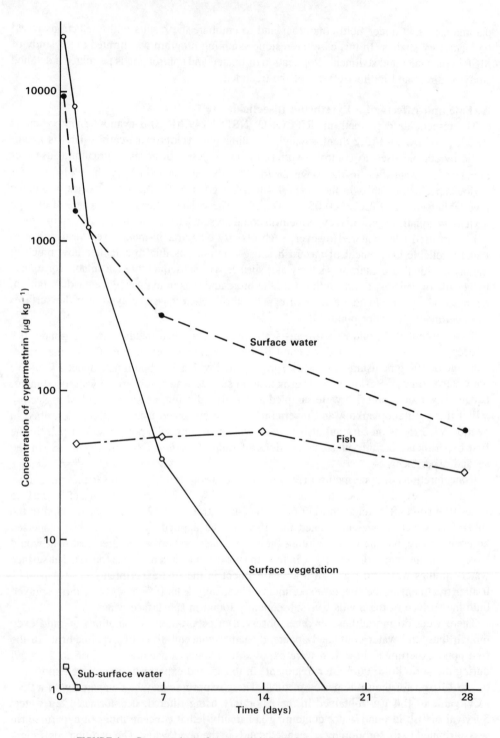

FIGURE 1.   Concentrations of cypermethrin in various kinds of pond samples.

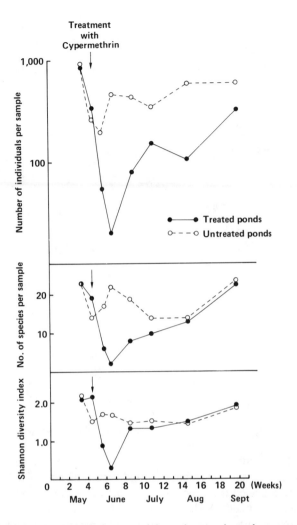

FIGURE 2. Effects of cypermethrin on the macroinvertebrate community of an experimental pond.

tests in which various species of aquatic insects and zooplankters were shown to be susceptible to concentrations exceeding 0.5 μg/l. About 8 weeks after treatment populations of daphnids and copepods, virtually eliminated by the treatment, were recovering, and 12 weeks after treatment their numbers were comparable to or greater than in a control pond. Numbers of individuals, species, and the diversity of the macroinvertebrate community were also severely affected but had recovered 8 weeks after treatment (Figure 2).

B. Fate and Effects of MEP

MEP was used as a reference compound to compare fate and effects of a toxic chemical in the field with predictions based on laboratory data.[20] On the basis of a literature survey it was concluded that biodegradation and sorption would be the most important processes for loss of MEP from the water. Assuming that biodegradation in sediment could be ignored, a mathematical model was used to calculate that the overall rate of loss would be 0.0024 to 0.018/d, corresponding to half-lives of 37 to 290 d. The results of a literature survey of data obtained from toxicity tests indicated that treatment of the outdoor ponds at a concentration of 100 μg/l would have adverse effects on populations of aquatic insects and Cladocera. No direct effects on populations of algae, Copepoda, or fish were expected at this dosage level.

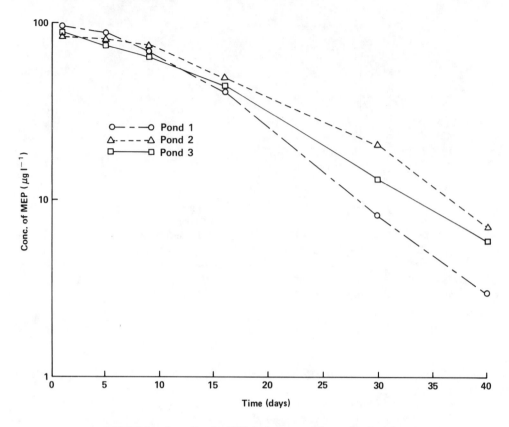

FIGURE 3.   Ct profiles for MEP in the water of three treated ponds.

MEP was applied beneath the surface of three ponds at a nominal concentration of 100 μg/l. Three similar ponds remained untreated and served as controls. Water samples for residue analysis were collected from each pond 1 h before treatment and at various intervals from 1 h to 40 d after treatment. Sediment samples for residue analysis were collected on only one occasion, 16 days after treatment, when it was expected that sediment concentrations would be greatest and therefore relatively easy to detect.

Water temperature, DO concentration, and pH were measured daily in each pond. Population densities of zooplankton and macroinvertebrates were monitored using the sampling techniques described in Section VI. A group of 25 juvenile rainbow trout (*Salmo gairdneri*) were individually weighed and released into each pond 2 weeks before treatment. They were removed 9 weeks after treatment, using an electrofishing technique, and reweighed to evaluate effects on growth.

The concentration-time (ct) profiles for MEP in the water of the three treated ponds are shown in Figure 3. It is apparent that there was very little difference in MEP concentrations between ponds. From the ct profiles it was calculated that the rate of loss from the water was 0.041 to 0.079/d, corresponding to half-lives of 8.7 to 17 d, very different from predictions. MEP was not found in any of the samples of pond sediment, and, therefore, sorption alone could not account for the difference between observed and predicted losses of MEP from the water. From these results it was deduced that biodegradation in sediment might be the key loss process.

Laboratory studies were then carried out in which it was demonstrated that (1) addition of plants or sediments to aquaria containing water and MEP resulted in a greatly accelerated rate of loss of MEP from the water and (2) in bacterial cultures the rate of loss of MEP was very fast, but it was almost totally inhibited when the bacteria were killed by the addition of sodium azide.[21] It was concluded that the key loss process for MEP from the outdoor ponds was biodegradation in sediments and, to a lesser extent, biodegradation on plants and other surfaces. It was then calculated that the rate of biodegradation in sediment (4.0/d) was very much faster than the rate of transport from water to sediment (0.02 to 0.05/d). Thus, the rate-limiting mechanism for loss of MEP from the water was transport to the sediments, a mechanism that is primarily dependent on the mixing characteristics of the system. This pond experiment, in conjunction with mathematical modeling techniques and laboratory experiments, yielded some useful insights into the loss processes affecting MEP and, more fundamentally, into the ways in which sorption and biodegradation can interact to affect the loss of organic chemicals in aquatic systems.

Direct effects of MEP on the aquatic fauna of the ponds were similar to predictions based on the results of laboratory toxicity tests. There was no effect on gastropods or oligochaetes, but there was an overall reduction in the numbers of aquatic insects for a period of 6 weeks after treatment. There was a marked decline in the numbers of mayfly larvae in the control ponds during the last 2 weeks of April (Figure 4), an effect attributable to emergence of the adults. Their numbers remained relatively low through May and June and then increased during July and August following the development of a second generation of larvae. In the treated ponds the numbers of mayfly larvae followed a similar seasonal pattern, but during the months of May and June their numbers were significantly less ($p < 0.05$) than in the control ponds. There was also a significant decrease ($p < 0.05$) in the numbers of dipterous larvae in treated ponds during May and June, and their numbers also increased very rapidly in July. Overall, effects on the aquatic insect fauna were severe for a period of 6 weeks following treatment. However, insect populations were able to recover very quickly when MEP concentrations in the water declined below threshold toxicity levels (0.1 to 1 μg/l for the most susceptible species). The concentration of MEP in pond water decreased from 90 to 5 μg/l between 0 and 40 d after treatment. Assuming no change in the rate of loss of MEP from the water, the concentration would have decreased to sublethal levels 60 to 80 d after treatment. Seventy days after treatment there was evidence of repopulation by Diptera, Ephemeroptera, and Cladocera. Ninety days after treatment populations of these invertebrates had all recovered from effects of the treatment.

The dominant organisms among the zooplankton belonged to three families: Daphniidae (Cladocera), Diaptomidae, and Cyclopidae (Copepoda). *Daphnia* populations were virtually eliminated by the MEP treatment (Figure 5a). Their numbers remained very low until 10 weeks after treatment, and thereafter their numbers increased very rapidly. *Cyclops* populations were also affected by the MEP treatment but to a lesser extent than the *Daphnia* populations.

Various indirect, or secondary, effects of the MEP treatment were observed. There was an increase in the density of populations of *Diaptomus* in treated, compared with control ponds (Figure 5b). This difference emerged about 3 weeks after treatment and was sustained until about 13 weeks after treatment. It appears that *Diaptomus* populations were able to increase in the absence of competition from *Daphnia* and predation from *Cyclops* and aquatic insect predators. Similar predator-prey interactions have been reported elsewhere in the literature.[22,23]

There was a significant effect ($p < 0.05$) on the growth of rainbow trout. The mean weight of fish removed from control ponds was 26.2 g compared with 20.6 g from treated ponds (Table 2). This effect cannot be attributed to the direct toxicity of MEP since it has been shown that the growth of rainbow trout is not affected when they are exposed to higher concentrations than occurred in the ponds.[10] The effect may therefore be attributed to an indirect effect on the food supply of the rainbow trout, i.e., the aquatic invertebrates.

In one of the treated ponds mortality of some of the rainbow trout was associated with a

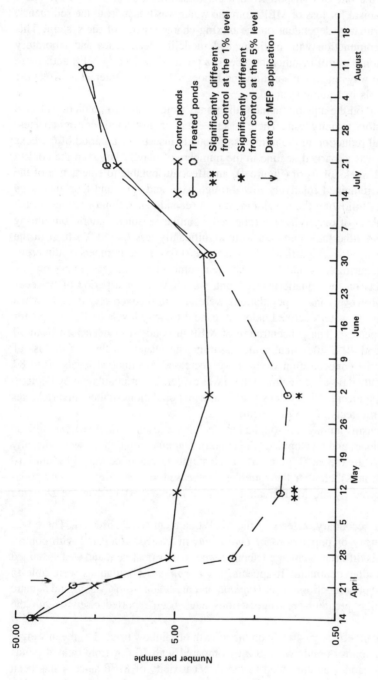

FIGURE 4.   Effects of MEP treatment on populations of mayfly larvae.

depression of DO concentrations (Figure 6) caused by the death and decay of a bloom of filamentous algae. This effect occurred in only one of the treated ponds, and, therefore, it cannot be attributed to the MEP treatment with any degree of confidence. However, the association between algal blooms, their death and decay, and the depression of DO concentrations is well documented.[22] In the pond where this effect was observed the growth of a filamentous alga, *Enteromorpha* sp., was much greater than in the other two treated ponds, where vascular plants were dominant. This species of alga is known to be a preferred food source of mayfly larvae and daphnids, and, thus, the death of these herbivores might well have led to the algal bloom, its death and decay, depression of DO concentrations, and fish mortalities.

IX. EXAMPLES OF POND EXPERIMENTS TO ASSESS THE HAZARD OF CHRONIC TOXICITY TO THE AQUATIC ENVIRONMENT

Chemicals that are manufactured on a large scale, are widely distributed, and persistent, may pose a hazard arising from chronic toxicity, i.e., the biological effects that may occur as a result of long-term exposure to relatively low concentrations. Assessment of a hazard of this kind depends on determining concentrations of a chemical that have no adverse effects on sensitive species and comparing these with environmentally relevant concentrations. In this section two examples, pentachlorophenol (PCP) and 3,4-dichloroaniline (DCA), are described to illustrate how pond experiments may be used to evaluate the potential hazard of chronic toxicity.

From a review of the literature the threshold concentration for toxicity of PCP to susceptible species of algae, mollusks, and fish is 0.05 to 0.10 mg/l. *Daphnia magna* is less susceptible, with a "no effect" concentration of 0.34 mg/l in a 21-d life cycle test. For various organisms the ratio of acute to chronic toxicity is very low, approximately 1.0, indicating that prolonged exposure to PCP is no more hazardous than relatively short (24- to 96-h) exposures. These data were used as a basis for choosing the dosage of PCP in a pond experiment. PCP was maintained in the water of three ponds for 30 d at a concentration estimated to be near to the threshold concentration for toxicity to fish, algae, and mollusks (0.05 to 0.10 mg/l).

The spectrum of aquatic toxicity for DCA is very different from that of PCP, and the ratio of acute to chronic toxicity is relatively high, 10 to 500 depending on species, indicating that prolonged exposure to DCA is potentially more hazardous than short exposures. Among seven species used in acute and chronic toxicity tests, *Daphnia magna* was much more susceptible than other species.[24,25] In a chronic toxicity test, effects on the birth rate of *D. magna* occurred at a much lower concentration (0.01 mg/l) than that causing mortality (0.10 mg/l). In a 96-h exposure using rainbow trout the LC_{50} was 2.0 to 3.0 mg/l, but in a 28-d exposure effects on growth occurred at a concentration of only 0.04 mg/l. These data were used as a basis for choosing the dosage of DCA in a pond experiment. DCA was maintained in the water of three ponds for 30 d at a "low" concentration (0.045 mg/l) and in three ponds at a "high" concentration (0.45 mg/l). Three ponds remained untreated and were used as controls. A stock of 25 juvenile rainbow trout was introduced into each pond shortly before treatment.

A. Fate and Effects of PCP

PCP is a broad-spectrum biocide that, together with its sodium salt, is used as a wood preservative, fungicide, herbicide, insecticide, and molluskicide. Because of its relatively large scale of manufacture and its many uses, it is widely distributed in the environment. Its chemistry, toxicology, and fate in the environment have been widely studied.[24,26,27]

A pond experiment was carried out with PCP to assess the hazard of its chronic toxicity in aquatic environments.[28] PCP was applied to three ponds in a series of nine treatments

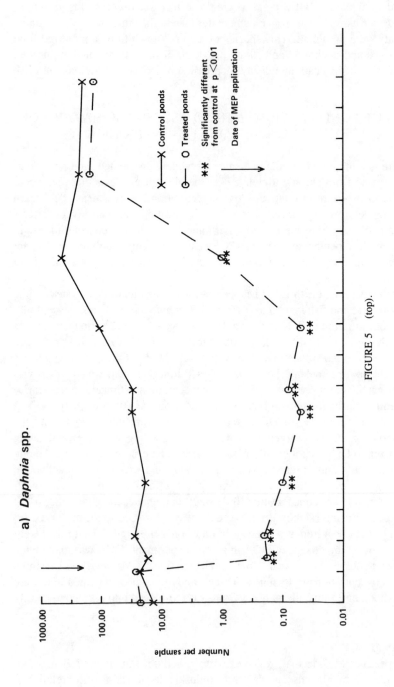

FIGURE 5 (top).

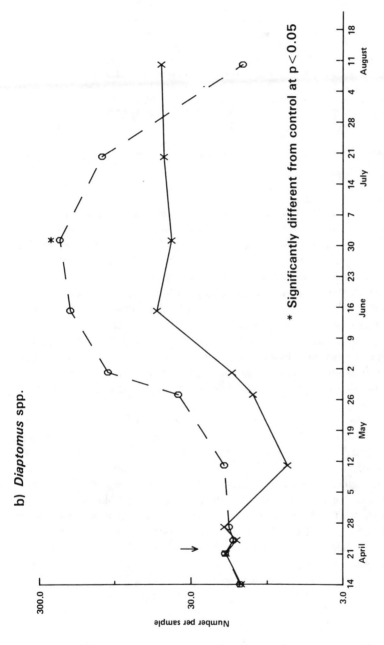

b) *Diaptomus* spp.

* Significantly different from control at p < 0.05

Number per sample

FIGURE 5. Effects of MEP treatment on zooplankton populations. (a) *Daphnia* spp. (susceptible species); (b) *Diaptomus* spp. (nonsusceptible species).

Table 2
EFFECT OF MEP TREATMENT ON THE
GROWTH OF RAINBOW TROUT

	2 weeks before treatment		9 weeks after treatment	
	Number of fish	Mean weight (g)	Number of fish	Mean weight (g)
Control ponds	25	4.72 (0.20)[a]	21	23.3 (1.27)
	25	4.42 (0.18)	21	26.9 (1.48)
	25	4.88 (0.20)	17	28.5 (1.62)
Treated ponds	25	4.65 (0.22)	18	21.0 (1.56)
	25	4.84 (0.19)	3	20.7 (3.29)
	25	4.82 (0.15)	20	20.1 (1.25)

Note: The effect is indirect, through the food chain.

[a] SE in parentheses.

carried out once every 2 to 4 d. The frequency of treatments and the concentration of PCP added at each treatment were dependent on the results of chemical analyses of samples of pond water. The aim of the treatment schedule was to maintain concentrations of PCP in the pond water within the range 0.05 to 0.10 mg/l for a period of 30 d to facilitate comparison of biological effects in the ponds with the results of chronic toxicity tests.

A process analysis for PCP in the ponds indicated that phototransformation would be the only important loss process.[28] The rate of this process was calculated using the mathematical model SOLAR,[29] together with an appropriate correction for cloud cover. Chemical-related inputs used for this calculation were the light absorption spectrum for PCP and the quantum yield of the light-activated reaction. Environment-related inputs were the light attenuation of the pond water, depth of ponds, season, latitude, and cloud cover. The rate of photo-transformation was thus calculated to be 0.23 to 0.46/d, corresponding to half-lives of 1.5 to 3.0 d.

The observed rate of loss of PCP from the ponds varied from 0.15 to 0.34/d, corresponding to half-lives of 2.0 to 4.7 d, in good agreement with predictions. Most of the variation in half-lives could be attributed to variation in the light attenuation of pond water caused by variable amounts of dissolved organic carbon, suspended solids, and unicellular algae. The rate of loss of PCP was greatest in pond 4 (Figure 7) where relatively clear water was associated with the presence of vascular plants and the absence of algal blooms. Results of this experiment demonstrated that phototransformation can be an important loss process for PCP, that the rate of this process can be predicted with reasonable confidence, and that environmental factors that affect light attenuation of natural waters may have a considerable effect on the rate of phototransformation.

Aquatic vegetation of two of the three treated ponds was dominated by the filamentous alga *Enteromorpha* sp. In the third pond the dominant vegetation consisted of vascular plants, *Potamogeton natans* and *Callitriche stagnalis,* and there were smaller amounts of the filamentous alga *Chara* sp. There was a marked effect of PCP on the filamentous algae about 3 weeks after the start of treatment. It was characterized by the death and rapid decay of *Enteromorpha* sp. in two of the ponds and of *Chara* sp. in the other pond. There was no effect on the vascular plants. Death and subsequent decay of *Enteromorpha* sp. in the treated ponds was associated with depression of DO concentrations and mortality of rainbow trout. No such effects were seen in control ponds.

There were no effects of the PCP treatment on population densities of various species of

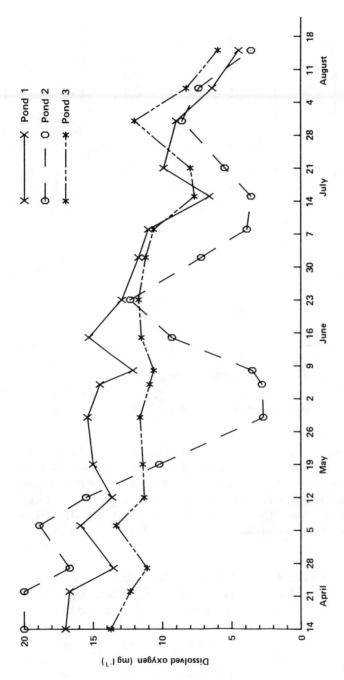

FIGURE 6. Seasonal variation of DO concentrations in ponds treated with MEP. (In pond 2 the dominant vegetation was filamentous algae. In ponds 1 and 3 vascular plants were dominant.)

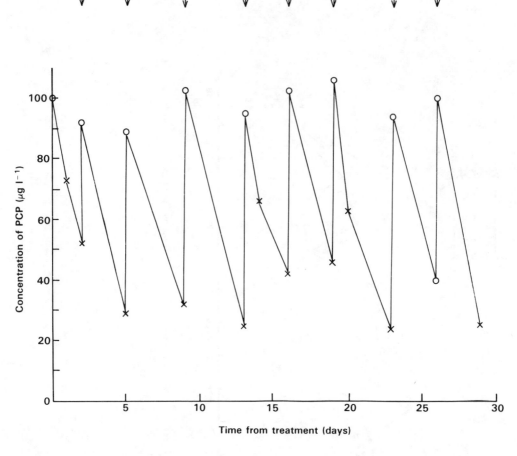

FIGURE 7.   Ct profile for PCP in a treated pond. ↓, dates of PCP treatments; ○, calculated PCP concentrations; ×, measured PCP concentrations.

zooplankters and aquatic insects, and there was no direct effect on fish survival or growth. These results were consistent with the results of chronic toxicity tests using invertebrates and fish. However, there was a significant reduction ($p < 0.05$) of the population density of gastropod snails in treated compared with untreated ponds. A possible contributory factor to this population decline was an effect of PCP on the viability of snail eggs. This effect was also consistent with the results of chronic toxicity tests using *Lymnaea stagnalis*.[28]

B. Fate and Effects of DCA

DCA is formed in the aquatic environment by hydrolysis of several widely used herbicides, and it is an intermediate in their manufacture. Together with other aromatic amines it has been found in various surface waters in Europe.[30] Its fate in the environment and its chronic toxicity to aquatic organisms have been studied by various workers.[24,25,31]

The experiment with DCA was carried out using nine experimental ponds and a randomized block design. Three ponds were treated with a "high" concentration of DCA (450 μg/l), three with a "low" concentration (45 μg/l), and three were untreated controls. Concentrations of DCA were maintained in the water for a period of 28 d.

Concentrations of DCA in the water were monitored using high-performance liquid chromatography, and the results were used as a guide to the frequency and magnitude of doses required to maintain the nominal concentrations in the pond water. The resulting ct profiles for DCA in the six treated ponds are shown in Figure 8.

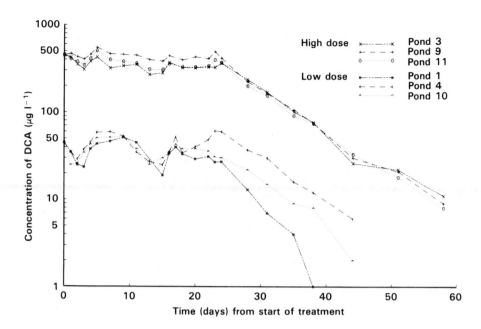

FIGURE 8. Ct profiles for DCA in six treated ponds.

A process analysis for DCA in the ponds indicated that phototransformation would be the dominant loss process. The rate of this process was calculated using the methods described for PCP in Section VIII.A. The calculated rate of loss was 0.13 to 0.22/d, corresponding with half-lives of 3.2 to 5.3 d. The observed rate of loss was 0.11 to 0.17/d, corresponding with half-lives of 4.1 to 6.3 d, in good agreement with calculations.

Population densities of species of zooplankters and aquatic invertebrates were monitored using previously described methods (Section VI). There was no effect on population densities of various species of macroinvertebrates, and there was no effect on indexes of community structure. On the other hand, population densities of various species of zooplankters were severely affected by DCA. *Daphnia longispina* and other Cladocera were the most susceptible species.

Detailed studies were carried out to evaluate effects of DCA on birthrates, death rates, and population growth rates of *D. longispina*. Birthrates were calculated from the numbers of young born to samples of adults confined for periods of 24 h in plexiglass cylinders. Death rates were calculated from the numbers of live and dead adults removed from the cylinders after 24 h. The cylinders were filled with pond water that was sieved, using a mesh size of 75 μm, to remove all *D. longispina* and their predators. A sample of zooplankton was collected and sieved, using a mesh size of 850 μm. The sample was transferred to a dish where dead or damaged specimens of *D. longispina* were removed. Predators of *D. longispina,* mainly *Chaoborus* sp. and *Cyclops* sp., were also removed, and the remaining sample, consisting mainly of live, adult *D. longispina,* was placed in a cylinder. After 24 h the contents of the cylinders were sieved, using a mesh size of 75 μm, and taken to the laboratory for examination and counting.

Birthrates, death rates, and population growth rates are given in Table 3, and effects on population densities are illustrated in Figure 9. In ponds treated with a low concentration of DCA the birthrate declined from 0.51 to 0.14 and the population growth rate from 0.43 to 0.07. These changes, particularly the change in birthrate, were reflected in an overall reduction in population density from 40 to 8/l. The population, with a much-reduced birthrate, was able to maintain itself at a lower density level against the combined effects of toxic

Table 3
MEAN BIRTHRATES (b), DEATH RATES (d), AND GROWTH RATES (r) OF *D. LONGISPINA* POPULATIONS

Day	Control			Low concentration			High concentration		
	b	d	r	b	d	r	b	d	r
−11	0.73	0.08	0.65	0.55	0.07	0.48	0.41	0.06	0.35
−5	0.43	0.09	0.34	0.28	0.14	0.14	0.20	0.08	0.12
0 (May 17, 1982)	0.76	0.02	0.74	0.80	0.04	0.76	0.46	0.04	0.42
3	0.39	0.06	0.33	0.25	0.04	0.21	0.15	0.08	0.07
7	0.43	0.02	0.41	0.28	0.04	0.24	0.05	0.12	−0.07
10	0.49	0.03	0.46	0.03	0.03	0	0.01	0.06	−0.05
14	0.58	0	0.58	0.05	0.05	0	0.03	0.18	−0.15
17	0.59	0.04	0.55	0.18	0.09	0.09			
21	0.57	0.05	0.52	0.09	0.05	0.04			
24	0.35	0.03	0.32	0.16	0.05	0.11			
28	0.49	0.02	0.47	0.15	0.08	0.07			
31	0.45	0.08	0.37	0.89	0.12	0.77			
35	0.32	0.03	0.29	0.58	0.04	0.54			
38	0.52	0.03	0.49	0.51	0.04	0.47			
52	0.41	0.02	0.39						
58	0.33	0.04	0.29				0.22	0.03	0.19
65	0.34	0.04	0.30				0.55	0.07	0.48
72	0.40	0.03	0.37				0.94	0.07	0.87
79	0.35	0.03	0.32				0.92	0.03	0.89

Note: Values are means (per day) for three ponds per treatment. Low and high concentrations were 45 and 450 µg/l DCA, respectively.

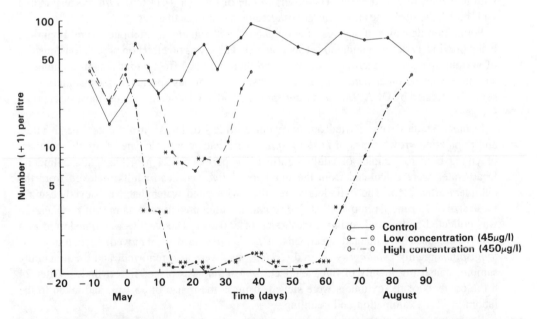

FIGURE 9.   Effect of DCA on density of *Daphnia longispina* populations. * significantly different from control at the 5% level; ** significantly different from control at the 1% level.

stress and predation. Very soon after the end of treatment, on day 31, the birthrate increased to 0.89 (from 0.15 on day 28), and the population density was quickly restored to pretreatment levels. In ponds treated with a high concentration of DCA, effects on the birthrate and population growth rate were seen soon after the start of treatment. These were reflected in an overall reduction in population density to near zero after 14 d. The density remained very low until 58 d after treatment, by which time the DCA concentration had decreased to 9 μg/l, below an estimated threshold concentration for toxicity of 14 μg/l. The birthrate then increased rapidly from 0.22 on day 58 to 0.94 on day 72, and the population density was quickly restored to pretreatment levels.

Effects of DCA on primary productivity were observed in ponds treated with low concentrations and high concentrations. These effects were dependent on the nature of the aquatic vegetation, i.e., on whether the plant community was dominated by filamentous algae or vascular plants. In two of the six treated ponds, treated with 45 and 450 μg/l of DCA, respectively, there was no indication of phytotoxicity. On the other hand, there were marked phytotoxic effects in the other four treated ponds, where filamentous algae were the dominant aquatic plants. These were first observed 20 d after the start of treatment in ponds treated with a high concentration of DCA and were characterized by black discoloration and death of the algal fronds, associated with depression of concentrations of DO and mortality of rainbow trout. Similar effects were observed 23 d after the start of the treatment in ponds treated with a low concentration of DCA.

There was a 10% depression of growth rate of rainbow trout in ponds treated with 45 μg/l, and there was a 66% depression of growth rate, in ponds treated with 450 μg/l. The nature of this response indicates a dose-response relationship, and it is similar to results obtained in the laboratory.

REFERENCES

1. **Caspers, V. N. and Hamburger, B.,** Aquatische Modellökosysteme als Indikatoren für Schadstoffbelastungen, *Z. Wasser Abwasser Forsch.,* 16, 205, 1983.
2. **Hall, D. J., Cooper, W. E., and Werner, E. E.,** An experimental approach to the production dynamics and structure of freshwater animal communities, *Limnol. Oceanogr.* 15, 839, 1970.
3. **Mauck, W. L., Mayer, F. L., Jr., and Holz, D. D.,** Simazine residue dynamics in small ponds, *Bull. Environ. Contam. Toxicol.,* 16, 1, 1976.
4. **Robinson-Wilson, E. F., Boyle, T. P., and Petty, J. D.,** Effects of increasing levels of primary production on pentachlorophenol residues in experimental pond ecosystems, in *Aquatic Toxicology and Hazard Assessment,* Bishop, W. E., Cardwell, R. D., and Heidolph, B. B., Eds., American Society for Testing and Materials, Philadelphia, 1983, 239.
5. **Sanders, H. O., Walsh, D. F., and Campbell, R. S.,** Abate: effects of the organophosphate insecticide on bluegills and invertebrates in ponds, Tech. Pap. No. 104, U.S. Fish and Wildlife Service, Washington, D.C., 1981.
6. **Giddings, J. M., Franco, P. J., Cushman, R. M., Hook, L. A., Southworth, G. R., and Stewart, A. J.,** Effects of chronic exposure to coal-derived oil on freshwater ecosystems. II. Experimental ponds, *Environ. Toxicol. Chem.,* 3, 465, 1984.
7. **Crossland, N. O. and Wolff, C. J. M.,** Outdoor ponds: their construction, management and use in experimental ecotoxicology, in *The Handbook of Environmental Chemistry,* Vol. 2D, Hutzinger, O., Ed., Springer-Verlag, Berlin, 1986, 1.
8. **Matthews, R. A., Buikema, A. L., Jr., Cairns, J., Jr., and Rodgers, J. H., Jr.,** Biological monitoring. IIA. Receiving system functional methods, relationships and indices, *Water Res.,* 16, 129, 1982.
9. **Sheehan, P. J.,** Functional changes in the ecosystem in *SCOPE 22: Effects of Pollutants at the Ecosystem Level,* Sheehan, P. J., Miller, D. R., Butler, G. C., and Bourdeau, P., Eds., John Wiley & Sons, Chichester, 1984, 101.
10. **Crossland, N. O.,** Fate and biological effects of methyl parathion in outdoor ponds and laboratory aquaria. II. Effects, *Ecotoxicol. Environ. Saf.,* 8, 482, 1984.

11. **Hurlbert, S. H., Mulla, M. S., Keith, J. O., Westlake, W. E., and Dusch, E.,** Biological effects and persistence of Dursban in freshwater ponds, *J. Econ. Entomol.*, 63, 43, 1970.

12. **Hardy, R., Mackie, P. R., Whittle, K. J., McIntyre, A. D., and Blackman, R. A. A.,** Occurrence of hydrocarbons in the surface film, subsurface water and sediment in the waters around the United Kingdom, *Rapp. P. V. Reun. Cons. Int. Explor. Mer.*, 171, 61, 1977.

13. **de Heer, H.,** Measurements and computations on the behaviour of the insecticides azinphosmethyl and dimethoate in ditches, *Agric. Res. Rep.*, Pudoc, Wageningen, 884.

14. **Williams, S., Ed.,** *Official Methods of Analysis of the Association of Official Analytical Chemists*, 14th ed., Association of Official Analytical Chemists, VA, 1984.

15. **Watts, R. R., Ed.,** Manual of Analytical Methods for the Analysis of Pesticides in Humans and Environmental Samples, No. 600/8-80-038, U.S. Environmental Protection Agency, Washington, D.C., 1980.

16. **Harris, J. C., Hayes, M. J., Levins, P. L., and Lindsay, D. B.,** Procedures for level 2 sampling and analysis of organic materials, No. 600/7-79-003, U.S. Environmental Protection Agency, Washington, D.C., 1979.

17. **Marcotte, A. L. and Bradley, M., Eds.,** FDA Pesticide Analytical Manual, U.S. Food and Drug Administration, Rockville, MD, 1986.

18. **Stephenson, R. R.,** Aquatic toxicology of cypermethrin. I. Acute toxicity to some freshwater fish and invertebrates in laboratory tests, *Aquat. Toxicol.*, 2, 175, 1982.

19. **Crossland, N. O.,** Aquatic toxicology of cypermethrin. II. Fate and biological effects in pond experiments, *Aquat. Toxicol.*, 2, 205, 1982.

20. **Crossland, N. O. and Bennett, D.,** Fate and biological effects of methyl parathion in outdoor ponds and laboratory aquaria. I. Fate, *Ecotoxicol. Environ. Saf.*, 8, 471, 1984.

21. **Crossland, N. O., Bennett, D., Wolff, C. J. M., and Swannell, R. P. J.,** Evaluation of models used to assess the fate of chemicals in aquatic systems, *Pestic. Sci.*, 17, 297, 1986.

22. **Hurlbert, S. H.,** Secondary effects of pesticides on aquatic ecosystems, *Res. Rev.*, 57, 81, 1985.

23. **Hurlbert, S. H., Mulla, M. S., and Willson, H. R.,** Effects of an organophosphorus insecticide on the phytoplankton, zooplankton and insect populations of freshwater ponds, *Ecol. Monogr.*, 42(3), 269, 1972.

24. **Adema, D. M. M. and Vink, G. J.,** A comparative study of the toxicity of 1,1,2-trichloroethane, dieldrin, pentachlorophenol and 3,4-dichloroaniline for marine and freshwater organisms, *Chemosphere*, 10, 533, 1981.

25. **Crossland, N. O. and Hillaby, J. M.,** Fate and effects of 3,4-dichloroaniline in the laboratory and in outdoor ponds. II. Chronic toxicity to *Daphnia* spp. and other invertebrates, *Environ. Toxicol. Chem.*, 4, 489, 1985.

26. **Ranga Rao, K.,** *Pentachlorophenol: Chemistry, Pharmacology and Environmental Toxicology*, Plenum Press, New York, 1978.

27. **Crosby, D. G.,** The environmental chemistry of pentachlorophenol, *Pure Appl. Chem.*, 53, 1051, 1981.

28. **Crossland, N. O. and Wolff, C. J. M.,** Fate and biological effects of pentachlorophenol in outdoor ponds, *Environ. Toxicol. Chem.*, 4, 73, 1985.

29. **Zepp, R. G. and Cline, D. M.,** Rates of direct photolysis in aquatic environments, *Environ. Sci. Technol.*, 11, 359, 1977.

30. **Wegman, R. C. C. and de Korte, G. A. L.,** Aromatic amines in surface waters of the Netherlands, *Water Res.*, 15, 391, 1981.

31. **Wolff, C. J. M. and Crossland, N. O.,** Fate and effects of 3,4-dichloroaniline in the laboratory and in outdoor ponds. I. Fate, *Environ. Toxicol. Chem.*, 4, 481, 1985.

Chapter 11

ARTIFICIAL STREAMS IN ECOTOXICOLOGICAL RESEARCH

R. J. Kosinski

TABLE OF CONTENTS

I. INTRODUCTION

The use of increasing amounts of toxic chemicals by society has created a need for information on the fate, transport, and effects of these substances in the environment. While historically the main source of this information has been toxicological tests on single species, more recently there has been a move toward use of multispecies tests in model ecosystems, including artificial streams.[1-5] The model ecosystem approach is seen as a bridge between the controlled environment of the laboratory and the uncontrolled environment of the field, with its numerous interactions and indirect effects.[6] Several authors[6-10] have described an "integrated" toxicological research program in which work elucidating the basic environmental chemistry of a toxicant is followed by single species tests and then multispecies tests to determine toxicities, transfer coefficients, bioconcentration factors, etc. The next step is construction of a mathematical model which predicts the impacts of the toxicant. Validation of this model is done first in replicated model ecosystems and then in the field. Giesy[11] argues that multispecies tests are strongest in this validation role. Finally, the validated model is used to assist long-term field monitoring.

Artificial streams may resemble small fragments of natural environments, but they have several inherent differences from them. These include small size, usually noninclusion of higher trophic levels, high surface/volume ratio, and often a lack of natural recruitment from a larger environment.[8,9,12] The designers of artificial streams also have to decide between conflicting goals for their systems:

1. *Tractability vs. realism.* As complexity of a system increases, so does the amount of data which must be gathered from it. Also, more complicated systems are less suitable for use in the calibration of impact assessment models.[13,14] For this reason, Sanders[9] recommends constructing artificial streams as "analogs" to natural streams, with only enough complexity to assure realistic reactions to the toxicant. On the other hand, a toxicological artificial-stream study is not an end in itself. It exists to model the environment, and if it diverges too far from it, its point is lost.[8]

2. *Replication of streams vs. the size of individual streams.* Inclusion of higher trophic levels in model ecosystems is a frequent suggestion,[6,9] but in order to sustain higher-level consumers with autochthonous productivity, the model must be large. Giesy and Odum[12] cite a case in which a pelagic microcosm of 10^4 l was necessary to support one mosquito fish. Also, small artificial streams cannot survive extensive destructive sampling. On the other hand, large streams are so expensive and require so much sampling effort that replication of streams within treatments tends to be low or absent. This is a serious lapse of good experimental design,[15] but it is very common in toxcological artificial-stream studies.

These two dilemmas, tractability/realism and size/replication, will be the major themes of this review. Emphasis will be on design — both the physical design of the streams and experimental design — in papers published after about 1970. The review does not include studies that use tanks with rapid water renewal but without a significant unidirectional current[17-19] nor studies that use flowing water chambers merely as chemical avoidance arenas for fish.[20-22] At the end several artificial-stream issues will be discussed, especially experimental design and the utility of artificial-stream studies as confirmers of single species tests and predictors of events in the field.

Shriner and Gregory[10] published a review of artificial streams in toxicology in 1984, but this review will be somewhat different from theirs. Also, Warren and Davis[16] published a much-cited artificial-stream review (both toxicological and nontoxicological) in 1971.

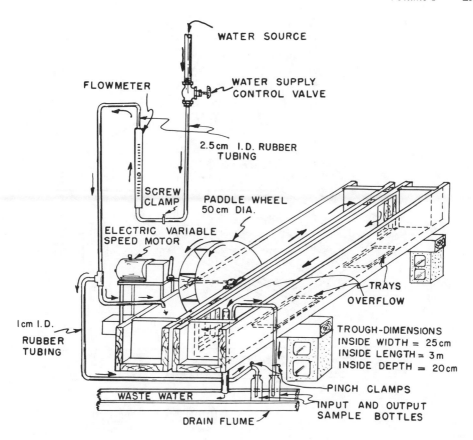

FIGURE 1. The "McIntire" artificial stream, probably the most commonly used small-stream design. (From McIntire, C. D., Garrison, R. L., Phinney, H. K., and Warren, C. E., *Limnol. Oceanogr.*, 9, 92, 1964. With permission.)

II. STREAM DESIGN

Past reviewers of artificial streams[10,23] have tended to classify studies by the degree to which the streams employed were open, flow-through systems or closed, recirculating types. A list of nontoxicological studies will be mentioned, and then the toxicological investigations will be classified into single-species tests, multispecies tests using small streams and usually only the lower trophic levels, and multispecies tests employing large streams and several trophic levels.

A. Nontoxicological Artificial-Stream Studies

An emphatically incomplete sampling of nontoxicological studies is listed in the references.[24-46] The work of McIntire et al.[34-38] is especially noteworthy for two reasons. First, the stream design used by McIntire and his co-workers has been used repeatedly for partially recirculating stream studies. A 3-m-long trough is divided longitudinally by a partition which does not reach the end walls of the trough. A paddle wheel then drives the water around the partition. The circulating water is gradually renewed by external input and losses through a standpipe (Figure 1).

Second, in his 1973 and 1975 papers,[37,38] McIntire discusses his construction of a computer model of periphyton dynamics using data from his artificial-stream experiments. The model made accurate predictions of the annual periphyton cycle in Berry Creek, OR, the stream from which the laboratory streams had been derived. Although McIntire's work is not

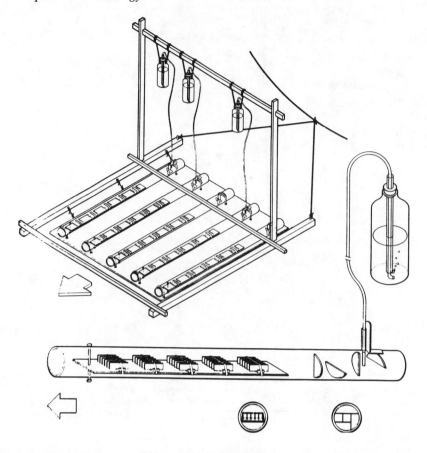

FIGURE 2.   The floating artificial streams of Petersen et al.,[41] used on a tundra river with a large water-level fluctuation. Arrows indicate the direction of the current. Nutrient (or toxicant) treatments can be administered from the Mariotte bottles. (From Petersen, B. J., Hobbie, J. E., Corliss, T. L., and Kriet, K., *Limnol. Oceanogr.*, 28, 583, 1983. With permission.)

toxicological, this is one of the few cases where the modeling component of the integrated assessment strategy has been carried to completion by an artificial-stream researcher.

The ingenious "streams" of Petersen et al.[41] should also be mentioned. The channels were a series of clear plastic tubes, 1.2 m long, which contained periphyton colonization slides and which were suspended on a floating rack in an Alaskan tundra river. The flow of the river forced water through the tubes, and different nutrient treatments were added from a series of Mariotte bottles on a rack above. The tendency of the river to flood was not a problem, since the rack rose and fell with the water level, always keeping the colonization slides just beneath the water surface (Figure 2).

B. Toxicological Studies

Table 1 lists 57 toxicological artificial-stream studies along with some details of the streams and the experimental procedures followed. The stream dimensions (l × w × d) are in meters. The length refers to the length of the path that the water takes through the channel. Thus, a 3-m-long McIntire-style stream would be listed as having a 6-m channel. If "description" is listed for the toxicant, it means that the paper described the system rather than reporting experimental results. "Cur." is the current speed in cm/s, with pool and riffle speeds indicated by P and R, respectively. "R & P" refers to the presence (Y) or absence (N) of both riffles and pools. Circulation ("Circ.") is given as FT (flow-through), R (totally recirculating, either with no outside water input or with input only to replace evaporation),

Table 1
SUMMARY OF TOXICOLOGICAL ARTIFICIAL-STREAM STUDIES

Toxicant	Tr.	Ch.	Dimensions	Cur.	R & P	Circ.	Time	Exp.	Biota	I/G/O	Design	Ref.
Anthracene	2	2					36	C	P	O		7
Acid, Al	3	3	6 × 0.215 × 0.24		N	FT	41 + 85	C	P + I	O	NRT	47
Heat	4	12	114 × 4.3 × 2		Y	FT	67 + 177	C	P	O	RT	48
Zn	3	3		0.23 P / 0.93 R	N	R	21 + 30	C	I	I + O	NRT	49
Heat	2	2	20 × 1.3 × 0.8	60	Y	R		C	F	O	NRT	50
NTA[a]	5	5	0.56 × 0.23		N	R	10	C	P	G	NRT	51
NTA[a] + CU	4	4	7.3 × 0.6 × 0.18	30—42	Y	FT or R	730			G	NRT	52
Acid + Al	3	3	2.1 × 0.6 × 0.56			R	28 + 28	C	I	I	NRT	53
Description		6	4 m long	9	N	R		I	P	O		
p-Cresol	2	2	1040 × 3.9 × 0.8	12.2 P / 25.0 R	Y	R	14	I	I + F	O	NRT	54
Sediments	6	12	6.6 m long			PR	180 + 90	P	I + F	G	RT	56
Zn	1	1	1 m long		N	FT or R			P	I	NRT	57
Heat	3	3	0.86 × 0.27 × 0.1		N	R		C	I	I	NRT	58
Chlorpyrifos	3	3	245 × 3.9 × 0.8		Y	FT	100	C + I	P	O	NRT	59
Treated sewage	4	4	200 × 0.35 × 0.2	15—25	N	FT	365 + 730	C	I + I + F	O	NRT	60
Co, Cu, Zn	5	8	75 × 0.2 × 0.2	8—12	N			C	P	I	NRT	61
Coal leachate	2	5	2.3 × 0.1	10—30	N	PR		C	P	I	RT	62
Volcanic ash			3.2 × 0.5 × 0.15	18	N	R		P	I	I	RT	63
Cd	3	6	91.5 × 0.61 × 0.31	1.3	Y	FT	365	C	B, Fn, P, M, I, F	O	RT	64
Cd	8	4	2 × 0.15 × 0.15		N	FT or R	14	C	I	I	NRT	65
Cd, Cu, Zn	4	6	3.5 × 0.7 × 0.3		N		30	C	I	I	NRT	66
Diflubenzuron	5	10	12.2 m long	30	Y	PR	90 + 150	C	B, Fn, P, I	I	RT	67
Description		6	6.7 × 0.3 or 0.6		Y	FT		C	P, I, M	G		68
PCP	4	4	152 × 3.9 × 0.8	1.4	Y	FT	84	C	B, P, I, F	O	NRT	69
NH₄Cl	12	12	6 × 0.25 × 0.2	24	N	PR		C	F	I	NRT	70
Sewage		12	5 × 0.15 × 0.2	40—74	N	FT	23	C	P	O		71
Cu	4	4	3.5 × 0.7 × 0.3	8	N		7 + 18	C + I	P	O	NRT	72
Atrazine, trifluralin, MSMA,[b] paraquat	8	32	2.4 × 0.12 × 0.1	3	N	PR	28 + 21	P	P	I	RT	73
Cl, chloramines	4	12	6.6 × 0.66 × 0.25	24	Y	PR	240 + 54	C	P + I	G	RT	75

Table 1 (continued)
SUMMARY OF TOXICOLOGICAL ARTIFICIAL-STREAM STUDIES

Toxicant	Tr.	Ch.	Dimensions	Cur.	R & P	Circ.	Time	Exp.	Biota	I/G/O	Design	Ref.
Atrazine, HCBP[c]	4	6	4 × 0.58 × 0.27		N	FT	365 + 90	C	P, I, M, F	I	NRT	76
TFM	2	6	8 × 0.6 × 0.25	2.4	Y	FT	300 + 3	P	P, I	I	RT	77
TFM			4.86 × 0.29 × 0.19	15	N	R	1 + 14	P	M	I		78
N + P	6	6	3.46 × 0.29	26	N	PR		C	P	O	NRT	79
N + P	4	4	20 × 0.5	45	N	FT	10	C	P	O	NRT	80
Atrazine	4	16	2.4 × 0.12 × 0.1	3	N	PR	35 + 7	P	P	I	RT	81
Permethrin, temephos			0.25 × 0.04 × 0.005	30	N	FT	1	C	I	O	NRT	82
Heat	2	2	520 × 3.9 × 0.8	1.4	Y	FT	365	C	I	O	NRT	83
PCP	4	4	488 × 3.9 × 0.8	1.4	Y	FT	84 or 107	C	B	O	NRT	84
PCP	3	3	180 × 3.4 × 0.8	1.4	Y	FT	88	C	B	O	NRT	85
Heat	4	12	112 × 4.3 × 2	0.23 P / 0.93 R	Y	FT		C	I	O	RT	
Cu, Cr, Ca(OCl)$_2$	6	6	4 × 0.35	9	N	FT	20	C	P	I		86
Dieldrin	3	6	6 × 0.3 × 0.18	28 or 0		PR	60 + 120	C	P	G		87
	4											
	6											
Temephos, chlorpyrifos	1		4 × 0.36	9 or 15	N	R		C	I	I		88
Kraft mill effluent	3	6	6.6 × 0.66 × 0.25	10	Y	PR		C	I, F	G	RT	89
Description	5		6 × 0.22 × 0.24		N	FT		C	P, I	O		90
Hg	3	6	91.5 × 0.61 × 0.31	1.3	Y	FT	150 + 365	C + I	P, I	O	RT	91
p-Cresol	2	2	1000 × 3.9 × 0.8	12.2	Y	R		I	P	O	NRT	92
p-Cresol	2	2	1000 × 3.9 × 0.8	12.2	Y	R			P	O	NRT	93
Sucrose	4	1.	152 × 1.5		N	FT	620	P	I, F	O	NRT	94
Treated sewage	3	3	300 × 1.25 × 0.4	10 R / 43 P	Y	FT		C	I	O	NRT	95
Heat	5	6	6.7 × 0.3 × 0.6	8.2	Y	FT	1095 + 365	C	P	G	NRT	96
Zn	4	4	6.1 × 0.3 × 0.2	0.056	N	FT	98	C	P + I	O	NRT	97
Description		4			N	FT		C	I	O		98
Heat	2	2	10 × 0.15	30	N	FT		C	P	O	NRT	99
Heat	4	12	112 × 4.3 × 2	0.23 P / 0.93 R	Y	FT	1095 + 185	C	P, I, Z, F	O	RT	100

Heat	4	12	112 × 4.3 × 2	0.23 P 0.93 R	Y	FT	1095 + 185	C	F	O	RT	101
Temephos, chlorphoxim	2	2	100 × 0.23	12	N	FT		I	P, I	O	NRT	102
PCP	4	4	520 × 3.9 × 0.8		Y	FT	7 + 20	C	P	O		103
Acid	3	3	122 × 3.9 × 0.8		Y	FT	119	C	I + F	O	NRT	104

Note: See text for explanation of column headings.

a Nitrilotriacetic acid.
b Monosodium methanarsonate.
c Hexachlorobiphenyl.

or PR (partially recirculating, with gradual replacement of the circulating water). "Time" is the length of the experiment in days, expressed as colonization period plus period after toxicant injection. If one figure is given, it is the whole period of experimental monitoring. Exposure ("Exp.") is coded as C (continuous — toxicant continuously introduced to maintain a constant concentration), I (intermittent — toxicant concentration maintained for a period and then flushed from the system), or P (pulsed — toxicant introduced once and then allowed to decay and be flushed from the system). "Biota" are the organisms studied, coded as B (bacteria), Fn (fungi), P (periphyton, mostly algal), Z (zooplankton), M (macrophytes), I (macroinvertebrates), and F (fish). Under "I/G/O," I indicates an indoor location with artificial lighting, G is a greenhouse location, and O is an outdoor location. Finally, "NRT" under "Design" means "nonreplicated treatments," the practice of assigning only one stream per treatment. "RT" indicates an experimental design with true replicates.

The following summary uses only data that appear in the table, and if one study is described by several papers, then the statistics include several entries for it. Although 35 toxicants appear in Table 1, the most studied ones are heat (9 papers), zinc and copper (5 each), and pentachlorophenol, or PCP (4). The number of channels ("Ch." in table) used per experiment ranges from 1 to 32, with a median of 4. The median number of treatments ("Tr." in table) per experiment is three, reflecting the large number of experiments that use only one channel per treatment. The channel lengths range from a fraction of a meter to over 1000 m (median about 7 m). Cited current speeds range from 0 to 74 cm/s (median 10 cm/s). The studies are almost equally divided between those that provide riffles and pools and those with only one "habitat"; 56% used flow-through streams, 20% recirculating, 19% partially recirculating, and 5% used both flow-through and a recirculating operation. The lengths of the experiments are approximate, especially with regard to colonization time, since some investigators use streams that have been flowing for years. The median colonization time given is about 50 d, and four studies cite colonization periods of 1 year or longer. Lengths of experiments range from a few hours to 2 years of monitoring. The great majority of the papers (84%) used continuous exposure to the toxicant. The most-studied artificial-stream biota are macroinvertebrates (58%) and periphyton (54%), followed by fish (23%). Outdoor streams were used by 59% of the papers, indoor streams with artificial lighting by 26%, a greenhouse arrangement by 14% and one study used streams in both an indoor and an outdoor setting. Finally, of the papers where it could be determined or was relevant, 70% used experimental designs with one stream per treatment.

With this overall background, noteworthy design and protocol features of the papers in Table 1 will now be reviewed.

1. Single-Species Studies

The streams used by the single-species studies run the gamut from general purpose to highly specialized. Variations of the McIntire design are most common, either without much modification[56,70,75,89] or with a more oval-tank shape and generally smaller size.[49,58,63,88] Short troughs, with water either recirculating or flowing through from a head box, were also used.[65,66,98]

An unusual design was used by Bisson and Davis[50] in a thermal study on chinook salmon. A single stream was actually two troughs, each 10 m long. The troughs were connected at both ends by pipes large enough for a juvenile salmon to pass through. The bottoms of the troughs were also contoured to provide both riffles and pools. A pump propelled water in a circular path from the downstream end of one trough to the upstream end of the other.

Graney et al.[65] were interested in substrate effects on the bioconcentration of cadmium by *Corbicula*, and researchers held the number of channels they used to a minimum by using four different substrate types in each system. In addition, they performed a series of 2×2 factorial experiments to determine the effect of temperature, feeding on contaminated algae, pH, and Cd concentration. To perform all these experiments simultaneously, evaluate the significance of all interactions, use only one substrate per stream, and use two streams per treatment would have required the use of 192 channels. The investigators only used four, which testifies both to their economical approach and to the extent of nonreplication, unestimated interactions, and possible nonindependence of treatments which their design required.

Finally, Muirhead-Thomson[81] provided an engaging example of a specialized stream. Blackfly larvae clinging to aquatic vegetation were placed in a 10-l aspirator bottle which was aerated near one wall. Soon the larvae congregated in a band on the side of the jar where the bubbles were rising. The jar was then slowly tipped onto its side and emptied. The blackflies were then placed in the path of a stream of water which entered from the port at the bottom of the aspirator bottle. This concentrated the blackflies into an even tighter band. The "stream" was only 4 cm wide and 5 mm deep, and toxicants could be added to it at will.

2. Multispecies Studies

In terms of stream design, these studies can be divided into those with smaller streams (generally under 10 m long) and those using large "mesocosm"-scale streams.

The designs of the smaller streams are quite diverse.[47,51-53,57,62,71,72-74,76-78,80,86,87,90,97] The most common stream design is similar to that described by Clark et al.[53] Six aluminum hatchery troughs 4 m long were coated with epoxy paint and lined with washed cobble from the New River, VA. Water was pumped from the river to an insulated mixing box at the head of the streams. From there water flowed through the streams at 9 cm/s and back into the river. The gradient and current in the streams could be varied by changing their slopes with automobile jacks. Toxicants were added from an insulated dosing box.

Some noteworthy features of other streams with this basic design include damming a natural stream to provide gravity feed of water to the head box of artificial streams,[47,90] the use of a refrigeration unit in an indoor stream,[52] the division of a standard hatchery trough into two segments with one segment used solely as a settling basin for river particulates,[76] and the heavily replicated (32 streams) factorial experimental design of Kosinski et al.[73,74] Gerhart et al.[62] reported very good replicability by a series of small streams in a growth-chamber-type enclosure.

Some small streams are quite elaborate. The "ecosystem" streams of Bott et al.[51] had a total channel length of 7.3 m and were made of polyvinyl chloride (PVC), but they provided both riffles and pools and even ledge-like "littoral" areas which became colonized with macrophytes. The Flowing Stream Laboratory of Aiken, SC[68,96] contained six PVC streams, each with a channel length of 6.7 m and a "mission" of investigating the effects of elevated water temperatures. However, the description by Harvey[68] indicates the requirements of even this relatively modest installation. These included a controlled-access location on an unpolluted creek that flows year-round and has a protected watershed, an intake structure, a pump house, a thermal head tank, and an air-conditioned greenhouse for the streams. Harvey also mentions that 9 to 12 months of colonization was required before the diversity of the artificial stream biota matched that of Three Runs Creek nearby.

Finally, this category also includes some specialized "streams" that bear little resemblance to the usual channels. Patrick[105-107] pumped water through small plastic boxes (0.46 m long) containing periphyton colonization slides. The community structure of the colonizing diatoms was found to be highly reproducible between boxes and reflected local ecological conditions. Bott et al.[51] used a variety of stream types, including a clear pipe and a flat, closed box 0.55 m long, containing coverslips for periphyton colonization. Cushing and Rose[57] probably used the most unusual "stream" ever devised. Periphyton was allowed to colonize the outside of a test tube, and then a photomultiplier was inserted into the tube. The whole apparatus was contained in a larger tube which channeled water over the periphyton. A similar apparatus, but with no periphyton, was on the same water circuit. When radioactive zinc was injected into the water stream, the periphyton took it up, and its radioactivity could be measured nondestructively and in real time. Correction for dissolved and glass-absorbed zinc was automatically provided by the periphyton-free apparatus.

The papers from the larger multispecies streams are dominated by a few large and elaborate installations which have been the source of multiple research reports. However, a group of papers with streams of less than mesocosm scale should be discussed first. Eichenburger[60,61] and Yasuno et al.[102] and their respective colleagues used streams which were quite long (200, 75, and 100 m), but narrow and shallow, and only for the study of periphyton and macroinvertebrates. Yasuno et al. constructed their streams with the first 70 m under the shade of a pine forest and the last 30 m in an open field.

The smallest true "mesocosm" streams were those used by Giesy et al.[64] and Sigmon et al.[91] at the Savannah River Plant, SC. These streams were 90 m long and 0.61 m wide, with riffles and pools, macrophytes, crayfish, and fish. The report on an ambitious study on cadmium toxicity[64] and a subsequent critique of the results[108] is lucid and enlightening reading for anyone interested in working with large artificial streams. Bowling et al.[108] reported that the original study attempted to include all major functional groups (without introducing unmanageable complexity) by including only one or two representatives of each group (two species of clams, two species of fish, etc). This study was not completely successful, since some of the introduced species could not survive in the streams. Also, the streams were too small to support the extensive destructive sampling necessary to determine cadmium-uptake kinetics by higher trophic levels. In addition, the streams were too complex for the determination of mechanisms and for calibrating models. The authors also mention that the study was troubled by seasonal change and the differential colonization of macrophytes in different channels. Finally, they cite sobering statistics on the resources and work the study consumed: a long list of materials including 36 metric tons of sand, 1500 mosquito fish, and 1800 *Corbicula*, plus 21 man-years and $300,000.

Other artificial streams dwarf those used by the workers above. Watton and Hawkes[95] used three earthen channels next to a river. Each was 300 m long, a maximum of 1.25 m wide, and each contained two riffle and two pool sections. Despite the size of this system, they achieved a current of 43 cm/s in the riffles and 10 cm/s in the pools. However, the channels were differentially colonized by macrophytes to the extent that the investigators had difficulty determining whether it was channel differences or their unreplicated sewage treatments which were causing differences in snail populations.

The Browns Ferry Experimental Channels in northern Alabama have a biothermal mission.[48,100,101] The installation contains 12 channels, each 112 m long and a remarkable 4.3 m wide. Each channel contains 530,000 l of water. Because of this large volume, there is a current of less than 1 cm/s even in the shallow areas. The water also cools appreciably on its way down the channels, and so the investigators tried to confine fish to the higher-temperature section by placing mesh barriers in two of the three channels per treatment. However, they concluded that this had been a mistake.[100] They also commented on high biomass variation between the channels. However, their streams apparently were large enough to support large numbers of introduced blugills and walleyes with autochthonous productivity. Another noteworthy point was the fact that the channels were monitored for 3 years before any thermal manipulation in order to obtain baseline data. Finally, despite the large size of these streams, all the Browns Ferry studies have been adequately replicated, an usual situation for which the investigators deserve credit.

The last truly artificial-stream system is the largest: the Monticello Ecological Research Station of the U.S. Environmental Protection Agency in Minnesota. This has a series of eight earthen channels, each 520 m long, with alternating 30-m riffle and pool sections

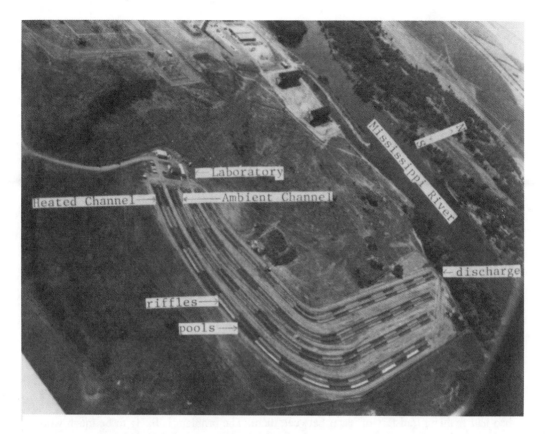

FIGURE 3. An aerial view of the eight channels at the Monticello Ecological Research Station, MN. Each channel is 520 m long. Note the size of the laboratory and parked cars at the head of the channels. (From Nordlie, K. J. and Arthur, J. W., *Environ. Pollut. Ser. A,* 25, 53, 1981. With permission.)

(Figure 3). Temporarily, two pairs of channels were linked to make two recirculating systems, each over 1000 m long. Unlike Browns Ferry, the Monticello channels have been used for a variety of toxicants: heat,[82] acid,[104] PCP,[69,83,84,103] *p*-cresol,[54,55,92,93] and chlorpyrifos.[59] Because of the size of the Monitcello channels and the literature productivity of the workers using them, the Monticello papers are an enlightening demonstration of the advantages and disadvantages of very large artificial streams. These points will be discussed below.

One final unusual example is the work of Warren et al.[94] on the administration of light and sucrose treatments to sequential sections of Berry Creek, OR. The creek could have been considered an artificial stream inasmuch as diversion dams and a channel bypass controlled the flow into the study channels. To prevent drift, screens were installed between the treatment sections and cleaned daily. Aside from nonreplication of treatments, this "sequential" design made the treatments nonindependent of one another and cannot be recommended.

III. ARTIFICIAL-STREAM ISSUES

To conclude this review, issues in artificial-stream research — experimental design, the reuse of contaminated channels by multiple experiments, and the utility of artificial-stream research to ecotoxicology — will be discussed. Another issue is the variables that should be measured in an artificial-stream study, but this has been extensively discussed elsewhere.[9,16,115]

A. Experimental Design

The one feature of artificial-stream research which was most striking during the preparation of this review was the contrast between the attention paid to experimental procedures by researchers vs. the remarkable lack of care with experimental design. Topics such as residue analysis, sampling of periphyton, and computer-controlled dosing would be discussed in elaborate detail in a study; meanwhile, the experimental design, *a priori*, would not allow any conclusions about the effect of the toxicant. The practice of using only one stream per treatment and then (in most studies) testing for a significant difference between treatments appeared in 70% of the studies in Table 1. Therefore, it warrants discussion. Many of the remarks below follow the arguments in the recent, but influential, paper by Hurlbert.[15]

In order to correctly estimate the significance of the difference between two treatments, it is necessary to have both unbiased estimates of the means of the treatments and also estimates of the variability between experimental units within a treatment.[110] In a typical artificial-stream study, the experimental units are streams, *not* samples within streams. To use one stream per treatment, no matter how extensively subsampled, is to assume that that one stream is an adequate estimate of the mean of all possible streams subjected to the same treatment. This is very improbable.

Due to the well-known "individuality of microcosms"[111] even streams treated alike will tend to develop differently. Examples of high interstream variation abound in the artificial-stream literature.[50,53,71,73,75,95,100,103] In the herbicide study by Kosinski,[73] on four occasions periphyton biomass and species composition variables were assessed within streams, between streams in the same treatment, and between treatments. Streams within a treatment were significantly different from one another in 28 out of 32 tests, and streams treated alike were significantly different in total biovolume in 4 out of 4 tests. This was true despite the fact that the streams were small, of simple design, had only been developing for 4 weeks, and had had frequent transfers of water between them. The problem is likely to be much worse in large artificial streams that have been developing a long time and which may have had different treatments in the past. An outstanding example was provided by one of the Monticello papers.[59] As the experiment started, 85% of all the macroinvertebrates in the one control stream were isopods, in the one pulse-dose stream 67% were chironomids, and in the one continuous-dose stream 69% were amphipods.

When a significance test is performed based on within-stream variability, the hypothesis being tested is that the streams are significantly different. This conclusion is entirely valid. However, whether the treatments or merely interstream variability caused the difference above is unresolved. In fact, we have excellent reason (cited above) to assume that the streams would have been significantly different even if they had received the same treatment. With few exceptions, in order to make statements about treatment effects, results from replicate streams within each treatment must be compiled.[15] As the number of replicates increases, confidence increases that the difference is due to the treatment rather than interstream variation. Without replicates, the interpreter of the data is left in the position of a researcher described by Day.[112] The unnamed investigator reported that $1/3$ of the mice in an experiment were cured by the tested drug, $1/3$ had remained unaffected; and that the third mouse had gotten away.

Surprisingly, the users of small streams are just as prone to employ nonreplicated designs as are the users of large streams. The percentage of studies with unreplicated treatments in Table 1 is almost identical for streams longer than and shorter than 10 m. Indeed, one of the highest numbers of streams per treatment is recorded for the very large Browns Ferry system. Nevertheless, the problem of replication can be especially acute for the user of large streams. Large artificial-stream systems are rare, are usually not heavily replicated, may have to be shared by several researchers, impose great sampling workloads, and cannot easily be cleaned out and restarted no matter how different the channels become over years of operation. For many purposes, large size may cause more problems than it solves.

On the other hand, in studies where destructive sampling of higher trophic-level organisms is involved, large systems are necessary; the bad results of overloading a small-stream system with fish and crayfish are described by Bowling et al.[108] In contrast, Wrenn et al.,[100] in a study on fish-spawning behavior, were able to compile data on 60 to 120 bluegill nests per treatment because of the large size of their streams. Also, long streams may be necessary where the degradation of a toxicant as it passes downstream is being assessed. Finally, it can be argued that large, complex systems with all trophic levels present are necessary to exhibit realistic system properties and community interactions.

There may be a statistical "safety net" for the large-stream investigator who cannot use replicate streams per treatment. Hedtke[113] pointed out that is possible to seek toxicological evidence by dose-response rather than by classical hypothesis testing. In hypothesis testing, power is greatest when there are few treatments and a maximum number of replicates per treatment. But for dose-response patterns, the most information is obtained when the number of treatments (and points on the response curve) is high, implying that replication per treatment is low.

This point should be considered. Yount and Richter[103] designated "positive" or "negative" responses of various taxonomic groups if their numbers increased or decreased monotonically over the series of four unreplicated PCP treatments. With N treatments and no real treatment effects, the probability of obtaining a monotonically decreasing response is $1/(N!)$. With four treatments this probability was 1/24, less than 0.05. A randomly derived, monotonically increasing response has the same probability. Thus, a significant ($p = 0.05$) treatment-related pattern could be declared without replication and without the assumptions of parametric statistics. The disadvantage is that a monotonic trend is required, and at least four treatments must be used for probability less than 0.05 for one experiment (the probability of a randomly derived monotonic decreasing response with three treatments is 0.17).

The use of upstream controls and downstream experimentals risks biasing means by longitudinal succession. Replication in time[102] is valid only insofar as the streams experience no mean difference between years, an unlikely situation. Similar dose-response patterns in successive tests are more convincing. With three treatments, the probability of two randomly derived, monotonically decreasing trends in a row is 0.03; with four treatments the corresponding probability is 0.002.

However, unless treatment differences are very large, if interstream variation is high the monotonic trends hoped for may be obscured by random variation. No matter what statistics are planned, the experimental design should include replication of streams within a treatment.

B. Reuse of Contaminated Channels

As well as can be determined, the reuse of streams which have already been contaminated by previous experiments is not discussed anywhere in the literature. Presumably, most users of small streams clean out their channels between experiments. Even Marker and Casey,[33] with a stream length of 53 m, replaced the top 40% of the substrate in the channels at the end of the first year. This is not possible with really large streams. At Browns Ferry, for example, shallow water areas are made by placing piles of crushed limestone in the channels. This limestone alone amounts to over 1100 m^3, enough to fill over 100 dump trucks!

The worst reuse situation seems to be at Monticello, which has the largest channels and also has used a variety of toxicants (at different times heat, acid, PCP, p-cresol, and chlorpyrifos).Reuse is seen to an unusual degree in a report by Cooper and Stout.[55] Working with the two recirculating Monticello channels in summer of 1980, one channel, A, was designated as the reference channel and the other, B, was dosed with p-cresol. Three weeks later, channel A was dosed, and B was used as a reference. The next summer, A was again the reference and B was dosed, and 16 d after that A was dosed and B was used as the reference.

Hedtke[113] indicated that the Monticello group was aware of the reuse problem, avoided persistent contaminants, monitored residues in the water and sediment, and felt assured that degradation and flushing had removed all traces of past toxicants. While this is quite possible, another impact of a toxicant is to leave behind a suite of resistant organisms.

Cooper and Stout[54] noted a startling decrease in the half-life of *p*-cresol after they started dosing their recirculating Monticello streams. Hedtke et al.[69] and Pignatello et al.[83] (also at Monticello) noted that bacterial degradation of PCP became progressively more effective as exposure continued. A toxicant input which killed fish at the beginning of the experiment no longer did so toward the end. Thus, even the most transient toxicants (including heat) might select for resistant organisms and influence future results.

The channel reuse problem is rendered less serious by long periods between experiments, lack of inducible cross-resistance between successive toxicants, and high population turnover of the test organisms. Nevertheless, it seems reasonable that workers who use previously treated channels should give channel histories in their site descriptions.

C. The Utility of Artificial-Stream Studies

Bowling et al.[108] maintained that model ecosystem studies have two purposes: they allow verification of the results of single-species tests in a field-like setting, and they reveal significant indirect and system-level effects which cannot be predicted from single-species tests alone. Assessing the evidence for these assertions is the most important question we can ask about toxicological artificial-stream studies.

Some investigators have found that single-species tests adequately predicted model stream response. Cooper and Stout[54,55] gave laboratory tests high marks for predicting both species survival and community responses to *p*-cresol. Hansen and Garton[67] found that laboratory-derived criteria left no artificial-stream species underprotected from diflubenzuron; in fact, the laboratory tests were up to ten times more sensitive than the stream tests. Hedtke and Arthur[69] found good agreement between single-species tests and stream results in a PCP study, but also found unpredicted deleterious effects at the "criterion" concentration. Finally, Maki[114] performed tests on the lampricide 3-trifluoromethyl-4-nitrophenol (TFM) in the laboratory, in artificial streams, and then in a channelized "natural" stream. He found excellent agreement between the test series both in overall effects and on the variables that were the most sensitive indicators of TFM impact.

On the other hand, Geckler et al.[115] found that variations in stream water quality and fish avoidance behavior caused poor agreement between laboratory and artificial-stream tests on copper. Unlike Hansen and Garton, Geckler et al. concluded that the single-species tests were not conservative. Larson et al.[75] reported problems with high variability between channels and varying concentration of the toxicant (chlorine compounds), determined that the artificial-stream communities had a higher chlorine tolerance than previous aquarium tests had indicated, and were discouraged about the probability of accurate extrapolation from laboratory to field. Finally, Manuel and Minshall[79] determined that while artificial-stream periphyton increased in response to nitrogen and phosphorus inputs, there was no response in a field test in the Big Wood River, ID because the river periphyton were limited by sloughing and grazing.

Predictably, the role of artificial streams as a bridge between the laboratory and the field remains in doubt. Many more studies like the ones above are needed. However, at present, the evidence seems to support the conclusion of Giesy:[11] "One will never be able to absolutely establish the accuracy of multispecies tests."

It is in the role of detecting unexpected indirect effects of toxicants that artificial-stream studies should really show their value, and many examples of indirect effects can be found. Perhaps the most dramatic was reported by Cooper and Stout.[54] They found that *p*-cresol so depressed the photosynthesis and enhanced the respiration of the flora of productive, recirculating Monticello channels that oxygen depletion caused widespread fish mortality. This drastic effect was not predicted from well-aerated laboratory tests. Several studies[51,55,83,84] have noted the increasing efficiency of degradation of a toxicant as exposure to it continues, sometimes effectively producing a no-effect toxicant input which declines with time. Algal growth seems to be a frequent cause and result of indirect effects in artificial-stream studies. When stimulated by treatments, it can prevent fish from feeding,[50,89] shade the bottom, and hinder sampling.[85] Treatments which destroy grazers also benefit algal growth.[67] On the other hand, Giesy et al.[64] and Bowling et al.[108] reported how destruction of crayfish by cadmium allowed increased growth of macrophytes, which in turn stripped nutrients from the water and caused a decline in algal populations. Finally, an indication of how the nature of an impact can depend on the whole community was given by Wrenn et al.[100] Before the introduction of fish, the Browns Ferry channel zooplankton had been dominated by clado-cerans. After fish were introduced, the zooplankton fauna shifted to copepods and ostracods, which would have rendered irrelevant zooplankton toxicity data based on cladocerans.

Are these indirect effects important enough to justify a large investment in artificial-stream research? The oxygen depletion by *p*-cresol probably would have been less intense in a flow-through natural stream. Increased toxicant degradation with time is extremely significant, and how long it persists in the absence of toxicant would be a worthwhile research topic. Shifts in community composition might be transitory or (in a system with multiple stable states) more long lasting. More research is needed here also. To summarize, it is known that model ecosystems can demonstrate indirect effects. What is needed is to know how durable and significant these effects are for the toxicological response of the system.

IV. CONCLUSION

Predicting the effects of toxicants on the environment is very difficult. Due to historical, legal, and economic factors, most of the data which can be used for this task are derived from single-species tests, but it is clear that indirect effects can make toxicants more or less dangerous than predicted. One of the best ways to investigate these effects is with model ecosystems like artificial streams. Aside from the environmental control that can be exerted over these models, their great advantage is that they can be replicated. They are a powerful tool, but only when used with correct experimental design.

Hurlbert[15] points out that while the statistical analysis of data and the interpretation of data can easily be redone if faulty, a trivial hypothesis, incorrect execution of procedures, or a bad experimental design can render data permanently unsalvageable. Experimental design has been the leading problem of the oceans of expensive, hard-won data in Table 1. The hypothesis of 70% of the entries, as read from the one-stream-per-treatment designs, is that streams differ from one another, not that the toxicant is causing streams to differ from one another. Users of small streams are just as likely to fail to replicate as users of large streams, so the economic impossibility of increasing the number of streams is not the problem in many cases. If the cause is policy, based on the idea that the best use of resources is to use one stream per treatment, then one may hope that the situation will be improving soon.

Experimental design issues aside, artificial streams have made valuable contributions to ecotoxicology. The literature in Table 1 show that the stream designs are often ingenious and that the amount of work put in per study is impressive. Many researchers are asking the right questions about the correlation of model-ecosystem research with the field and with traditional single-species toxicology. Some studies on this question are encouraging, sug-gesting that the single-species/model-ecosystem approach is giving us reliable results. Others raise doubts. Certainly, more studies are needed.

The two best steps that could be taken to strengthen future toxicological artificial-stream research are the replication of streams within treatments and more attempts to perform an integrated, comparative series of single-species, artificial-stream, and field experiments.

REFERENCES

1. **Cairns, J., Jr.,** Multispecies toxicity testing, *Bull. Ecol. Soc. Am.,* 65, 301, 1984.
2. **Cairns, J., Jr.,** Multispecies toxicity testing, *Environ. Toxicol. Chem.,* 3, 1, 1984.
3. **Conway, R. A.,** Introduction to environmental risk analysis, in *Environmental Risk Analysis for Chemicals,* Conway, R. A., Ed., D Van Nostrand-Rheinhold, New York, 1982, 1.
4. **Horning, W. B., II and Weber, C. I.,** Short-Term Methods for Estimating the Chronic Toxicity of Effluents and Receiving Waters to Freshwater Organisms, EPA-600/4-85/014, U.S. Environmental Protection Agency, Cincinnati, OH, 1985.
5. **Kimerle, R. A.,** Has the water quality criteria concept outlived its usefulness?, *Environ. Toxicol. Chem.,* 5, 113, 1986.
6. National Research Council, *Testing for the Effect of Chemicals on Ecosystems,* National Academy Press, Washington, D.C., 1981.
7. **Bowling, J. W., Haddock, J. D., and Allred, P. M.,** Disposition of Anthracene in Water and Aufwuchs Matrices of a Large, Outdoor Channel Microcosm: A Data Set for Mathematical Simulation Models, EPA-600/S3-84-036, U.S. Environmental Protection Agency, Washington, D.C., 1984.
8. **Pritchard, P. H.,** Model ecosystems, in *Environmental Risk Analysis for Chemicals,* Conway, R. A., Ed., D Van Nostrand-Rheinhold, New York, 1982, 257.
9. **Sanders, F. S.,** Guidance for the Use of Artificial Streams to Assess the Effects of Toxicants in Freshwater Systems, ORNL/TM-7185, Oak Ridge National Laboratory, Oak Ridge, TN, 1982.
10. **Shriner, C. and Gregory, T.,** Use of artificial streams for toxicological research, *Crit. Rev. Toxicol.,* 13, 253, 1984.
11. **Giesy, J. P., Jr.,** Multispecies tests: research needs to assess the effects of chemicals on aquatic life, in *Aquatic Toxicology and Hazard Assessment: Eighth Symposium,* Bahner, R. C. and Hansen, D. J., Eds., American Society for Testing and Materials, Philadelphia, 1985, 67.
12. **Giesy, J. P., Jr. and Odum, E. P.,** Microcosmology: introductory comments, in *Microcosms in Ecological Research,* Giesy, J. P., Jr., Ed., Technical Information Center, U.S. Department of Energy, Oak Ridge, TN, 1980, 1.
13. **Draggen, S.,** The microcosm as a tool for estimation of environmental transport of toxic materials, *Int. J. Environ. Stud.,* 10, 65, 1976.
14. **Suter, G. W., II, Barnthouse, L. W., Breck, J. E., Gardner, R. H., and O'Neill, R. V.,** Extrapolating from the Laboratory to the Field: How Uncertain Are You?, CONF-8304101-1, Oak Ridge National Laboratory, Oak Ridge, TN, 1983.
15. **Hurlbert, S. H.,** Pseudoreplication and the design of ecological field experiments, *Ecol. Monogr.,* 54, 187, 1984.
16. **Warren, C. E. and Davis, G. E.,** Laboratory stream research: objectives, possibilities and constraints, *Annu. Rev. Ecol. Syst.,* 2, 111, 1971.
17. **Chapman, G. A.,** Toxicities of cadmium, copper and zinc to four juvenile stages of chinook salmon and steelheads, *Trans. Am. Fish. Soc.,* 107, 841, 1978.
18. **Dodson, J. J. and Mayfield, C. I.,** Modification of the rheotropic response of rainbow trout *(Salmo gairdneri)* by sublethal doses of the aquatic herbicides diquat and simazine, *Environ. Pollut.,* 18, 147, 1979.
19. **Finlayson, B. J. and Verrue, K. M.,** Toxicities of copper, zinc and cadmium mixtures to four juvenile stages of chinook salmon, *Trans. Am. Fish. Soc.,* 111, 645, 1982.
20. **Cherry, D. S., Hoehn, R. C., Waldo, S. S., Willis, D. H., Cairns, J., Jr., and Dickson, K. L.,** Field-laboratory determined avoidances of the spotfin shiner and the bluntnose minnow to chlorinated discharges, *Water Res. Bull.,* 13, 1047, 1977.
21. **Giattina, J. D., Garton, R. R., and Stevens, D. G.,** Avoidance of copper and nickel by rainbow trout as monitored by a computer-based acquisition system, *Trans. Am. Fish. Soc.,* 111, 491, 1982.
22. **Stott, B. and Buckley, B. R.,** Avoidance experiments with homing shoals of minnows, *Phoxinus phoxinus,* in a laboratory stream channel, *J. Fish Biol.,* 14, 135, 1979.
23. **Hammons, A. S., Ed.,** Ecotoxicological Test Systems, proc. of a series of workshops, EPA-560/6-81-004, U.S. Environmental Protection Agency, Washington, D.C., 1981.
24. **Brusven, M. A.,** A closed system Plexiglass stream for studying insect-fish-substrate relationships, *Prog. Fish Cult.,* 35, 87, 1973.

25. **Cummins, K. W., Klug, J. J., Wetzel, R. G., Petersen, R. C., Suberkropp, K. F., Manny, B. A., Wuycheck, J. C., and Howard, F. O.,** Organic enrichment with leaf leachate in experimental streams, *BioScience,* 22, 719, 1972.

26. **Davis, G. E. and Warren, C. E.,** Trophic relations of a sculpin in laboratory stream communities, *J. Wildl. Manage.,* 29, 846, 1965.

27. **Eichenburger, E. and Schlatter, F.,** The effect of herbivorous insects on the production of benthic algal vegetation in outdoor channels, *Verh. Int. Ver. Theor. Angew. Limnol.,* 20, 1806, 1978.

28. **Feltmate, B. W., Baker, R. L., and Pointing, P. J.,** Distribution of the stonefly nymph *Paragnetina media* (Plecoptera: Perlidae): influence of prey, predators, current speed and substrate composition, *Can. J. Fish. Aquat. Sci.,* 43, 1582, 1986.

29. **Irvine, J. R. and Northcote, T. G.,** Selection by young rainbow trout *(Salmo gairdneri)* in simulated stream environments for live and dead prey of different sizes, *Can. J. Fish. Aquat. Sci.,* 40, 1745, 1983.

30. **Ladle, M., Cooling, D. A., Welton, J. S., and Bass, J. A. B.,** Studies on Chironomidae in experimental recirculating stream systems. II. The growth, development and production of a spring generation of *Orthocladius calvus, Freshwater Biol.,* 15, 243, 1985.

31. **Lawson, P. W.,** A simple air-powered pump for laboratory streams, *Freshwater Invertebr. Biol.,* 1, 48, 1982.

32. **Mackay, R.,** A miniature laboratory stream powered by air bubbles, *Hydrobiologia,* 83, 383, 1981.

33. **Marker, A. F. H. and Casey, H.,** The population and production dynamics of benthic algae in an artificial recirculating hard-water stream, *Philos. Trans. R. Soc. London Ser. B,* 298, 265, 1982.

34. **McIntire, C.D., Garrison, R. L., Phinney, H. K., and Warren, C. E.,** Primary production in laboratory streams, *Limnol. Oceanogr.,* 9, 92, 1964.

35. **McIntire, C. D.,** Some factors affecting respiration of periphyton communities in lotic environments, *Ecology,* 47, 918, 1966.

36. **McIntire, C. D.,** Structural characteristics of benthic algal communities in laboratory streams, *Ecology,* 49, 520, 1968.

37. **McIntire, C. D.,** Periphyton dynamics in laboratory streams: a simulation model and its implications, *Ecol. Monogr.,* 43, 399, 1973.

38. **McIntire, C. D.,** Periphyton assemblages in laboratory streams, in *River Ecology,* Whitton, B. A., Ed., University of California Press, Los Angeles, 1975, 403.

39. **Mueller-Haeckel, A.,** Experiments zum Bervegungsverhalten von einzelligen Fliesswasseralgen, *Hydrobiologia,* 41, 221, 1973.

40. **Mulholland, P. J., Elwood, J. W., Newbold, J. D., and Ferren, L. A.,** Effect of a leaf-shredding invertebrate on organic matter dynamics and phosphorus spiralling in heterotrophic laboratory streams, *Oecologia (Berlin),* 66, 199, 1985.

41. **Petersen, B. J., Hobbie, J. E., Corliss, T. L., and Kriet, K.,** A continuous flow periphyton bioassay: tests of nutrient limitation in a tundra stream, *Limnol. Oceanogr.,* 28, 583, 1983.

42. **Ryer, C. H., Wetmore, J. A., and Gooch, J. L.,** An artificial stream design for lotic invertebrates, *Am. Midl. Nat.,* 101, 447, 1979.

43. **Sechnick, C. W. and Carline, R. F.,** Habitat selection by smallmouth bass in response to physical characteristics of a simulated stream, *Trans. Am. Fish. Soc.,* 115, 314, 1986.

44. **Sheldon, S. P. and Taylor, M. K.,** Community photosynthesis and respiration in experimental streams, *Hydrobiologia,* 87, 3, 1982.

45. **Sumner, W. T. and McIntire, C. D.,** Grazer-periphyton interactions in laboratory streams, *Arch. Hydrobiol.,* 93, 135, 1982.

46. **Wetzel, R. G. and Manny, B. A.,** Decomposition of dissolved organic carbon and nitrogen compounds from leaves in an experimental hard-water stream, *Limnol. Oceanogr.,* 17, 927, 1972.

47. **Allard, M. and Moreau, G.,** Short-term effect on the metabolism of lotic benthic communities following experimental acidification, *Can. J. Fish. Aquat. Sci.,* 42, 1676, 1985.

48. **Armitage, B.,** Effects of temperature on periphyton biomass and community composition in the Browns Ferry Experimental Channels, in *Microcosms in Ecological Research,* Giesy, J. P., Jr., Ed., Technical Information Center, U.S. Department of Energy, Oak Ridge, TN, 1980, 668.

49. **Belanger, S. E., Farris, J. L., Cherry, D. S., and Cairns, J., Jr.,** Growth of Asiatic clams *(Corbicula* sp.) during and after long-term zinc exposure in field-located and laboratory artificial streams, *Arch. Environ. Contam. Toxicol.,* 15, 427, 1986.

50. **Bisson, P. A. and Davis, G. E.,** Production of juvenile chinook salmon, *Oncorhynchys tshawytscha,* in a heated model stream, *Fish. Bull.,* 74, 763, 1976.

51. **Bott, T. L., Preslan, J., Finlay, J., and Brunker, R.,** The use of flowing-water streams to study microbial degradation of leaf litter and nitrolotriacetic acid (NTA), *Dev. Ind. Microbiol.,* 18, 171, 1977.

52. **Burton, T. M. and Allan, J. W.,** Influence of pH, aluminum and organic matter on stream invertebrates, *Can. J. Fish. Aquat. Sci.,* 43, 1285, 1986.

53. **Clark, J., Rodgers, J. H., Jr., Dickson, K. L., and Cairns, J., Jr.,** Using artificial streams to evaluate perturbation effects on aufwuchs structure and function, *Water Res. Bull.,* 16, 100, 1980.
54. **Cooper, W. E. and Stout, R. J.,** Assessment of transport and fate of toxic materials in an exprimental stream ecosystem, in *Modeling the Fate of Chemicals in the Aquatic Environment,* Dickson, K. L., Maki, A. W., and Cairns, J., Jr., Eds., Ann Arbor Science Publishers, Woburn, MA, 1982, 347.
55. **Cooper, W. E. and Stout, R. J.,** The Monticello Experiment — A Case Study, EPA-600/D-84-067, U.S. Environmental Protection Agency, Duluth, MN, 1984.
56. **Crouse, M. R., Callahan, C. A., Malueg, K. W., and Dominguez, S. E.,** Effects of fine sediments on growth of juvenile coho salmon in laboratory streams, *Trans. Am. Fish. Soc.,* 110, 281, 1981.
57. **Cushing, C. E. and Rose, F. L.,** Cycling of zinc-65 by Columbia River periphyton in a closed lotic microcosm, *Limnol. Oceanogr.,* 15, 762, 1971.
58. **de Kozlowski, S. J. and Bunting, D. L., II,** A laboratory study on the thermal tolerance of four southeastern stream insect species (Trichoptera, Ephemeroptera), *Hydrobiolgia,* 79, 141, 1981.
59. **Eaton, J., Arthur, J., Hermanutz, R., Kiefer, R., Mueller, L., Anderson, R., Erikson, R., Nordling, B., Rogers, J., and Pritchard, H.,** Biological effects of continuous and intermittent dosing of outdoor experimental streams with chlorpyrifos, in *Aquatic Toxicology and Hazard Assessment: 8th Symp.,* Bahner, R. C. and Hansen, D. J., Eds., American Society for Testing and Materials, Philadelphia, 1985, 85.
60. **Eichenburger, E.,** Ökologische Untersuchungen an Modelfleissengewässern. I. Die Jahreszeitliche Verteilung der bestandesbildenden pflanzlichen Organismen bei verscheidner Abwasserbelastung, *Schweiz, Z. Hydrol.,* 29, 1, 1967.
61. **Eichenburger, E., Schlatter, F., Weilenmann, H., and Wuhrmann, K.,** Toxic and eutrophying effects of cobalt, copper and zinc on algal benthic communities in rivers, *Verh. Int. Ver. Limnol.,* 21, 1131, 1981.
62. **Gerhart, D. Z., Anderson, S. M., and Richter, J.,** Toxicity bioassays with periphyton communities: design of experimental streams, *Water Res.,* 11, 567, 1977.
63. **Gersich, F. M. and Brusven, M. A.,** Volcanic ash accumulation and ash-voiding mechanisms of aquatic insects, *J. Kans. Entomol. Soc.,* 55, 290, 1982.
64. **Giesy, J. P., Jr., Kania, H. J., Bowling, J. W., Knight, R. L., Mashburn, S., and Clarkin, S.,** Fate and Biological Effect of Cadmium Introduced into Channel Microcosms, EPA-600/3-79-039, U.S. Environmental Protection Agency, Athens, GA, 1979.
65. **Graney, R. L., Jr., Cherry, D. S., and Cairns, J., Jr.,** The influence of substrate, pH, diet and temperature upon cadmium accumulation in the Asiatic clam *(Corbicula fluminea)* in laboratory artificial streams, *Water Res.,* 18, 833, 1984.
66. **Graney, R. L., Jr., Cherry, D. S., and Cairns, J., Jr.,** Heavy metal indicator potential of the Asiatic clam *(Corbicula fluminea)* in artificial stream systems, *Hydrobiologia,* 102, 81, 1983.
67. **Hansen, S. R. and Garton, R. R.,** Ability of standard toxicity tests to predict the effects of the insecticide, diflubenzuron, on laboratory stream communities, *Can. J. Fish. Aquat. Sci.,* 39, 1273, 1982.
68. **Harvey, R. S.,** A flowing stream laboratory for studying the effects of water temperature on the ecology of stream organisms, *Assoc. Southeast. Biol. Bull.,* 20, 3, 1973.
69. **Hedtke, S. F. and Arthur, J. W.,** Evaluation of a Site-Specific Criterion Using Outdoor Experimental Streams, EPA-600/D-85-041, U.S. Environmental Protection Agency, Monticello, MN, 1985.
70. **Hedtke, J. L. and Norris, L. A.,** Effect of ammonium chloride on predatory consumption rates of brook trout *(Salvelinus fontinalis)* on juvenile chinook salmon *(Oncorhynchus tshawytscha)* in laboratory streams, *Bull. Environ. Contam. Toxicol.,* 24, 81, 1980.
71. **Hoffman, R. W. and Horne, A. J.,** On-site flume studies for assessment of impacts on stream aufwuchs communities, in *Microcosms in Ecological Research,* Giesy, J. P., Jr., Ed., Technical Information Center, U.S. Department of Energy, Oak Ridge, TN, 1980, 610.
72. **Kaufman, L. H.,** Stream Aufwuchs accumulation: disturbance frequency and stress resistance and resilience, *Oecologia (Berlin),* 52, 57, 1982.
73. **Kosinski, R. J.,** The effect of terrestrial herbicides on the community structure of stream periphyton, *Environ. Pollut. Ser. A,* 36, 165, 1984.
74. **Kosinski, R. J. and Merkle, M. G.,** The effect of four terrestrial herbicides on the productivity of artificial stream algal communities, *J. Environ. Qual.,* 13, 75, 1984.
75. **Larson, G. L., Warren, C. E., Hutchins, F. E., Lamperti, L. P., Schlesinger, D. A., and Seim, W. K.,** Toxicity of Residual Chlorine Compounds to Aquatic Organisms, EPA-600/3-78-023, U.S. Environmental Protection Agency, Washington, D.C., 1978.
76. **Lynch, T. R., Johnson, H. E., and Adams, W. J.,** Impact of atrazine and hexachlorobiphenyl on the structure and function of model stream ecosystems, *Environ. Toxicol. Chem.,* 4, 399, 1985.
77. **Maki, A. W. and Johnson, H. E.,** Evaluation of a toxicant on the metabolism of model stream communities, *J. Fish. Res. Board Can.,* 33, 2740, 1976.
78. **Maki, A. W. and Johnson, H. E.,** The influence of larval lampricide (TFM: 3-trifluoromethyl-4-nitrophenol) on growth and production of two species of aquatic macrophytes, *Elodea canadensis* (Michx.) Planchon and *Myriophyllum spicatum* (L.), *Bull. Environ. Contam. Toxicol.,* 17, 57, 1977.

79. **Manuel, C. Y. and Minshall, G. W.,** Limitations on the use of microcosms for predicting algal response to nutrient enrichment in lotic systems, in *Microcosms in Ecological Research,* Giesy, J. P., Jr., Ed., Technical Information Center, U.S. Department of Energy, Oak Ridge, TN, 1980, 645.

80. **Moorhead, D. L. and Kosinski, R. J.,** Effect of atrazine on the productivity of artificial stream algal communities, *Bull. Environ. Contam. Toxicol.,* 37, 330, 1986.

81. **Muirhead-Thomson, R. C.,** Comparative tolerance levels of blackfly *(Simulium)* larvae to permethrin (NRDC 143) and Temephos, *Mosq. News,* 37, 172, 1977.

82. **Nordlie, K. J. and Arthur, J. W.,** Effects of elevated water temperature on insect emergence in outdoor experimental channels, *Environ. Pollut. Ser. A,* 25, 53, 1981.

83. **Pignatello, J. J., Martinson, M. M., Steiert, J. G., Carlson, R. E., and Crawford, R. L.,** Biodegradation and photolysis of pentachlorophenol in artificial freshwater streams, *Appl. Environ. Microbiol.,* 46, 1024, 1983.

84. **Pignatello, J. J., Johnson, L. K., Martinson, M. M., Carlson, R. E., and Crawford, R. L.,** Response of the microflora in outdoor exprimental streams to pentachlorophenol: compartmental contributions, *Appl. Environ. Microbiol.,* 50, 127, 1985.

85. **Rodgers, E. B.,** Effects of elevated temperatures on macroinvertebrate populations in the Browns Ferry experimental ecosystems, in *Microcosms in Ecological Research,* Giesy, J. P., Jr., Ed., Technical Information Center, U.S. Department of Energy, Oak Ridge, TN, 1980, 684.

86. **Rodgers, J. H., Jr., Clark, J. R., Dickson, K. L., and Cairns, J., Jr.,** Nontaxonomic analysis of structure and function of aufwuchs communities in lotic microcosms, in *Microcosms in Ecological Research,* Giesy, J. P., Jr., Ed., Technical Information Center, U.S. Department of Energy, Oak Ridge, TN, 1980, 625.

87. **Rose, F. H. and McIntire, C. D.,** Accumulation of dieldrin by benthic algae in laboratory streams, *Hydrobiologia,* 35, 481, 1970.

88. **Ruber, E. and Kocor, R.,** The measurement of upstream migration in a laboratory stream as an index of potential side-effects of Temephos and Chlorpyrifos on *Gammarus fasciatus* (Amphipoda: Crustacea), *Mosq. News,* 36, 424, 1976.

89. **Seim, W. K., Lichatowich, J. A., Ellis, R. H., and Davis, G. E.,** Effects of Kraft mill effluents on juvenile salmon production in laboratory streams, *Water Res.,* 11, 189, 1977.

90. **Sérodes, J. B., Moreau, G., and Allard, M.,** Dispositif expériméntal pour l'étude de divers impacts sur la faune benthique d'un cours d'eau, *Water Res.,* 18, 95, 1984.

91. **Sigmon, C. F., Kania, H. J., and Beyers, R. J.,** Reductions in biomass and diversity resulting from exposure to mercury in artificial streams, *J. Fish. Res. Board Can.,* 34, 493, 1977.

92. **Stout, R. J. and Cooper, W. E.,** Effect of p-cresol on leaf decomposition and invertebrate colonization in experimental outdoor streams, *Can. J. Fish. Aquat. Sci.,* 40, 1647, 1983.

93. **Stout, R. J. and Kilham, S. S.,** Effects of p-cresol on photosynthetic and respiratory rates of a filamentous green alga *(Spirogyra), Bull. Environ. Contam. Toxicol.,* 30, 1, 1983.

94. **Warren, C. E., Wales, J. H., Davis, G. E., and Doudoroff, P.,** Trout production in an experimental stream enriched with sucrose, *J. Wildl. Manage.,* 28, 617, 1964.

95. **Watton, A. J. and Hawkes, H. A.,** Studies on the effects of sewage effluent on gastropod populations in experimental streams, *Water Res.,* 18, 1235, 1984.

96. **Wilde, E. W. and Tilly, L. J.,** Structural characteristics of algal communities in thermally altered artificial streams, *Hydrobiologia,* 76, 57, 1981.

97. **Williams, L. G. and Mount, D. L.,** Influence of zinc on periphyton communities, *Am. J. Bot.,* 52, 26, 1965.

98. **Wilton, D. P. and Travis, B. V.,** An improved method for simulated stream tests of blackfly larvae, *Mosq. News,* 25, 118, 1965.

99. **Wojtalik, T. A. and Waters, T. F.,** Some effects of heated water on the drift of two species of stream invertebrates, *Trans. Am. Fish. Soc.,* 99, 782, 1970.

100. **Wrenn, W. B., Armitage, B. J., Rodgers, E. B., and Forsythe, T. D.,** Browns Ferry Biothermal Res. Ser. II. Effects of Temperature on Bluegill and Walleye, and Periphyton, Macroinvertebrates, and Zooplankton Communities in Experimental Ecosystems, EPA-600/3-79-092, U.S. Environmental Protection Agency, Duluth, MN, 1979.

101. **Wrenn, W. B. and Grannemann, K. L.,** Effect of temperature on bluegill reproduction and young-of-the-year standing stocks in experimental ecosystems, in *Microcosms in Ecological Research,* Giesy, J. P., Jr., Ed., Technical Information Center, U.S. Department of Energy, Oak Ridge, TN, 1980, 703.

102. **Yasuno, M., Sugaya, Y., and Iwakuma, T.,** Effects of insecticides on the benthic community in a model stream, *Environ. Pollut. Ser. A,* 38, 31, 1985.

103. **Yount, J. D. and Richter, J. E.,** Effects of pentachlorophenol on periphyton communities in outdoor experimental streams, *Arch. Environ. Contam. Toxicol.,* 15, 51, 1986.

104. **Zischke, J. A., Arthur, J. W., Nordlie, K. J., Hermanutz, R. O., Standen, D. A., and Henry, T. P.,** Acidification effects on macroinvertebrates and fathead minnows *(Pimephales promelas)* in outdoor experimental channels, *Water Res.,* 17, 47, 1982.

105. **Patrick, R.,** The structure of diatom communities in similar ecological conditions, *Am. Nat.,* 102, 173, 1968.

106. **Patrick, R.,** The effects of increasing light and temperature on the structure of diatom communities, *Limnol. Oceanogr.,* 16, 405, 1971.

107. **Patrick, R.,** Diatoms as bioassay organisms, in *Bioassay Techniques and Environmental Chemistry,* Glass, G. E., Ed., Ann Arbor Science Publishers, Woburn, MA, 1973, 139.

108. **Bowling, J. W., Giesy, J. P., Jr., Kania, H. J., and Knight, R. L.,** Large-scale microcosms for assessing fates and effects of trace contaminants, in *Microcosms in Ecological Research,* Giesy, J. P., Jr., Ed., Technical Information Center, U.S. Department of Energy, Oak Ridge, TN, 1980, 224.

109. **Patrick, R., Crum, R. B., and Coles, J.,** Temperature and manganese as determining factors in the presence of diatoms or bluegreen floras in streams, *Proc. Natl. Acad. Sci. U.S.A.,* 64, 472, 1969.

110. **Steel, R. G. D. and Torrie, J. H.,** *Principles and Procedures of Statistics,* McGraw-Hill, New York, 1960, chap. 3.

111. **Whittaker, R. H.,** Experiments with radiophosphorus tracer in aquarium microcosms, *Ecol. Mongr.,* 31, 157, 1961.

112. **Day, R. A.,** *How to Write and Publish a Scientific Paper,* 1st ed., ISI Press, Philadelphia, 1979, 31.

113. **Hedtke, S. F.,** personal communication.

114. **Maki, A. W.,** Evaluation of toxicant effects on structure and function of model stream communities: correlations with natural stream effects, in *Microcosms in Ecological Research,* Giesy, J. P., Jr., Ed., Technical Information Center, U.S. Department of Energy, Oak Ridge, TN, 1980, 583.

115. **Geckler, J. R., Horning, W. B., Neiheisel, T. M., Pickering, Q. H., and Robinson, E. L.,** Validity of Laboratory Tests for Predicting Copper Toxicity in Streams, EPA-600/3-76-116, U.S. Environmental Protection Agency, Duluth, MN, 1976.

Index

INDEX

pollutants and, 62, 177

B

Bacteria, see also individual species
  artificial stream studies and, 304
  as bioaccumulator organisms, 61
  biodegradation of persistent toxic organic
      chemicals by, 239
  colonization of fine particulate organic matter by, 9
  colonization of rithron debris by, 11
  decomposition by in lake ecosystems 27, 31—33
  enclosure technique and, 254
  organic ligand excretion by, 103
  outdoor pond studies and, 274—275, 285
  sediment bioavailability to, 133
  transfer of contaminants within trophic networks
      and, 62
Bacterial cell walls, ecotoxicological importance of
      as membrane barriers, 53—54
Balanced microcosms, used to assess pollutant
      effects on populations, 250
Benthic invertebrates
  mercury in, 204—205, 207—208, 211—212,
      214—215
  sediment bioavailability to, 125, 138—141
Benzenes
  in Laurentian Great Lakes, 220, 227, 229
  sorption isotherms and, 117
Beryllium, determination of, 85
BHC, see Lindane
Bieler See, trace metals in, 193
Bioaccumulation
  nutrient-water interactions and, 219—221, 239—
      241, 243, 245
  of toxicants in organisms, 35, 52—58, 61—62
Bioamplification, mechanism of, 64—65
Biocenoses, contamination of and effects on
      structure and functioning of ecosystems, 35,
      37—39, 45—69
Bioconcentration factor, for bioaccumulation of
      toxicants, 61
Biodegradation reactions
  dependence of on microbial degradation, 51
  regulation of in lake ecosystems, 21—22, 29—32
Biological barriers, ecotoxicological importance of,
      35, 55, 70
Biological integration levels, in freshwater
      ecosystems, 37 38
Biomagnification, see Bioamplification
Biomass, pollution effect on, 151, 176—177
Biotic factors, in natural systems, 36, 40, 69
Biotic index, table standard of determination of, 172
Biotoxification, concept of, 58
Biotransformation, of contaminants, 35, 52, 58—59
Biotopes, contaminant distribution and chemical
      transformations in, 35
  sediments, principal ecotoxicological processes in,
      48—52
  "water column", principal ecotoxicological
      processes in, 45—48

Biouptake, concept of, 52, 58
Bird scarers, as part of enclosure technique
      methodology, 253, 258
Bismuth, determination of, 79, 85, 87
Bisson and Davis artificial stream, 304
Blood, role of in overall mechanism of bioaccumula-
      tion, 59, 61
Blue-green algae, in lake ecosystems, 29
Bosmina
  aragoni, 178
  longirostris, 258
Bott ecosystem streams, 305
Bottom-up control, of contaminant fate, 238
Brahmaputra River, statistics pertaining to, 5
Brillouin's index of water pollution, Margaleff's
      variant, 171
Bryophytes, as bioaccumulator organisms, 61
Burial, sediment-water interactions and, 219, 230,
      233—235

C

Cadmium
  acid rain impact on solubilization of, 179
  adsorption of, 102—103, 113, 115, 129—130, 136,
      143, 164, 186, 189, 193—198
  artificial stream studies and, 301, 305—306, 311
  as contaminant, 43, 45, 47, 57—58, 61, 64
  determination of, 78—80, 82, 84—87
  enclosure technique testing and, 256, 264
  speciation of, 97, 102—104
Calanoid copepods, in lake ecosystems, 31
Calcium, microbial processing and, 8
Calcium carbonate, adsorption of, 186, 189
Callitriche stagnalis, 290
Carbamates
  as contaminants, 43
  resistance to, 67
Carbohydrate proteins, biochemical nature of
      dissolved organic matter and, 9, see also
      Glycoproteins
Carbon
  enclosure technique testing and, 269
  discharge of oxides of into atmosphere as
      pollutants, 42
  in lake ecosystems, 26, 29—30
  outdoor pond studies and, 290
  oxidation of organic, 10—11
  partitioning and, 232
  in river continuum, 13—14
Carbon dioxide
  enclosure technique testing and, 258
  in lake ecosystems, 26—27, 29
  outdoor pond studies and, 279
  as pollutant, 153
Carbonates
  complexation capacity of in natural waters, 48
  as sediments, 134
Carbon tetrachloride
  as global contaminant, 155—157
  sorption isotherms and, 116